Fractal Functions Communication with Fractional Calculus

A. Gowrisankar · T. M. C. Priyanka ·
Santo Banerjee

Fractal Functions Communication with Fractional Calculus

Springer

A. Gowrisankar
Department of Mathematics
School of Advanced Sciences
Vellore Institute of Technology
Vellore, Tamil Nadu, India

T. M. C. Priyanka
Department of Mathematics
Guizhou University
Guiyang, Guizhou, China

Santo Banerjee
Dipartimento di Scienze Matematiche
Politecnico di Torino
Turin, Italy

ISBN 978-3-031-96992-8 ISBN 978-3-031-96993-5 (eBook)
https://doi.org/10.1007/978-3-031-96993-5

This Springer imprint is published by the registered company Springer Nature Switzerland AG
The registered company address is: Gewerbestrasse 11, 6330 Cham, Switzerland

Preface

Fractal functions are incredibly fascinating functions due to their self-similar structure and relatively simple construction procedure. The concept of fractal *sets* is perfectly taken to the theme of fractal *functions* with similar construction procedure via *fractal interpolation functions*. These fractal interpolation functions can be utilized to explore naturally existing functions, which share irregularity and self-similarity in their graphs along with non-integer dimension. The non-integer dimension theory beginning from the "Hausdorff dimension" has evolved a lot in various phenomena and seems to have promising applications in many science and engineering problems. Meanwhile, despite the progress made in the study of fractal interpolation functions, there is still much to learn and apply to real-world challenges. One of the interesting studies is the discussion of fractional calculus methods on fractal functions. This study primarily focuses on the fractal scaling parameter and fractional derivative/integral order, not to the surprise both are mostly non-integers! This book mainly covers the types of fractal interpolation functions and their ordinary as well as fractional calculus investigation.

The following is a brief description of the framework of this book. Chapter 1 introduces what are fractals and narrates how the theory of fractal geometry was emerged in late 1970's. In fact, fractals are not something artificially designed for mathematical curiosity after 1970, the objects possessing self-similarity and non-integer Hausdorff dimension are consolidated under the name "fractals", to name a few, fern, lightning, Cantor set and Sierpinski gasket. It means that there are some notable pioneers of fractals, who designed fractal structures and made significant contributions before the name "fractal" was coined. This chapter recalls some of such contributors at first look. Thereafter, presents the construction of fractals using the iterated function theory, which is the fundamental construction procedure for the fractal interpolation functions discussed in the Chap. 2.

The two most important topics of this book are fractal (interpolation) functions and methods of fractional calculus. Chapter 2 discusses the former topic—*fractal interpolation functions*—where Barnsley's construction of fractal function from the theory of iterated function system is discussed in detail. In addition, the approach of defining fractal functions as fixed points of Read–Bajraktarević operator is also presented.

This chapter gives a concreteness of what factor or function reasons behind the classification of fractal interpolation functions like affine and non-affine fractal functions. Types of fractal functions including linear fractal function, quadratic fractal function, α-fractal function and hidden variable fractal function are discussed with numerical illustrations. Furthermore, the fractal dimension is estimated with different choice of scaling parameters generating different fractal functions.

Chapter 3 is devoted to highlight the significance of *scaling parameter* in the definition of fractal functions. It is the primary parameter distinguishes the fractal functions from the classical interpolation techniques and its range lies in the interval $(0, 1)$. Construction of fractal interpolation functions follow the approach of construction of fractals, in both the places, the scaling parameter acts as a contraction factor. Thus, enables fractals and fractal functions to possess non-integer fractal dimension and self-similar nature at any scale. In addition, this chapter highlights the importance of choosing variable scaling parameter and optimized scaling parameter with numerical illustrations.

In Chap. 4, the classical integral of types of fractal interpolation functions, namely, linear fractal interpolation function, α-fractal function, and hidden variable fractal interpolation function are discussed. The definite integral and indefinite integral for the classical fractal functions are investigated. The integral of fractal functions are again fractal functions to a different set of interpolation data if the integrals are predefined at the initial point or end point of the given data. In addition, the integral of fractal functions with continuous scaling functions is discussed which generalizes the integral of fractal functions with constant scaling factors.

Chapter 5 gives a precise introduction to the *fractional calculus*, the second important topic of this book. From here, the story behind the communication of fractal functions with fractional calculus begins! Some of the popular methods like Riemann–Liouville calculus, Weyl–Marchaud fractional derivative and Caputo-fractional derivative are discussed, only which will be dealt in the subsequent discussions. This chapter is framed in such a way that the stuff is essential and may be sufficient to the readers who proceed with the following chapters to understand the relationship between fractional calculus and fractal functions.

Chapter 6 presents the Riemann–Liouville fractional integral of non-affine fractal function, namely, α-fractal function. It also explores a fractional operator that maps the given continuous function to its Riemann–Liouville fractional integral. The Riemann–Liouville fractional integral is explored for α-fractal function by choosing vertical scaling factor as a constant as well as a continuous function defined on the closed interval of interpolation. Further, the boundedness and linearity of the fractional operator of α-fractal function are investigated. Finally, the semigroup property for the collection of fractional operators defined on the space of continuous functions is discussed.

In Chap. 7, the fractional derivative of various fractal interpolation functions like affine, non-affine fractal functions, and fractal functions in higher dimensions are discussed. It is ensured that the Weyl–Marchaud fractional derivative of fractal functions possess fractal properties and obey essential conditions of fractal functions whenever the order of the fractional derivative meets the necessary conditions. The important relation between the fractional functions and fractional calculus methods lies in the stated necessary condition which restricts the range of fractional order with respect to the scaling factor in order to generate new fractal functions. Finally, the beauty behind the connection between fractional derivative and fractal functions is enhanced by estimating the fractal dimension of fractal functions in terms of fractional derivative order.

Chapter 8 is devoted to investigate the fractional derivative of special class of fractal functions which are *differentiable* in nature, which is in contrast to the fractal functions discussed in Chaps. 2–4. The Caputo fractional derivative of a real continuous function g distinguishes from the other fractional derivative methods like Riemann–Liouville and Weyl–Marchaud, with the demand for the existence of at least the first-order derivative of g. This attribute leads to the investigation of Caputo fractional derivative of α-fractal splines rather than just a continuous non-differentiable α-fractal function. A bounded linear operator corresponding to the Caputo fractional derivative of fractal version is explored. In addition, a new family of fractal perturbations is presented in association with the fractional derivative. This chapter ends with the discussion of numerical approach to determine the exact Caputo fractional derivative of fractal functions in terms of Legendre polynomials.

Chapter 9 attracts every fractal and non-fractal audience with its interesting application of fractal interpolation on epidemiology and climate dynamics. As a first application, the positive cases of Omicron, one of the severe COVID-19 variants, are analyzed using linear fractal function to predict the upcoming scenario. Next, using the scheme of hidden variable fractal function, the temperature data with respect to time is extended to the higher dimensional space by adding CO_2 emissions as third hidden variables. The preprocessed temperature data is employed for determining future climate changes as a direct consequence of varying land and sea temperatures. Finally, tailoring fractal interpolation for pre-processing the mpox cases, the irregular and fractal patterns in the trend of mpox transmission is highlighted. This application is combined with deep learning model to forecast the future mpox spread to alert the outbreak.

Now, the readers can understand how the book is organized, in short, Chap. 1 introduces fractals, from Chap. 2 to 4, types of fractal functions and their classical integral are discussed. Providing a brief introduction to fractional calculus in Chap. 5, Chaps. 6–8 discusses how fractional calculus methods are investigated over fractal functions, the book ends by presenting the application part in Chap. 9. With this clear organization, the book aims to provide readers with an understanding of the present stage of research on "the fractional calculus on fractal functions" and encourage to extend the studies for additional knowledge and applications.

Vellore, India — A. Gowrisankar
Guiyang, China — T. M. C. Priyanka
Turin, Italy — Santo Banerjee

Contents

Chapter 1
Fractal Geometry—A Brief History

1.1 Introduction

The word 'fractal' is coined by the French-American mathematician Benoit B Mandelbrot in 1975, which is derived from the Latin adjective *fractus* associated with the verb frengere meaning to break. Now, it's half century since its birth! The book "*Fractals: Form, Chance, and Dimension*" published in 1977 is the translation of his French book "*Les Objects Fractals: Forme, Hasard et Dimension*" which is the first publication revealed his ideas of fractals and has attracted many to use fractals in their own field. Though there is a large contribution and many developments in almost all branches of science and engineering, the essence of fractals depends on the geometry. In the book "*Fractal Geometry of Nature*" [1], Mandelbrot's famous phrase

> Clouds are not spheres, mountains are not cones, coastlines are not circles, and bark is not smooth, nor does lightning travel in a straight line

evidences his perception that many devised theories of mathematics is unrelated with nature. He claims that many patterns of nature are irregular and fragmented which cannot be investigated using Euclidean geometry. For the proper understanding of irregularity or fragmentation, dimension cannot always be defined as the number of co-ordinates. In this way, before defining fractal, he has slowly stated the difficulties in the Euclidean geometry via the shapes and dimension. So fractal geometry provides a general framework to study the morphology and properties of irregular sets and functions. The work "How long is the coastline of Britain?" [2] is the seed for combining the two most important concepts of *self-similarity and non-integer dimension* with fractals. Mandelbrot defined 'a fractal is a set for which the Hausdorff Besicovitch dimension strictly exceeds its topological dimension', also, he stated 'a fractal is a shape made of parts similar to the whole in some way'. Even there is no complete mathematical definition to describe what exactly is a fractal, ultimately his idea was to bring objects with irregular and fragmented patterns under one family

A. Gowrisankar et al., *Fractal Functions Communication with Fractional Calculus*,
https://doi.org/10.1007/978-3-031-96993-5_1

called *fractals*. Fractals have their own dimension called the *fractal dimension* satisfying the first definition. The Hausdorff Besicovitch dimension is often fractional dimension, but the fractal dimension of an object may be an integer which is not greater than the Euclidean dimension (dimension of the space where the object is embedded) but strictly greater than the topological dimension. Hence, what we would like to clarify is that a fractal set may have an integer or a non-integer dimension in general.

The second definition of fractal which is based on self-similarity can be better understood with a real-life example, if we zoom into the lightning, we find that its ruggedness is repeated on finer and finer scales, and under rescaling looks essentially the same; the lightning is a fractal curve. Similarly, we can say many examples of fractals found in nature, for instance, coastline, cauliflower, fern leaf, branches of a tree, structure of a neuron and a wall crack. These are the examples of random fractals. There are special mathematical fractal sets, which can be constructed using mathematical ingredients to study the naturally existing irregular functions/signals, called the *deterministic fractals*. Random fractals are *approximately self-similar* whereas deterministic fractals are *exactly self-similar*. In late 1980s, K. J. Falconer personally felt that defining a fractal set seems to be as hard as that of defining the life of a living being, since the above definitions of Mandelbrot for fractals are unsatisfactory as they exclude a number of sets which clearly ought to be considered as fractals. In his book "*Fractal Geometry - Mathematical Illustration*" [3], from mathematician's point of view, he listed out certain characteristic properties for a set to be a fractal. They are fine structure, irregularity, recursive description, in addition to the dimension and self-similar property. The book "*Fractals*" by Jens Feder [4] is another interesting book which has beautifully highlighted the Mandelbrot's inspired concept of *multifractal measure*, currently active in many applications of physical and computer sciences. The book of J. M. Fraser, "*Assouad Dimension and Fractal Geometry*" [5] is a recent contribution. This book is dedicated to a thorough analysis of the Assouad dimension and some of its natural variants claiming that dimension theory is the heart of fractal geometry.

1.2 Fractal Dimension

In most of the useful books of fractal geometry, the one common and important concept discussed is the 'Fractal Dimension', since it is the mathematically quantitative representation to understand the abstract fractal characteristics like irregularity and detailed structuring. Among the variety of fractal dimensions including box-counting dimension, information dimension, correlation dimension and so on, the definition of Hausdorff dimension is the oldest and probably the most important. It was introduced by Hausdorff in [6], greatly developed by Besicovitch; it is based on measures and can be defined for any set. Its calculation or estimation by computational methods is hard, however it is essential to familiar with the concepts of Hausdorff measure and dimension for understanding the mathematics of fractals. It is defined in terms of Hausdorff measure, which can be viewed as an extension of Lebesgue measure to non-integer dimensions. If U is any non-empty subset of n-dimensional Euclidean space $\mathbb{R}^n$, the diameter of U is defined as

$$|U| = \sup\{|x - y| : x, y \in U\},$$

i.e., the greatest distance apart of any pair of points in U, where $|.|$ refers the usual Euclidean distance. If $\{U_i\}$ is a countable (or finite) collection of sets of diameter at most δ that cover F, i.e.,

$$K \subset \bigcup_{i=1}^{\infty} U_i$$

with $0 < |U_i| \leq \delta$ for each i, we say that $\{U_i\}$ is a δ-cover of K. Suppose that K is a subset of $\mathbb{R}^n$ and s is a non-negative number. For any $\delta > 0$ we define

$$\mathcal{H}_\delta^s(K) = \inf\left\{\sum_{i=1}^{\infty} |U_i|^s : \{U_i\} \text{ is a } \delta - \text{cover of } K\right\}.$$

As δ decreases, the class of permissible covers of K is reduced. Therefore, the infimum $\mathcal{H}_\delta^s(K)$ increases, and so approaches a limit as $\delta \to 0$. Thus,

$$\mathcal{H}^s(K) = \lim_{\delta \to 0} \mathcal{H}_\delta^s(K).$$

This limit exists for any subset K of $\mathbb{R}^n$, though the limiting value can be 0 or ∞. We call $\mathcal{H}^s(K)$ as the s-dimensional Hausdorff measure of K. Then, the Hausdorff Dimension or Hausdorff-Besicovitch Dimension of K is defined as,

$$\dim_H(K) = \inf\{s : \mathcal{H}^s(K) = 0\} = \sup\{s : \mathcal{H}^s(K) = \infty\},$$

so that

$$\mathcal{H}^s(K) = \begin{cases} \infty, & \text{if } s < \dim_H(K), \\ 0, & \text{if } s > \dim_H(K). \end{cases}$$

If $s = \dim_H(K)$, then $\mathcal{H}^s(K)$ may be zero or infinite, or may $0 < \mathcal{H}^s(K) < \infty$.

Hausdorff dimension of a given set K involves taking the infimum over covers consisting of balls of radius less than or equal to a given $\delta > 0$, this procedure makes the explicit computation difficult. A slight simplification is obtained by considering only covers by the balls of radius equal to δ. This gives rise to the concepts of box-dimension [7]. Let $K \in \mathcal{H}(X)$ and $N(\delta)$ denotes the smallest number of closed balls of radius $\delta > 0$ required to cover K. If

$$D = dim_B = \lim_{\delta \to 0} \left\{ \frac{\log N(\delta)}{\log(1/\delta)} \right\} \tag{1.1}$$

exists, then D is called the box-dimension or fractal dimension of K.

1.3 Pioneers of Fractal Geometry

When we intend to study 'Fractal Geometry', some classic examples of fractals that come in mind are Cantor set, Sierpinski triangle, von Koch curve and Wierstrass function. The surprising fact is that which are not constructed or defined after 1975, means after Mandelbrot coined the word 'fractal'. Unlike in physics, the word 'invention' is little hard in mathematics, since most of the ideas or concepts or definitions are a development or generalization of an existing theme. The fractal properties like non-integer dimension, self-similarity, nowhere differentiability have been individually discussed long ago 1970s. Here, let us see some pioneers and discuss their fascinating works which are unknowingly contributed to Fractal Geometry.

In the years 1879–1884, Georg Cantor, a German mathematician, wrote a series of papers entitled "Uber unendliche, lineare Punktmannichfaltigkeiten" [8], that contained the first systematic treatment of the point set topology of the real line. In 1883, by giving the definition for the *perfect set*, he has provided a perfect example for it, the Cantor set! However, there is no substantive evidence or indication about how he came upon the Cantor set and Cantor function [9]. The Cantor set is a subset of the real line $\mathbb{R}$. Let us start with the closed interval $C_0 = [0, 1]$. Then the set C_1 is obtained by removing the middle open third $(1/3, 2/3)$ from $[0, 1]$, leaving the two intervals $[0, 1/3]$ and $[2/3, 1]$. The next set C_2 is defined by removing the middle third of each of the two intervals of C_1. This leaves C_2 as the union of four intervals $[0, 1/9], [2/9, 1/3], [2/3, 7/9], [8/9, 1]$. Continuing this process, the triadic Cantor set is obtained as the 'limit' C of the sequence C_k of sets. The sets decrease: $C_0 \supseteq C_1 \supseteq C_2 \supseteq \cdots C_k \supseteq \cdots$. Therefore, the 'limit' can be defined as the intersection of the sets,

$$C = \bigcap_{k \in \mathbb{N}} C_k$$

which is the Cantor set. Mandelbrot has referred it as the 'Cantor dust' and the term "dust" refers to the fact that the set is zero-dimensional. The graphical representation of the Cantor set is given in Fig. 1.1a. The Cantor set is self-similar and it has infinitely many points but its length is zero. This fact gives a clue that its fractal dimension must be strictly less than 1. With $N = 2$ and $\delta = 1/3$, the fractal dimension of Cantor dust is estimated as

$$D = \frac{\log 2}{\log 3} = 0.6309,$$

its topological dimension is 0 and thus, it is a fractal set.

The von Koch curve, one of the earliest and standard fractal curves, was originally described by a Swedish mathematician Helge von Koch in [10]. It is constructed using only one of the three sides of the original triangle and a three Koch curves iterative construction make a Koch snowflake. The Koch curve is constructed using an iterative procedure beginning with the initiator of the set as the unit line segment. The unit line segment is divided into three, and the middle third is replaced with two equal segments, both one-third in length, which form an equilateral triangle; this step is the generator of the curve. At the next step, the middle third is removed from each of the four segments and each is replaced with two equal segments as before. This process is repeated an infinite number of times to produce the Koch curve. The Koch curve is displayed in Fig. 1.1c. The self-similarity of the set is evident: each sub-segment is an exact replica of the original curve. This curve's area vanishes, on the other hand, at each stage its total length increases in a ratio of $4/3$, hence the limit curve is of infinite length. The Hausdorff-Besicovitch dimension $\log 4/\log 3$ for the Koch curve exceeds its topological dimension 1, therefore, the Koch curve is a fractal set with the fractal dimension $\log 4/\log 3$ [4].

In 1915, a prolific Polish mathematician Waclaw Sierpiński, has devised the first example of a curve having almost all points as the branching points. This curve is

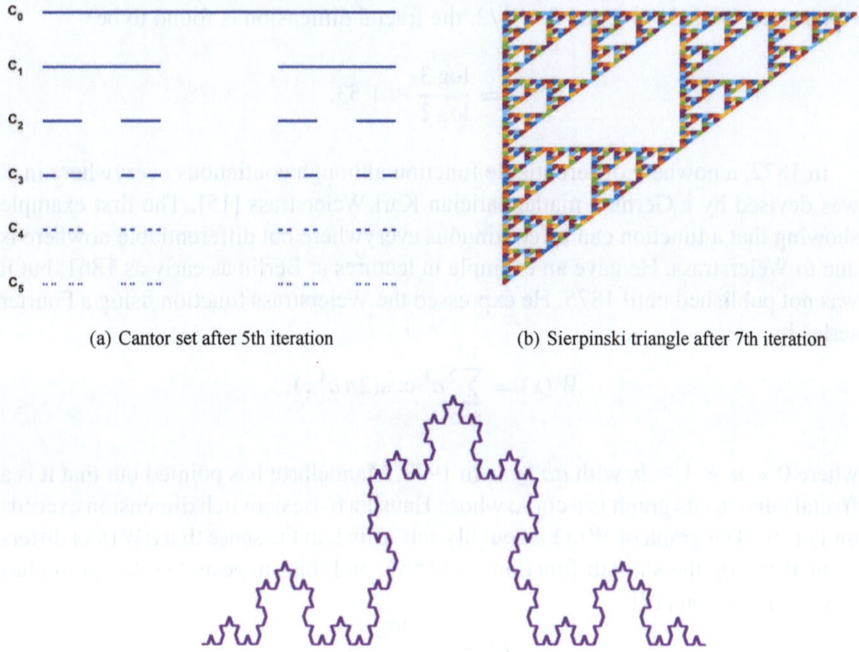

(a) Cantor set after 5th iteration

(b) Sierpinski triangle after 7th iteration

(c) von Koch curve after 4th iteration

Fig. 1.1 Examples of deterministic fractal sets: **a** Cantor set, **b** Sierpinski triangle and **c** von Koch curve

currently known as the Sierpiński gasket or Sierpiński triangle. It can be thought of as a Cantor-like dusts spread out in two dimensions. Images of this shape started to appear long before the seminal paper by Sierpiński [11]. During 11th century, one of the most ancient illustrations was found in a medieval mosaic in central Italy [12], similar drawing forms were also found in Chinese mathematics books in 13th century [13]. Throughout the history, many deterministic and probabilistic ways have been suggested to study the Sierpinski triangle for its simple construction and decorative applications [14]. Let S_0 be the equilateral triangle with side length 1. By joining the midpoints of each side, four smaller triangles can be obtained with side length $1/2$. Remove the middle open triangle leaving the boundary to obtain the set S_1. Repeating this procedure in each of the remaining three triangles, nine smaller triangles can be obtained with side length $1/4$, named S_2. In this manner, a decreasing sequence of subsets S_k are obtained as $S_0 \supseteq S_1 \supseteq S_2 \supseteq \cdots S_k \supseteq \cdots$ and its 'limit' S is the intersection of sets,

$$S = \bigcap_{k \in \mathbb{N}} S_k$$

which is the Sierpiński gasket. The graphical representation of the Sierpinski triangle is illustrated in Fig. 1.1b. This gasket has area zero while its topological dimension equals one. With $N = 3$ and $\delta = 1/2$, the fractal dimension is found to be

$$D = \frac{\log 3}{\log 2} \approx 1.53.$$

In 1872, a nowhere differentiable function although continuous everywhere in $\mathbb{R}$ was devised by a German mathematician Karl Weierstrass [15]. The first example showing that a function can be continuous everywhere but differentiable nowhere is due to Weierstrass. He gave an example in lectures at Berlin as early as 1861, but it was not published until 1875. He expressed the Weierstrass function using a Fourier series by

$$W(x) = \sum_{k=0}^{\infty} a^k \cos(2\pi b^k x),$$

where $0 < a < 1 < b$, with $ab \geq 1$. In 1977, Mandelbrot has pointed out that it is a fractal curve as its graph is a curve whose Hausdorff-Besicovitch dimension exceeds unity [16]. The graph of $W(x)$ is roughly self-affine, in the sense that $aW(bx)$ differs from $W(x)$ by the smooth function $\cos(2\pi x)$, and this suggests that the graph has fractal dimension [17]

$$D = 2 + \frac{\log a}{\log b}.$$

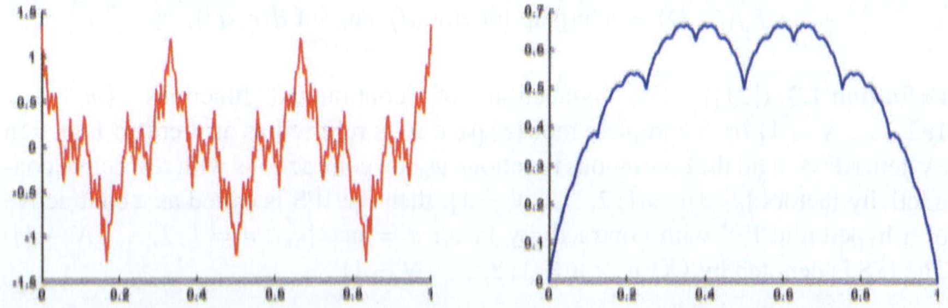

(a) Weierstrass function (b) Blancmange-Takagi function

Fig. 1.2 Examples of fractal curves: **a** Weierstrass function and **b** Blancmange-Takagi function

For instance, if $a = 0.5$ and $b = 3$; in this case

$$D = 2 - \frac{\log 2}{\log 3} \approx 1.37.$$

With non-integer fractal dimension, the Weierstrass function is highly irregular and self-similar at any scale. Takagi function is another example of a fractal curve first described in 1903 by Teiji Takagi [18]. This continuous nowhere differentiable function is sometimes called the Blancmange function as it resembles the blancmange dessert. The Weierstrass and Blancmange-Takagi functions are graphically illustrated in Fig. 1.2a and b, respectively.

1.4 Construction of Fractals

In 1981, Hutchinson [19] has theoretically framed the idea of constructing deterministic fractal by iterated function system and later, it has been popularized by Barnsley [20]. The iterated function system is the most widely used mathematical recipe for constructing fractals in a metric space with specified strict self-similar property.

Definition 1.1 ([21]) Let (X, d) be a complete metric space. A map $T : (X, d) \to (X, d)$ is said to be a Banach contraction map if there exists a constant $\kappa \in [0, 1)$ such that

$$d(Tx, Ty) \le \kappa d(x, y), \ \forall\, x, y \in X.$$

Here, κ is known as contractivity factor of T.

Definition 1.2 ([21]) The Hausdorff metric h_d between the two non-empty compact subsets C and D of X is defined as

$$h_d(C, D) = \max\{\sup_{c\in C}\inf_{d\in D} d(c, d), \sup_{d\in D}\inf_{c\in C} d(c, d)\}.$$

Definition 1.3 ([21]) The collection of continuous functions $\{w_n : n = 1, 2, \ldots, N-1\}$ on a complete metric space X is referred as an Iterated Function System (IFS). If all the continuous functions w_n are contractions with respective contractivity factors $\{\kappa_n : n = 1, 2, \ldots, N-1\}$, then the IFS is called as a contractive or a hyperbolic IFS with contractivity factor $\kappa = \max\{\kappa_n : n = 1, 2, \ldots, N-1\}$. The IFS is denoted by $\{X; w_n : n = 1, 2, \ldots, N-1\}$.

Let $\mathcal{H}(X)$ be the collection of all non-empty compact subsets of X. The pair $(\mathcal{H}(X), h_d)$ turns out to be a metric space with respect to Hausdorff metric h_d and referred as the *Hausdorff space or fractal space* where the deterministic fractals are generated. The completeness of the underlying space (X, d) implies the completeness of $(\mathcal{H}(X), h_d)$, similarly the space $(\mathcal{H}(X), h_d)$ is compact whenever (X, d) is compact [22]. Now, define $W : \mathcal{H}(X) \to \mathcal{H}(X)$ by

$$W(A) = \bigcup_{n=1}^{N-1} w_n(A), \text{ for } A \in \mathcal{H}(X),$$

where $w_n(A) = \{w_n(x) : x \in A\}$. Any set $G \in \mathcal{H}(X)$ satisfying $W(G) = G$ is called as an attractor for the IFS [20] and the IFS always admits at least one attractor. If w_i are contractions for some $0 \le s < 1$ satisfying

$$d(w_n(x), w_n(y)) \le s d(x, y), \ \forall\, x, y \in X,$$

then the IFS is termed hyperbolic IFS. In this case W is a contraction mapping on a complete metric space $(\mathcal{H}(X), h_d)$ and consequently admits a unique attractor by the Banach fixed point theorem,

$$G = \lim_{m\to\infty} W^{\circ m}(S), \ S \in \mathcal{H}(X).$$

Precisely, a finite collection of contraction mappings on a complete metric space constitutes a hyperbolic IFS and it admits a unique fixed point, referred as an *attractor or fractal*.

Some examples like Cantor set, Sierpinski triangle and von Koch curve, discussed above are perfect examples of deterministic fractals. Their associated IFSs are provided in the following. The IFS consisting of the following two functions: $w_1(x) = \frac{x}{3}$, $w_2(x) = \frac{x+2}{3}$ defined on $[0, 1]$, generates the Cantor set with the contraction factor $\kappa = 1/3$. Similarly, the three functions: $w_1(x, y) = (\frac{x}{2}, \frac{y}{2})$, $w_2(x, y) = (\frac{x+1}{2}, \frac{y}{2})$, $w_3(x, y) = (\frac{2x+1}{4}, \frac{2y+\sqrt{3}}{4})$ defined on $[0, 1] \times [0, 1]$ provide the IFS that generates the Sierpinski triangle. Here, the three functions w_i, $i = 1, 2, 3$ have contraction factor $1/2$. For constructing the von Koch curve, consider the IFS on $[-1, 1] \times [-1, 1]$ comprising the following four functions: $w_1(x, y) = (\frac{x-2}{3}, \frac{y}{3})$,

$w_2(x, y) = (\frac{x-\sqrt{3}y-1}{6}, \frac{\sqrt{3}x+y+\sqrt{3}}{6})$, $w_3(x, y) = (\frac{x+\sqrt{3}y+1}{6}, \frac{-\sqrt{3}x+y+\sqrt{3}}{6})$, $w_4(x, y) = (\frac{x+2}{3}, \frac{y}{3})$. The functions w_i, $i = 1, 2, 3, 4$ have contraction factor $1/3$.

References

1. B.B. Mandelbrot, *The Fractal Geometry of Nature* (W.H, Freeman and Company, 1983)
2. B. Mandelbrot, How long is the coast of Britain? statistical self-similarity and fractional dimension. Science **156**(3775), 636–638 (1967)
3. K.J. Falconer, *Techniques in Fractal Geometry*, vol. 3 (Wiley Chichester, 1997)
4. J. Feder, *Fractals* (Springer Science & Business Media, 2013)
5. J.M. Fraser, *Assouad Dimension and Fractal Geometry*, vol. 222 (Cambridge University Press, 2020)
6. Felix Hausdorff, Dimension und äußeres maß. Mathematische Annalen **79**(1), 157–179 (1918)
7. P.R. Massopust, *Fractal Functions, Fractal Surfaces, and Wavelets* (Academic Press, 2014)
8. Georg Cantor, Über unendliche, lineare punktmannichfaltigkeiten. Mathematische Annalen **20**(1), 113–121 (1882)
9. J.F. Fleron, A note on the history of the cantor set and cantor function. Math. Mag. **67**(2), 136–140 (1994)
10. H.V. Koch, Sur une courbe continue sans tangente, obtenue par une construction géométrique élémentaire. Arkiv for Matematik, Astronomi och Fysik **1**, 681–704 (1904)
11. Warclaw Sierpinski, Sur une courbe dont tout point est un point de ramification. CR Acad. Sci. **160**, 302–305 (1915)
12. E. Conversano, L.T. Lalli, Sierpinski triangles in stone on medieval floors in Rome. J. Appl. Math. **4**, 114–122 (2011)
13. Paola Magrone, Sierpinski's curve: a (beautiful) paradigm of recursion. Slov. Časopis Pre Geom. Graf **17**, 17–28 (2020)
14. M.-Á. Martínez-Cruz, J. Patiño-Ortiz, M. Patiño-Ortiz, A.S. Balankin, Some insights into the sierpiński triangle paradox. Fractal Fract. **8**(11), 655 (2024)
15. K. Weierstrass, Uber continuirliche functionen eines reellen arguments, die fur keinen worth des letzteren einen bestimmten differentailqutienten besitzen, akademievortrag. Math. Werke 71–74 (1872)
16. M.V. Berry, Z.V. Lewis, J.F. Nye, On the weierstrass-mandelbrot fractal function. Proc. R. Soc. London. A. Math. Phys. Sci. **370**(1743), 459–484 (1980)
17. Brian Hunt, The Hausdorff dimension of graphs of Weierstrass functions. Proc. Am. Math. Soc. **126**(3), 791–800 (1998)
18. T. Takagi, A simple example of the continuous function without derivative, in *Collected Papers*, pp. 5–6 (Springer, Berlin 1990)
19. J.E. Hutchinson, Fractals and self similarity. Indiana Univ. Math. J. **30**(5), 713–747 (1981)
20. M.F. Barnsley, *Fractals Everywhere* (Academic Press, 1993)
21. M.F. Barnsley, Fractal functions and interpolation. Constr. Approx. **2**, 303–329 (1986)
22. J. Henrikson, Completeness and total boundedness of the Hausdorff metric. MIT Undergrad. J. Math. **1**(69–80), 10 (1999)

Chapter 2
Fractal Functions—An Overview

In this chapter, the fractal functions constructed as attractors of iteration function systems, referred as *fractal interpolation functions*, are discussed in detail. In Chap. 1, some examples of fractal functions such as Weierstrass function and Takagi function are discussed, they are called fractal functions due to their non-differentiable nature and non-integer fractal dimension. However, they are not constructed using the IFS theory. In 1986, Barnsley [1] has developed the generalized interpolation functions with fractal characteristics and has beautifully narrated their greater efficiency in comparison with the classical (Euclidean) interpolation functions. In general, interpolation is a technique to estimate the missing value of a function $f : [a, b] \to \mathbb{R}$ at any given point $x \in [a, b]$ such that whose graph passes through all the given finite number of points $a = x_1 < x_2 < \cdots < x_N = b$. Suppose an experiment gives the data points $\{(0, 0.1), (0.25, 1.1), (0.5, -0.4), (0.75, 0.8), (1, 0.3)\}$ to be investigated. Then, the traditional method for analyzing the data begins with the graphical representation. For instance, a linear interpolation and polynomial fit with degree 4 are used for the approximation and the process is illustrated in Fig. 2.1a and b.

In general, there are many significant real-world phenomena such as climate patterns, EEG signals, fluctuations of gold price and stock price showing high irregularity, where the classical methods fail to depict their fine geometric patterns. For instance, if a climate data is collected $\{(x_n, y_n) : n = 1, 2, \ldots, N\}$, where N is a positive integer greater than 2, x_i denote the time and y_i denote the temperature, as a function of time. Then, a straight line segment or a polynomial of as low degree as possible do not give a useful approximation to the graph of collected data. In order to remedy this kind of approximation or interpolation of wild experimental data, fractal interpolation functions are developed with fractal characters and their fractal dimension aids to measure the roughness of the given data. Commencing with the discussion of classical fractal interpolation function, various types of fractal interpolation functions are encountered in this chapter with appropriate examples. For more information on fractal interpolation functions and their fractal dimension, reader may refer [1–8].

A. Gowrisankar et al., *Fractal Functions Communication with Fractional Calculus*,
https://doi.org/10.1007/978-3-031-96993-5_2

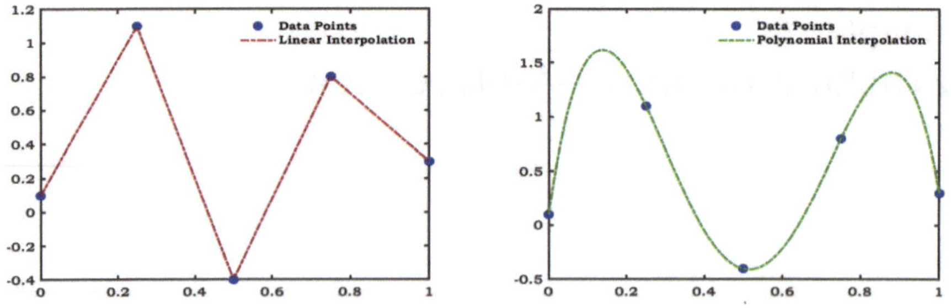

(a) Linear interpolation (b) Polynomial interpolation

Fig. 2.1 Graphical illustration of **a** linear interpolation and **b** polynomial fit of the given data points

2.1 Foundation of Fractal Interpolation Function

Let $\{(x_n, y_n) \in \mathbb{R}^2 : n = 1, 2, \ldots, N\}$ be the given set of data, where $x_1 < x_2 < \cdots < x_N$ with difference $x_{n+1} - x_n$ not necessarily equal, for all $n = 1, 2, \ldots, N-1$. Let $I = [x_1, x_N] \subset \mathbb{R}$ and $I_n = [x_n, x_{n+1}]$, for $n = 1, 2, \ldots, N-1$. Fractal interpolation looks for a continuous function $f : [x_1, x_N] \to \mathbb{R}$ satisfying the equation

$$f(x_n) = y_n, \ \forall\, n = 1, 2, \ldots, N,$$

and whose graph is obtained as an attractor of an IFS. The fundamental ingredients to construct a hyperbolic IFS are

(*i*) a complete metric space, taken as $X := I \times \mathbb{R}$ with Euclidean metric or its equivalent metric, and
(*ii*) a finite number of contraction mappings, taken as $w_n : X \to I_n \times \mathbb{R}$, $n = 1, 2, \ldots, N-1$.

Hereafter, we only use hyperbolic IFS hence, from here onward we drop the adjective 'hyperbolic' and the IFS reserves the notation $\{X; w_n : n = 1, 2, \ldots, N-1\}$, where the map w_n is described by

$$w_n(x, y) = (L_n(x), F_n(x, y)), \ \forall\, n = 1, 2, \ldots, N-1$$

with $L_n : I \to I_n$ and $F_n : X \to \mathbb{R}$. To contract the x-axis, the maps L_n are defined to be contraction homeomorphisms satisfying

$$\begin{aligned} |L_n(a_1) - L_n(a_2)| \le r|a_1 - a_2|, \ 0 \le r < 1, \ a_1, a_2 \in I, \\ L_n(x_1) = x_n, \ L_n(x_N) = x_{n+1}, \ n = 1, 2, \ldots, N-1. \end{aligned} \tag{2.1}$$

Meanwhile, to contract the y-axis the maps F_n are defined to be contractions in the second argument satisfying

$$|F_n(b, c_1) - F_n(b, c_2)| \le s|c_1 - c_2|,\ 0 \le s < 1,\ b \in I,\ c_1, c_2 \in \mathbb{R},$$
$$F_n(x_1, y_1) = y_n,\ F_n(x_N, y_N) = y_{n+1},\ n = 1, 2, \ldots, N-1. \tag{2.2}$$

The general form of the maps L_n and F_n are provided as

$$L_n(x) = a_n x + b_n,$$
$$F_n(x, y) = \alpha_n y + q_n(x),\ n = 1, 2, \ldots, N-1. \tag{2.3}$$

Here a_n and b_n are estimated using the endpoint conditions of L_n provided in Eq. (2.1),

$$a_n = \frac{x_{n+1} - x_n}{x_N - x_1},\quad b_n = \frac{x_N x_n - x_1 x_{n+1}}{x_N - x_1},\ \forall\, n = 1, 2, \ldots, N-1.$$

The free parameter $\alpha = (\alpha_1, \alpha_2, \ldots, \alpha_{N-1})$ is the scale vector satisfying $|\alpha_n| < 1$, as it scales the graph in vertical direction, it is referred as vertical scaling factor. In particular, when $\alpha_n = 0$ for all $n = 1, 2, \ldots, N-1$, then the obtained function is a piecewise linear interpolation function, connecting each two consecutive points with a straight line. The function q_n is a suitable continuous function and its choice must obey the Eq. (2.2).

In association with the IFS

$$\{X; w_n : n = 1, 2, \ldots, N-1\} \text{ or } \{(L_n(x), F_n(x, y)) : n = 1, 2, \ldots, N-1\}, \tag{2.4}$$

define the map $W : \mathcal{H}(X) \to \mathcal{H}(X)$ as

$$W(C) = \bigcup_{n=1}^{N-1} w_n(C),$$

for any $C \in \mathcal{H}(X)$, where $\mathcal{H}(X)$ denotes the collection of all non-empty compact subsets of X. Since, W is a contraction on the complete Hausdorff metric space $\mathcal{H}(X)$, then by the Banach contraction principle it has a unique fixed point $\mathcal{G}_f$ such that $\mathcal{G}_f = W(\mathcal{G}_f)$. Here, Barnsley has taken the ideology of employing this map W as a finite union of compact sets on a complete metric space for obtaining an invariant set $\mathcal{G}_f$ which is introduced by Hutchinson in the article [9]. This attractor $\mathcal{G}_f := \{(x, f(x)) : x \in I\}$ is the graph of the continuous function $f : I \to \mathbb{R}$ satisfying $f(x_n) = y_n$ for all $n = 1, 2, \ldots, N$. Barnsley [1] refer this function f as the *Fractal Interpolation Function* (FIF) or *fractal function* interpolating the given data set $\{(x_n, y_n) \in \mathbb{R}^2 : n = 1, 2, \ldots, N\}$.

To further discuss and analyze the properties of FIF f, it is obtained as a unique fixed point of the Read-Bajraktarević ($\mathcal{RB}$) operator. This operator is first investigated by Read [10] and later by Bajraktarević [11] in the domain of functional analysis. Then, Dubuc [12] and Bedford [13] have utilized this operator for generating fractal functions and in recent years, $\mathcal{RB}$ operator has been receiving significant attention for constructing types of FIFs in fractal literature. Now, let us briefly discuss the $\mathcal{RB}$ operator theory for generating FIF. Let $C^*(I)$ be the collection of functions $g : I \to \mathbb{R}$ which are necessarily continuous and satisfying $g(x_1) = y_1,\ g(x_N) = y_N$. The sup norm defined by $\|g\|_\infty = \max\{|g(x)| : x \in I\}$ completes $C^*(I)$. Now the $\mathcal{RB}$ operator $\mathcal{T} : C^*(I) \to C^*(I)$ is defined by

$$(\mathcal{T}g)(x) = F_n(L_n^{-1}(x), g(L_n^{-1}(x))),\ \forall\, x \in I_n,\ n = 1, 2, \ldots, N-1.$$

Employing the contractivity conditions of F_n, the operator $\mathcal{T}$ can be shown as a contraction. Then, it admits a unique fixed point, say f obeying

$$f(x) = F_n(L_n^{-1}(x), f(L_n^{-1}(x)),$$

(or)

$$f(L_n(x)) = F_n(x, f(x)) = \alpha_n f(x) + q_n(x),\ \forall\, x \in I_n,\ n = 1, 2, \ldots, N-1. \tag{2.5}$$

The continuity of f is guaranteed with the definition of $\mathcal{RB}$ operator on the space of continuous functions and this function f is our required fractal interpolation function associated with the IFS $\{X; w_n : n = 1, 2, \ldots, N-1\}$. Based on the particular choices of q_n, different types of fractal interpolation functions can be generated, for instance, if q_n is a linear polynomial, then the associated FIF is a linear FIF and if q_n is a quadratic polynomial, then it generates a quadratic FIF.

2.2 Affine Fractal Functions

Let $L_n(x) = a_n x + b_n$. If $q_n(x)$ in Eq. (2.3) is chosen to be a linear polynomial of the form $c_n x + d_n$, then $F_n(x, y) = \alpha_n y + (c_n x + d_n)$. The join-up conditions in Eqs. (2.1) and (2.2) estimate the parameters c_n and d_n as

$$\begin{aligned} c_n &= \frac{y_{n+1} - y_n}{x_N - x_1} - \alpha_n \frac{y_N - y_1}{x_N - x_1}, \\ d_n &= \frac{x_N y_n - x_1 y_{n+1}}{x_N - x_1} - \alpha_n \frac{x_N y_1 - x_1 y_N}{x_N - x_1},\ \forall\, n = 1, 2, \ldots, N-1. \end{aligned} \tag{2.6}$$

Now, the iterated function system expressed in Eq. (2.4) generates an *affine or linear fractal interpolation function* satisfying the functional equation given by

$$f(L_n(x)) = \alpha_n f(x) + c_n x + d_n,\ \forall\, x \in I_n,\ n = 1, 2, \ldots, N-1.$$

Example 2.1 In this example, we demonstrate the role of scale vector in the linear fractal interpolation functions which interpolates the same data set with different choice of scale vectors. For $n = 5$, let $\{(0, 0.1), (0.25, 1.1), (0.5, -0.4), (0.75, 0.8),$

$(1, 0.3)\}$ be the given data set. To interpolate the given data, linear fractal interpolation functions are generated with different scale vectors $(0, 0, 0, 0)$, $(0.1, 0.1, 0.1, 0.1)$, $(0.2, 0.2, 0.2, 0.2)$, $(0.4, 0.4, 0.4, 0.4)$. The choice of q_n as a linear polynomial $q_n(x) = c_n x + d_n$ in the Eq. (2.3), generates linear FIFs. For each distinct scale vector α, the values of parameters c_n and d_n can be determined using the Eq. (2.6).

In Fig. 2.2, the red dots represent the data points and blue curve represents the graph of FIF. From Fig. 2.2a, when $\alpha = (0, 0, 0, 0)$, graph of linear fractal interpolation is obtained as the attractor of following IFS:

$$
\begin{aligned}
L_1(x) &= 0.25x;\ F_1(x, y) = x + 0.1\\
L_2(x) &= 0.25x + 0.25;\ F_2(x, y) = -1.5x + 1.1\\
L_3(x) &= 0.25x + 0.5;\ F_3(x, y) = 1.2x - 0.4\\
L_4(x) &= 0.25x + 0.75;\ F_4(x, y) = -0.5x + 0.8
\end{aligned}
$$

and it coincides with the linear interpolation function demonstrated in Fig. 2.1a. On the other side, when the vertical scalings are gradually increased, the irregularity of the function also increases between each subintervals. Comparing the Fig. 2.2b, c and d, the steadily increasing non-smoothness can be observed while preserving the self-similar nature. For the same data set, now the scaling factors are chosen as $(0.8, 0.6, 0.4, 0.2)$ and $(-0.2, 0.7, -0.4, 0.5)$. That is, different scaling factor α_n for each sub-interval I_n satisfying $|\alpha_n| < 1$, for all n, which is contrary to the above discussed case of same scaling factor for the entire interval. The corresponding linear FIF with $\alpha = (0.8, 0.6, 0.4, 0.2)$ is obtained as an attractor of following IFS (2.7)

$$
\begin{aligned}
L_1(x) &= 0.25x;\ F_1(x, y) = 0.8y + 0.84x + 0.02\\
L_2(x) &= 0.25x + 0.25;\ F_2(x, y) = 0.6y - 1.62x + 1.04\\
L_3(x) &= 0.25x + 0.5;\ F_3(x, y) = 0.4y + 1.12x - 0.44\\
L_4(x) &= 0.25x + 0.75;\ F_4(x, y) = 0.2y - 0.54x + 0.78
\end{aligned}
\tag{2.7}
$$

and the linear FIF with $\alpha = (-0.2, 0.7, -0.4, 0.5)$ is obtained as an attractor of the IFS (2.8) expressed by

$$
\begin{aligned}
L_1(x) &= 0.25x;\ F_1(x, y) = -0.2y + 1.04x + 0.12\\
L_2(x) &= 0.25x + 0.25;\ F_2(x, y) = 0.7y - 1.64x + 1.03\\
L_3(x) &= 0.25x + 0.5;\ F_3(x, y) = -0.4y + 1.28x - 0.36\\
L_4(x) &= 0.25x + 0.75;\ F_4(x, y) = 0.5y - 0.6x + 0.75.
\end{aligned}
\tag{2.8}
$$

Their graphically illustrations are provided in Fig. 2.3a and b. Comparing Figs. 2.2 and 2.3, the shape of the graphs of linear FIFs vary for each different selection of scale vector showcasing the influence of scale vector.

In general, interpolation process increases the number of data points within the prescribed domain while determining the continuous function to interpolate

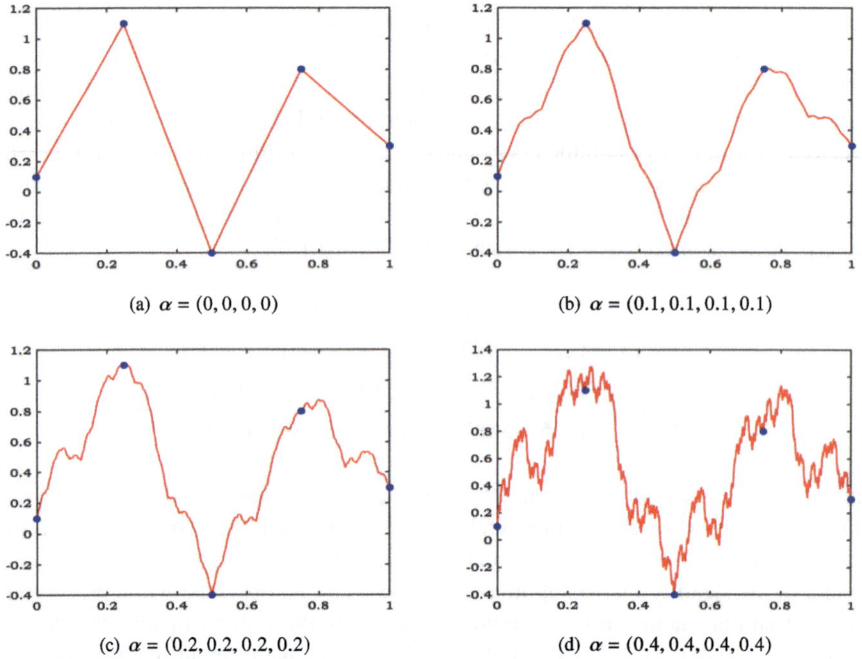

Fig. 2.2 Graphical illustration of linear fractal interpolation functions with different choice of scale vectors: **a** $\alpha_n = 0$, **b** $\alpha_n = 0.1$, **c** $\alpha_n = 0.2$ and **d** $\alpha_n = 0.4$, for all $n = 1, 2, 3, 4$

the known data set. Here, in the case of fractal interpolation, the new data points generation specifically depends on the iteration. Suppose $N + 1$ number of sample points are provided. In the first iteration, the contraction maps L_n yield $N - 1$ points in each of the N subintervals. As a result, at the end of first iteration,

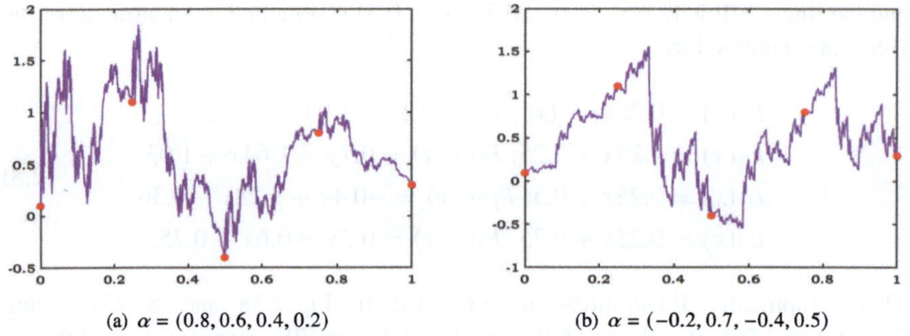

Fig. 2.3 Graphical illustration of quadratic fractal interpolation functions with two different choice of scale vectors: **a** $(0.8, 0.6, 0.4, 0.2)$ and **b** $(-0.2, 0.7, -0.4, 0.5)$

$N(N-1)+N+1=N^2+1$ new points are generated. Similarly, at the end of second iteration $N(N^2-1)+N+1=N^3+1$ distinct points are obtained. By induction method, at the end of nth iteration, the fractal interpolation function passes through a total number of $N^{n+1}+1$ new data points. It follows that the magnitude of the data set maximizes with every successive iteration on affine transformations. This process is clearly illustrated with the following example. Consider the data set $\{(0,0),(0.35,-0.2),(0.7,0.1),(1.05,-0.1)\}$. The linear FIF passing through the given data with scaling factors $\alpha_n=(0.5,0.4,0.3)$ is graphically demonstrated in Fig. 2.4. In Fig. 2.4a, blue lines represent the separation of 3 subintervals $[0,0.35]$, $[0.35,0.7]$, $[0.7,1.05]$ with 3 scaling factors for the provided 4 data points. The red lines indicate the number of data points generated after the first iteration within each 3 subintervals. Similarly, the number of subintervals divided and number of data points generated after second iteration are represented in black dotted lines in Fig. 2.4b.

2.3 Non-affine Fractal Functions

A non-affine fractal interpolation function can be generated by choosing $q_n(x)$ in Eq. (2.3) to be a polynomial or a continuous function with order greater than 1. If $L_n(x)=a_nx+b_n$ and $q_n(x)$ in Eq. (2.3) is chosen to be a quadratic polynomial of the form $x^2+e_nx+f_n$, then $F_n(x,y)=\alpha_ny+(x^2+e_nx+f_n)$. The iterated function system expressed in Eq. (2.4) generates a *quadratic fractal interpolation function* satisfying the functional equation given by

$$g(L_n(x))=\alpha_n g(x)+x^2+e_nx+f_n,\ \forall\, x\in I_n,\ n=1,2,\ldots,N-1.$$

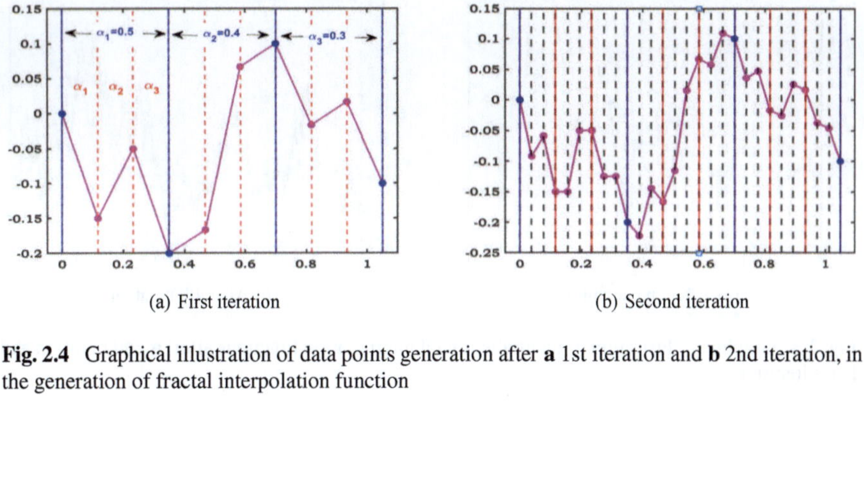

(a) First iteration

(b) Second iteration

Fig. 2.4 Graphical illustration of data points generation after **a** 1st iteration and **b** 2nd iteration, in the generation of fractal interpolation function

Example 2.2 Let $\{(0,0), (1/5, 1/4), (2/5, 2/5), (3/5, 1/2), (4/5, 3/4), (1, 0)\}$ be the given interpolation data with vertical scaling factors $\alpha_1 = 0.7$, $\alpha_2 = 0.4$, $\alpha_3 = -0.4$, $\alpha_4 = 0.5$, $\alpha_5 = 0.6$. Then, the quadratic fractal interpolation function corresponding to the above data determined by the IFS consists of following functions:

$$L_1(x) = 0.2x;\ F_1(x, y) = 0.7y + x^2 - 0.75x$$
$$L_2(x) = 0.2x + 0.2;\ F_2(x, y) = 0.4y + x^2 - 0.85x + 0.25$$
$$L_3(x) = 0.2x + 0.4;\ F_3(x, y) = -0.4y + x^2 - 0.9x + 0.4$$
$$L_4(x) = 0.2x + 0.6;\ F_4(x, y) = 0.5y + x^2 - 0.75x + 0.5$$
$$L_5(x) = 0.2x + 0.8;\ F_5(x, y) = 0.6y + x^2 - 1.75x + 0.75.$$

The graphical representation of the attractor (quadratic fractal interpolation function) after second and fourth iteration, associated with the above IFS, is depicted in Fig. 2.5. The graph in Fig. 2.5b looks dense than the graph in Fig. 2.5a, it is due to the more irregular nature as iteration increases.

In the article [1], while proposing the affine fractal interpolation, Barnsley has provided a lead for introducing a special kind of non-affine fractal interpolation called the α-fractal interpolation with the special choice of q_n as

$$q_n(x) = \phi(L_n(x)) - \alpha_n b(x). \tag{2.9}$$

where $\phi(x)$ and $b(x)$ are continuous functions respectively satisfying

$$\phi(x_n) = y_n,\ \forall\, n = 1, 2, \ldots, N,$$

and

$$b(x_1) = \phi(x_1),\ b(x_N) = \phi(x_N) \text{ and } b \neq \phi. \tag{2.10}$$

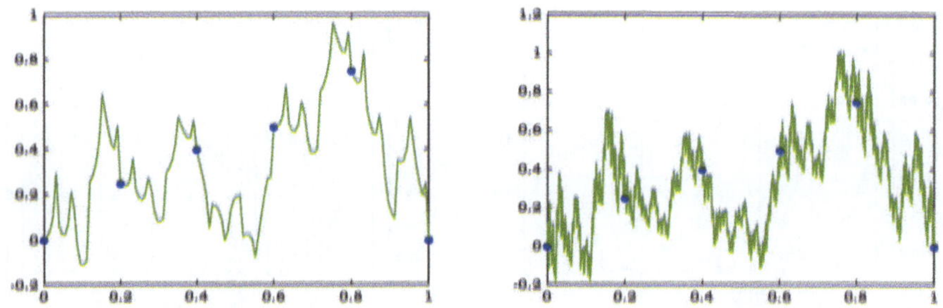

(a) After 2nd iteration (b) After 4th iteration

Fig. 2.5 Graphical illustration of quadratic fractal interpolation functions after: **a** 2nd iteration and **b** 4th iteration

The function $b(x)$ is called the base function as it shrinks the fractal graph horizontally and it can be chosen as any polynomials or splines obeying Eq. (2.10). In continuation to the remark of Barnsley [1], Navascués [3] has consolidated the construction of non-affine fractal interpolation function denoted by ϕ^α, as a *perturbation* of given continuous function ϕ. The IFS for generating ϕ^α consists of

- a partition of the original function's (ϕ) domain
- a scale vector associated with the partition
- a base function depending on the endpoints of ϕ.

Consider the partition $\Delta : a = x_1 < x_2 < \cdots < x_N = b$. To approximate the prescribed continuous function $g : [a, b] \to \mathbb{R}$, a family of fractal functions ϕ^α is invoked using

$$q_n(x) = \phi(L_n(x)) - \alpha_n b(x),$$

here the base function satisfies the constraints as provided in Eq. (2.10) with respect to ϕ. Then, the functional equation of ϕ^α is expressed by

$$\phi^\alpha(x) = \alpha_n \phi^\alpha \circ L_n^{-1}(x) + q_n \circ L_n^{-1}(x), \ \forall\, x \in I_n.$$

From the expression of q_n in Eq. (2.3),

$$\phi^\alpha(x) = \phi(x) + \alpha_n(\phi^\alpha - b) \circ L_n^{-1}(x), \ \forall\, x \in I_n. \tag{2.11}$$

The function ϕ^α satisfying Eq. (2.11) is referred as the α- *fractal interpolation function* of ϕ, refer [3]. If the base function is taken as a composition of prescribed function g with the continuously increasing function c satisfying $c(x_1) = x_1$ and $c(x_N) = x_N$, it follows that

$$\phi^\alpha(x) = \phi(x) + \alpha_n(\phi^\alpha - \phi \circ c) \circ L_n^{-1}(x), \ \forall\, x \in I_n.$$

The fractal perturbation ϕ^α mimics the behavior of g when each α_n is taken to be zero.

Proposition 2.1 ([14, 15]) *The α-fractal function ϕ^α corresponding to the given function ϕ obeys the inequality,*

$$\|\phi^\alpha - \phi\|_\infty \le \frac{|\alpha|_\infty}{1 - |\alpha|_\infty} \|\phi - b\|_\infty, \tag{2.12}$$

where $|\alpha|_\infty = \max_{1 \le n \le N-1}\{|\alpha_n|\}$. Also, ϕ^α interpolates g at each x, (i.e.),

$$\phi^\alpha(x_n) = \phi(x_n), \ \forall\, n = 1, 2, \ldots, N.$$

Consider the base function $b = \mathscr{L}\phi$, where

$$\mathscr{L} : C(I) \to C(I)$$

is a bounded linear operator with the sup norm $\|\phi\|_\infty = \max\{|\phi(x)| : x \in I\}$ on $C(I)$ satisfying $\mathscr{L}\phi(x_1) = x_1,\ \mathscr{L}\phi(x_n) = x_n,\ \mathscr{L} \neq Identity$. The operator norm is defined by $\|\mathscr{L}\| := \sup\{\|\mathscr{L}\phi\| : \|\phi\|_\infty \leq 1\}$.

Remark 2.1 If the operator $\mathscr{L}$ is taken as $\mathscr{L}\phi(x) = \phi \circ c(x)$ as discussed in [8], where c is an increasing continuous function satisfying $c(x_1) = x_1$ and $c(x_n) = x_n$, then

$$\|\mathscr{L}\phi\|_\infty = \|\phi \circ c\|_\infty = \|\phi\|_\infty$$

with $\|\mathscr{L}\| = 1$. Suppose the base function b is chosen as $\phi \circ c$, then from the above proposition, it follows that

$$\|\phi^\alpha - \phi\|_\infty \leq \frac{2|\alpha|_\infty}{1 - |\alpha|_\infty}\|\phi\|_\infty. \tag{2.13}$$

Further, in the article [3], Navascués has connected the fractal interpolation theory with the operator theory by proposing an α-fractal operator on the space of continuous functions. The α-fractal operator, with respect to the partition Δ and the continuous function c, is defined by

$$\begin{aligned} \mathcal{F}^\alpha_{\Delta,c} : C(I) &\to C(I) \\ \phi &\hookrightarrow \phi^\alpha. \end{aligned}$$

In a wider sense, the operator provides a family of fractal interpolation functions (ϕ^α) to generalize each continuous function (ϕ) defined in a real, closed and bounded interval.

Note 2.1 The operator $\mathcal{F}^\alpha$ converges to the identity when α approaches to zero. The bounds of the norm of operator $\mathcal{F}^\alpha$ [3] is expressed in terms of scale vector α as

$$1 \leq \|\mathcal{F}^\alpha\| \leq \frac{1 + |\alpha|_\infty}{1 - |\alpha|_\infty},$$

where $|\alpha|_\infty = \max\{|\alpha_n| < 1 : n = 1, 2, \ldots, N-1\}$.

In sequel to the fractal perturbation discussed in [3], a series of works are scrutinized in relation with the α-fractal function by Navascués, see for instance [8, 16].

Example 2.3 For the given function $\phi(x) = \sqrt{x}$, let us take the following set of interpolation data $\{(0, 0), (1/4, 1/2), (1/2, 1/\sqrt{2}), (3/4, \sqrt{3}/2), (1, 1)\}$. Choose the scaling factors $\alpha_n = (0.5, -0.3, 0.1, -0.1)$ such that $|\alpha_n| < 1$. To illustrate the effects of base function in α-FIFs, here we take three different base functions $b(x)$ as given below:
(a) $b(x) = x$, line passing through the end points $(0, 0)$ and $(1, 1)$,
(b) $b(x) = (\phi \circ c)(x)$, where $c(x) = (e^{\mu x} - 1)/(e^\mu - 1)$, $\mu = 1$,

(c) $b(x) = u(x)\phi(x)$, where $u(x) = \cos(\pi x/2) + \sin(\pi x/2)$,
(d) $b(x) = \sin(\pi x/2)$.

Observe that all the above three described base functions have been chosen as continuous functions and clearly which satisfy the stated underlying constraint in Eq. (2.10) with respect to $\phi(x)$. The graphs of α-FIF ϕ^α corresponding to the given $\phi(x)$ are portrayed in Fig. 2.6a, b, c and d. The red curve is the original continuous function ϕ and blue curve represents the interpolant ϕ^α. The difference in the shapes of the graphs and varying irregularity is due to the different selection of base functions. This graphical illustration shows that depending on the prescribed continuous function $\phi(x)$, the base function can be chosen appropriately obeying the condition Eq. (2.10) for better approximation despite the scaling factors selection.

Remark 2.2 While constructing an IFS with maps L_n and F_n as defined in Eq. (2.3), suppose $q_n(x)$ is chosen to be a rational function of the form,

$$q_n(x) = \frac{U_{n,M}}{V_{n,N}}.$$

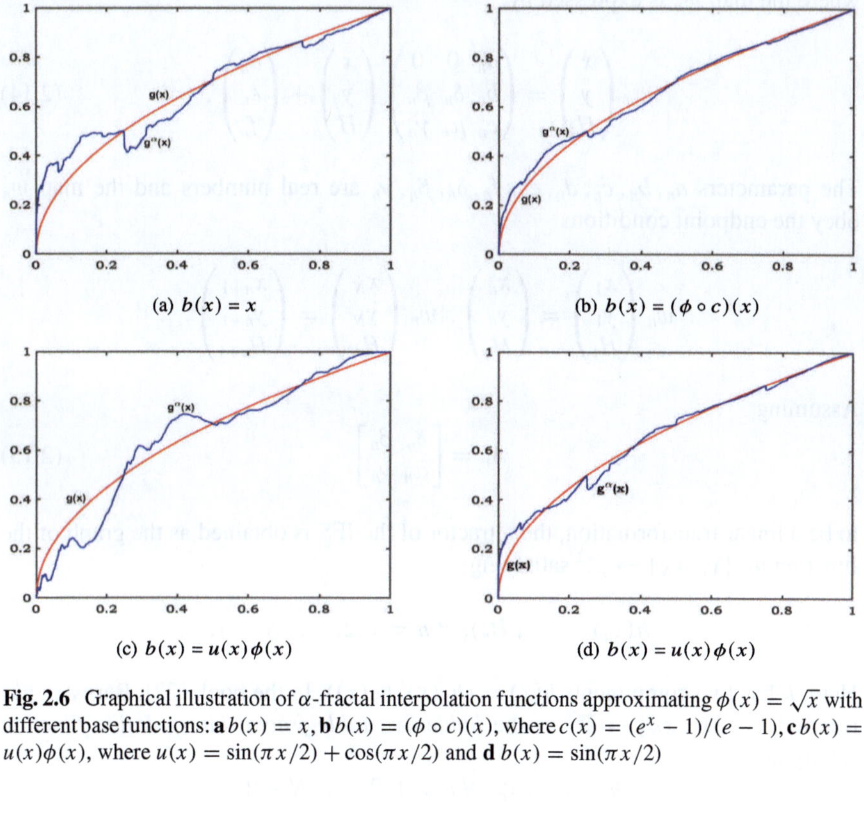

Fig. 2.6 Graphical illustration of α-fractal interpolation functions approximating $\phi(x) = \sqrt{x}$ with different base functions: **a** $b(x) = x$, **b** $b(x) = (\phi \circ c)(x)$, where $c(x) = (e^x - 1)/(e - 1)$, **c** $b(x) = u(x)\phi(x)$, where $u(x) = \sin(\pi x/2) + \cos(\pi x/2)$ and **d** $b(x) = \sin(\pi x/2)$

Here $U_{n,M}$ is a polynomial of degree M and $V_{n,N}$ is a polynomial of degree N such that $V_{n,N}(x) \neq 0,\ \forall x \in I$. Then, the resulting FIF is referred as rational fractal interpolation function (RFIF) [17, 18]. For more works on rational fractal interpolation functions, refer [19, 20].

2.4 Fractal Functions in Higher Dimension

In sequel to the above discussed affine fractal interpolation function, Barnsley and Massopust [21] have introduced the *hidden variable fractal interpolation function* (HFIF). The HFIF contains many hidden variables to offer more flexibility to the interpolant. The generalized fractal interpolation HFIF is constructed by defining an IFS on higher dimensional space $\mathbb{R}^3$ and project its graph into $\mathbb{R}^2$. The data set in $\mathbb{R}^2$ is generalized by working in $\mathbb{R}^3$ with the addition of "hidden" variables providing additional degrees of freedom. Consider an IFS

$$\{\mathbb{R}^3; w_n : n = 1, 2, \ldots, N-1\},$$

where the map w_n is expressed by

$$w_n \begin{pmatrix} x \\ y \\ H \end{pmatrix} = \begin{pmatrix} a_n & 0 & 0 \\ b_n & \delta_n & \beta_n \\ c_n & \mu_n & \gamma_n \end{pmatrix} \begin{pmatrix} x \\ y \\ H \end{pmatrix} + \begin{pmatrix} d_n \\ e_n \\ f_n \end{pmatrix}. \tag{2.14}$$

The parameters $a_n, b_n, c_n, d_n, e_n, f_n, \delta_n, \beta_n, \gamma_n$ are real numbers and the map w_n obey the endpoint conditions

$$w_n \begin{pmatrix} x_1 \\ y_1 \\ H_1 \end{pmatrix} = \begin{pmatrix} x_n \\ y_n \\ H_n \end{pmatrix},\ w_n \begin{pmatrix} x_N \\ y_N \\ H_N \end{pmatrix} = \begin{pmatrix} x_{n+1} \\ y_{n+1} \\ H_{n+1} \end{pmatrix}.$$

Assuming

$$A_n = \begin{bmatrix} \delta_n & \beta_n \\ \mu_n & \gamma_n \end{bmatrix} \tag{2.15}$$

to be a linear transformation, the attractor of the IFS is obtained as the graph of the function $h : [x_1, x_N] \to \mathbb{R}^2$ satisfying

$$h(x_n) = (y_n, H_n),\ \forall n = 1, 2, \ldots, N-1.$$

Here, h has two components $h(x) = (h_1(x), h_2(x))$. In the book [22], Barnsley has referred $h_1 : [x_1, x_N] \to \mathbb{R}^2$ as the *hidden variable fractal interpolation function* satisfying

$$h_1(x_n) = y_n,\ \forall n = 1, 2, \ldots, N-1.$$

He added that the graph of h_1 is 'not self-similar or self-affine or self-anything' since it is not the union of affine transformations of itself. There is no special interpretation on h_2 as both h_1 and h_2 behave in same manner from the affine transformation given in Eq. (2.14).

Later in the year 2007, Chand and Kapoor [5] have developed the *coalescence hidden variable fractal interpolation function* (CHFIF) as a self-affine function under certain conditions, in contrast to Barnsley's statement of 'the graph of HFIF is not self-similar or self-affine or self-anything'. CHFIF is constructed as the graph of the function h interpolating $\{(x_n, y_n) : n = 1, 2, \ldots, N\}$ such that it is the projection of an invariant subset of a higher dimensional space $\mathbb{R}^3$. Let $\{(x_n, y_n, z_n) \in I \times \mathbb{R}^2 : n = 1, 2, \ldots, N\}$ be the generalized data set, where $\{z_n : n = 1, 2, \ldots, N\}$ are real parameters (hidden variables). For $n = 1, 2, \ldots, N-1$, let $L_n : I \to I_n$ be the contraction homeomorphism satisfying (2.1), and the function $F_n : I \times \mathbb{R}^2 \to \mathbb{R}^2$ be defined as

$$F_n(x, y, z) = A_n(y, z)^t + \big(p_n(x), q_n(x)\big)^t.$$

where t denotes the transpose, A_n is an upper-triangular matrix $\begin{bmatrix} \delta_n & \beta_n \\ 0 & \gamma_n \end{bmatrix}$ and the functions $p_n(x)$ and $q_n(x)$ are continuous with at least two unknown co-efficients. Note that in the previously assumed A_n in Eq. (2.15), the element a_{21} is μ_n which is not necessarily zero, whereas here $a_{21} = 0$. Further, δ_n is a free variable satisfying $|\delta_n| < 1$ and β_n is a constrained free variable obeying $|\beta_n| + |\gamma_n| < 1$. Then the attractor $\mathcal{G}_h$ of the generalized IFS

$$\{\mathbb{R}^3; (L_n(x), F_n(x, y, z)) : n = 1, 2, \ldots, N-1\} \tag{2.16}$$

is the graph of the continuous vector valued function h obeying $h(x_n) = (y_n, z_n)$. The first component h_1 satisfies $h_1(x_n) = y_n$ and the second component h_2 satisfies $h_2(x_n) = z_n$. In the article [5] by Chand et al., the function h_1 is called as a *coalescence hidden variable fractal interpolation* which interpolates the data $\{(x_n, y_n) : n = 1, 2, \ldots, N\}$. The function h_1 is the projection of the attractor $\mathcal{G}_h$ and it is not always the union of affine transformations of itself. Therefore, in general, CHFIFs are non-self-affine like the Barnsley's hidden variable fractal interpolation function. However, as a particular case, when one chooses

$$z_n = y_n, \quad \delta_n + \beta_n = \gamma_n, \quad p_n = q_n, \tag{2.17}$$

then the associated CHFIF h_1 is obtained as the projection of the attractor of IFS (2.16) and coincides with the self-affine fractal function h_2. The $\mathcal{RB}$ theory provides the functional equation representations for h_1 and h_2 as follows,

$$\begin{aligned} h_1(L_n(x)) &= \delta_n h_1(x) + \beta_n h_2(x) + p_n(x), \\ h_2(L_n(x)) &= \gamma_n h_2(x) + q_n(x), \ \forall\, n = 1, 2, \ldots, N-1, \end{aligned}$$

here p_n and q_n are continuous functions. The CHFIFs can be effectively used to capture both self-affine and non-self-affine nature present in the real-world phenomena simultaneously.

Remark 2.3 In the construction of CHFIF, suppose one chooses A_n as a lower triangular matrix (i.e.,)

$$A_n = \begin{bmatrix} \delta_n & 0 \\ \beta_n & \gamma_n \end{bmatrix}$$

instead of an upper triangular matrix, then h_1 is a self-affine fractal interpolation function and h_2 is a coalescence hidden variable fractal interpolation function. Also, under the conditions, $z_n = y_n$ and $\beta_n + \gamma_n = \delta_n$, the function h_2 behaves like the self-affine fractal function h_1.

Example 2.4 Let$(x_n, y_n) = \{(0, 0), (0.25, 0.1), (0.5, -0.4), (0.75, -1.2), (1, 0.3)\}$ be the provided data set in $\mathbb{R}^2$. With the inclusion of hidden variables $z_n = \{0, 0.1, 1.1, 0.5, -0.3\}$, the provided data is extended to the higher dimensional space $\mathbb{R}^3$ and the generalized data set is

$$\{(0, 0, 0), (0.25, 0.1, 0.1), (0.5, -0.4, 1.1), (0.75, -1.2, 0.5), (1, 0.3, -0.3)\}. \tag{2.18}$$

The scaling factors in the upper triangular matrix are taken as $\delta = \{0.2, 0.3, 0.5, 0.6\}$, $\beta = \{0.3, 0.4, 0.2, 0.3\}$, $\gamma = \{0.5, 0.1, 0.4, 0.4\}$ such that the constraints $|\delta_n| < 1$ and $|\beta_n| + |\gamma_n| < 1$ are satisfied. The graphs of coalescence hidden variable fractal function h_1 interpolating (x_n, y_n) and self-affine fractal function h_2 interpolating (x_n, z_n) are generated using the below functions:

$$L_1(x) = 0.25x;\ F_1^1(x, y, z) = 0.2y + 0.3z + 0.13x,\ F_1^2(x, z) = 0.5z + 0.25x$$
$$L_2(x) = 0.25x + 0.2;\ F_2^1(x, y, z) = 0.3y + 0.4z - 0.47x + 0.1,\ F_2^2(x, z) = 0.1z + 1.03x + 0.1$$
$$L_3(x) = 0.25x + 0.25;\ F_3^1(x, y, z) = 0.5y + 0.2z - 0.89x - 0.4,\ F_3^2(x, z) = 0.4z - 0.48x + 1.1$$
$$L_4(x) = 0.25x + 0.5;\ F_4^1(x, y, z) = 0.6y + 0.3z + 1.41x - 1.2,\ F_4^2(x, z) = 0.4z - 0.68x + 0.5.$$

Their graphs are portrayed in Fig. 2.7a and b. Though both the graphs look irregular, the nature of self-similarity varies depending on the scaling factors associated with the functions h_1 and h_2. Now, to illustrate the special case as discussed in Eq. (2.17), consider a new generalized set of data

$$\{(0, 0, 0), (0.25, 0.1, 0.1), (0.5, -0.4, -0.4), (0.75, -1.2, -1.2), (1, 0.3, 0.3)\} \tag{2.19}$$

such that $z_n = y_n$. With the same scaling factors discussed above $\delta = \{0.2, 0.3, 0.5, 0.6\}$, $\beta = \{0.3, 0.4, 0.2, 0.3\}$, $\gamma = \{0.5, 0.1, 0.4, 0.4\}$, the graphs of h_1 and h_2 do not coincide since the scaling condition is not satisfied, observe the Fig. 2.7c. On the other hand, when the scaling factors are chosen as

$$\delta_n = (0.2, 0.3, 0.5, 0.6),\ \beta_n = (0.3, 0.4, 0.2, 0.3),\ \gamma_n = (0.5, 0.7, 0.7, 0.9)$$

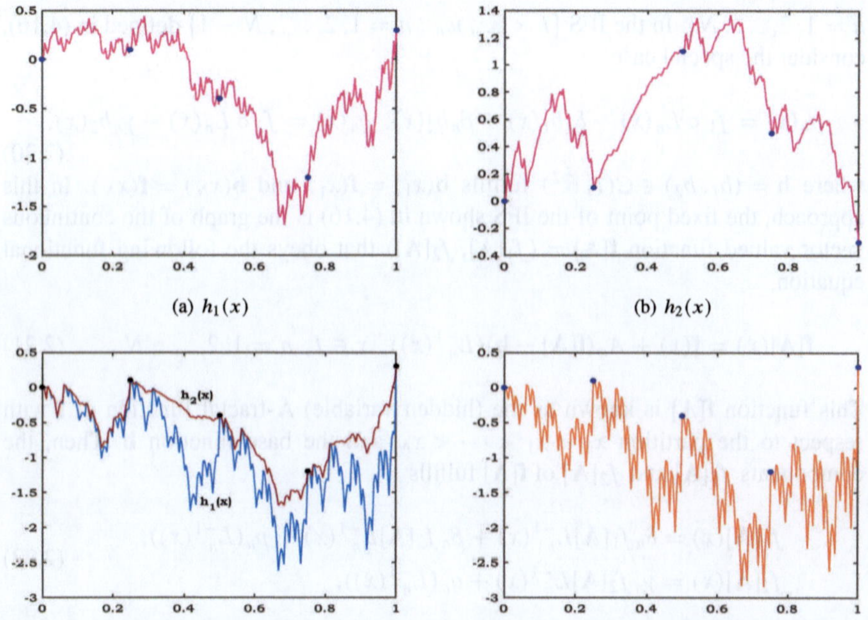

(a) $h_1(x)$ (b) $h_2(x)$

(c) $h_1(x) \neq h_2(x)$ (d) $h_1(x) = h_2(x)$

Fig. 2.7 Graphical illustration of **a** non-self-affine coalescence hidden variable fractal interpolation function $h_1(x)$, **b** self-affine fractal interpolation function $h_2(x)$ associated with the data set $\{(0, 0, 0), (0.25, 0.1, 0.1), (0.5, -0.4, 1.1), (0.75, -1.2, 0.5), (1, 0.3, -0.3)\}$, **c** $h_1(x) \neq h_2(x)$ associated with the data set $\{(0, 0, 0), (0.25, 0.1, 0.1), (0.5, -0.4, -0.4), (0.75, -1.2, -1.2), (1, 0.3, 0.3)\}$ such that $z_n = y_n$ but $\delta_n + \beta_n \neq \gamma_n$ and **d** $h_1(x) = h_2(x)$ associated with the data set $\{(0, 0, 0), (0.25, 0.1, 0.1), (0.5, -0.4, -0.4), (0.75, -1.2, -1.2), (1, 0.3, 0.3)\}$ such that $z_n = y_n$ and $\delta_n + \beta_n = \gamma_n$

such that $\delta_n + \beta_n = \gamma_n$, for all $n = 1, 2, 3, 4$. Then both the functions p_n and q_n are taken as same linear polynomials, the graph of h_1 coincides with the graph of h_2 as illustrated in Fig. 2.7d. The choice of data set (2.18) and its associated scalings do not obey the Eq. (2.17), hence the graphs of h_1 in Fig. 2.7a and h_2 in Fig. 2.7b are not same. On the other hand, Fig. 2.7d is the graphical illustration of h_1 which exactly matches with the graph of h_2 since the choice of its data (2.19) and its corresponding scalings agree the Eq. (2.17).

Now let us see the construction of an A-fractal function corresponding to continuous $\mathbb{R}^2$-valued function. A-fractal function is constructed by blending the α-fractal function and the hidden variable fractal interpolation function. A continuous function $\mathbf{f} : I \to \mathbb{R}^2$ is uttered in order to elicit a whole family of fractal functions $\mathbf{f}[\mathbf{A}]$ parametrized by a specific block matrix $\mathbf{A} = [A_n], n = 1, 2, \ldots, N-1$ with $A_n = \begin{bmatrix} \delta_n & \beta_n \\ 0 & \gamma_n \end{bmatrix}$, take a note of the fact $\mathbf{f}[\mathbf{A}] = \mathbf{f}$ when $\mathbf{A} = \mathbf{0}$. Consider the partition $x_1 < x_2 < \ldots < x_N$ of I and the data set $D = \{(x_n, f_1(x_n), f_2(x_n)) :$

$n = 1, 2, \ldots, N\}$. In the IFS $\{I \times \mathbb{R}^2; w_n : n = 1, 2, \ldots, N-1\}$ defined in (4.16), consider the special case,

$$p_n(x) = f_1 \circ L_n(x) - \delta_n b_1(x) - \beta_n b_2(x), \; q_n(x) = f_2 \circ L_n(x) - \gamma_n b_2(x), \tag{2.20}$$

where $\mathbf{b} = (b_1, b_2) \in C(I, \mathbb{R}^2)$ fulfills $\mathbf{b}(x_1) = \mathbf{f}(x_1)$ and $\mathbf{b}(x_N) = \mathbf{f}(x_N)$. In this approach, the fixed point of the IFS shown in (4.16) is the graph of the continuous vector valued function $\mathbf{f}[\mathbf{A}] = (f_1[\mathbf{A}], f_2[\mathbf{A}])$ that obeys the following functional equation,

$$\mathbf{f}[\mathbf{A}](x) = \mathbf{f}(x) + A_n(\mathbf{f}[\mathbf{A}] - \mathbf{b})(L_n^{-1}(x)), \; x \in I, \; n = 1, 2, \ldots, N. \tag{2.21}$$

This function $\mathbf{f}[\mathbf{A}]$ is known as the (hidden variable) $\mathbf{A}$-fractal function of $\mathbf{f}$ with respect to the partition $x_1 < x_1 < \cdots < x_N$ and the base function $\mathbf{b}$. Then, the components $f_1[\mathbf{A}]$ and $f_2[\mathbf{A}]$ of $\mathbf{f}[\mathbf{A}]$ fulfills

$$\begin{aligned} f_1[\mathbf{A}](x) &= \delta_n f_1[\mathbf{A}]L_n^{-1}(x) + \beta_n f_2[\mathbf{A}]L_n^{-1}(x) + p_n(L_n^{-1}(x)), \\ f_2[\mathbf{A}](x) &= \gamma_n f_2[\mathbf{A}]L_n^{-1}(x) + q_n(L_n^{-1}(x)), \end{aligned} \tag{2.22}$$

The functions $f_1[\mathbf{A}]$ and $f_2[\mathbf{A}]$ are non self-referential and self-referential functions respectively. For any choice of the block matrix $\mathbf{A}$ and for any choice of the base function $\mathbf{b}$ satisfying the above mentioned conditions, it is observed that $\mathbf{f}[\mathbf{A}](x_n) = \mathbf{f}(x_n), \; \forall \, n = 1, 2, \ldots, N$. Hence, the function $\mathbf{f}[\mathbf{A}]$ can be called as the fractal generalization of the continuous function $\mathbf{f}$.

When the parameters in the block matrix are taken as continuous functions from $[0, 1]$ to $\mathbb{R}$, then the Eq. (2.22) becomes,

$$\begin{aligned} f_1[\mathbf{A}](x) &= \alpha_n(L_n^{-1}(x)) f_1[\mathbf{A}]L_n^{-1}(x) + \beta_n(L_n^{-1}(x)) f_2[\mathbf{A}]L_n^{-1}(x) + p_n(L_n^{-1}(x)), \\ f_2[\mathbf{A}](x) &= \gamma_n(L_n^{-1}(x)) f_2[\mathbf{A}]L_n^{-1}(x) + q_n(L_n^{-1}(x)). \end{aligned} \tag{2.23}$$

The first and second equations in (2.23) are the functional equations of non self-referential fractal function $f_1[\mathbf{A}]$ and self-referential fractal function $f_2[\mathbf{A}]$.

2.5 Fractal Dimension of Fractal Functions

Fractal dimension is an index to quantify how densely an object occupies the space where it lies and to measure how rough/irregular the arbitrary functions. It is practically attached to the real-world phenomena, for example, the fractal dimension of the coastline of Great Britain is determined to be 1.2 (see [22]). The most commonly used fractal dimension is the *box counting dimension*, simply called as box-dimension, for more details refer [23]. The other famous fractal dimension is the Hausdorff dimension, however it is subtle and more complex than the box-dimension. It is generally not used for the experimental procedures in determining the fractal dimension of physical sets. In [1], Barnsley has experimentally employed box-dimension to determine the fractal dimension of fractal interpolation functions based on their scaling factor α_n. Let f be the fractal interpolation function associated with the following affine transformations

$$w_n\begin{pmatrix}x\\y\end{pmatrix}=\begin{pmatrix}a_n & 0\\ c_n & \alpha_n\end{pmatrix}\begin{pmatrix}x\\y\end{pmatrix}+\begin{pmatrix}b_n\\d_n\end{pmatrix},\ n=1,2,\dots,N-1.$$

If $\gamma=\sum_{n=1}^{N-1}|\alpha_n|>1$ and the interpolation points are not collinear, then the fractal dimension of the attractor $\mathcal{G}_f$ is the unique solution D of

$$\sum_{n=1}^{N-1}|\alpha_n|a_n^{D-1}=1.$$

Otherwise the fractal dimension of $\mathcal{G}_f$ is one, here $\mathcal{G}_f$ represents the graph of the function f. Suppose the interpolation points are provided with equal spacing and $a_n=1/N$, then it follows that

$$D=1+\frac{\log(\gamma)}{\log(N)}. \tag{2.24}$$

The above delightful formula clarifies that the fractal dimension of fractal interpolation function is less than two, meanwhile greater than one since $\gamma>1$. However, depending on γ, it is possible to increase D arbitrarily closer to 2 or can vary it smoothly down to 1. The value of fractal dimension D is independent of the y_n values apart from the non-collinear condition.

Note 2.2 Suppose the interpolation data points are collinear and the graph of the function f is a line segment connecting the initial point (x_1, y_1) to the end point (x_N, y_N), then its fractal dimension is 1. Similarly, the fractal dimension becomes one when $\gamma\le 1$, for more information refer [22].

Example 2.5 In this example, let us estimate the fractal dimension of linear FIFs interpolating the data set $\{(0, 0.1), (0.25, 1.1), (0.5, -0.4), (0.75, 0.8), (1, 0.3)\}$. We take three different set of scale vectors $\alpha^1=(0,0,0,0)$, $\alpha^2=(0.1,0.1,0.1,0.1)$, $\alpha^3=(0.4,0.4,0.4,0.4)$ and $\alpha^4=(0.8,0.6,0.4,0.2)$. The graphs of linear FIFs, say f_1, f_2, f_3 and f_4 with scalings $\alpha^1,\alpha^2,\alpha^3$ and α^4 are respectively demonstrated in Figs. 2.2a, b, d and 2.3a. Note that the x_n's are equidistant, then using the formula provided in Eq. (2.24), the fractal dimensions of linear FIFs are estimated as follows:

$$D_{f_1}=1,\ \text{since } \gamma<1,$$
$$D_{f_2}=1,\ \text{since } \gamma<1,$$

$$D_{f_3} = 1 + \frac{\log(1.6)}{\log(5)} \approx 1.2920,$$
$$D_{f_4} = 1 + \frac{\log(2)}{\log(5)} \approx 1.4307.$$

The estimations demonstrate that the fractal dimensions can vary for FIFs depending on the choice of scaling factors, though they interpolate the same set of data points.

References

1. M.F. Barnsley, Fractal functions and interpolation. Constr. Approx. **2**, 303–329 (1986)
2. P.R. Massopust, *Fractal Functions, Fractal Surfaces, and Wavelets* (Academic Press, 2014)
3. M.A. Navascués, Fractal polynomial interpolation. Zeitschrift für Analysis und ihre Anwendungen **24**(2), 401–418 (2005)
4. M.F. Barnsley, J. Elton, D. Hardin, P. Massopust, Hidden variable fractal interpolation functions. SIAM J. Math. Anal. **20**(5), 1218–1242 (1989)
5. A.K.B. Chand, G.P. Kapoor, Smoothness analysis of coalescence hidden variable fractal interpolation functions. Int. J. Nonlinear Sci. **3**(1), 15–26 (2007)
6. P.R. Massopust, Vector-valued fractal interpolation functions and their box dimension. Aequationes Math. **42**, 1–22 (1991)
7. M.A. Navascués, A fractal approximation to periodicity. Fractals **14**(04), 315–325 (2006)
8. M.A. Navascués, Fractal approximation. Complex Anal. Oper. Theory **4**, 953–974 (2010)
9. J.E. Hutchinson, Fractals and self similarity. Indiana Univ. Math. J. **30**(5), 713–747 (1981)
10. A.H. Read, XXIV. The solution of a functional equation. Proc. R. Soc. Edinb. Sect. A: Math. **63**(4), 336–345 (1952)
11. M. Bajraktarevic, Sur une équation fonctionnelle. Glasnik Mat.-Fiz. I Astr. **12**(3), 201–205 (1957)
12. S. Dubuc, Interpolation through an iterative scheme. J. Math. Anal. Appl. **114**(1), 185–204 (1986)
13. T.J. Bedford, F.M. Dekking, M.S. Keane, Fractal image coding techniques and contraction operators (1992)
14. M.A. Navascués, M.V. Sebastián, Fractal splines. Monografías del Seminario Matemático García de Galdeano **33**, 161–168 (2006)
15. M.A. Navascués, M.V. Sebastián, Smooth fractal interpolation. J. Inequalities Appl. **1–20**, 2006 (2006)
16. M.A. Navascués, A fractal approximation to periodicity. Fractals **14**(04), 315–325 (2006)
17. A.K.B. Chand, N. Vijender, Monotonicity preserving rational quadratic fractal interpolation functions. Adv. Numer. Anal. (2014)
18. A.K.B. Chand, S.K. Katiyar, G. Saravana Kumar, A new class of rational fractal function for curve fitting. In: Proceeding of Computer Aided Engineering CAE (2013), pp. 78–93
19. K.R. Tyada, A.K.B. Chand, M. Sajid, Shape preserving rational cubic trigonometric fractal interpolation functions. Math. Comput. Simul. **190**, 866–891 (2021)
20. P. Viswanathan, A.K.B. Chand, A C^1-rational cubic fractal interpolation function: convergence and associated parameter identification problem. Acta Appl. Math. **136**(1), 19–41 (2015)
21. M.F. Barnsley, J. Elton, D. Hardin, P. Massopust, Hidden variable fractal interpolation functions. SIAM J. Math. Anal. **20**(5), 1218–1242 (1989)
22. M.F. Barnsley, *Fractals Everywhere* (Academic Press, 2014)
23. K.J. Falconer, *Techniques in Fractal Geometry*, vol. 3 (Wiley Chichester, 1997)

Chapter 3
Fractal Scaling Parameters

In a nutshell, an iterated function system made up of a finite number of iterated mappings on a complete metric space produces a fractal interpolation function. The characteristics and shape of the fractal functions are significantly influenced by the vertical scaling factors, which are significant set of free parameters in the iterated mappings. Not only in the case of affine fractal function but also in the non-affine cases, the vertical scaling factors uniquely determine the corresponding fractal function when the interpolation points are prescribed in advance. In the previous Chapter, various types of fractal interpolation functions are discussed but all are limited within the case of constant vertical scaling factors (α_n, constant parameters). Using constant parameters in iterated function system, one can make each iterated mapping possess the same vertical compression ratios on an identical subinterval that belongs to a partition of some closed interval defining the fractal function. This means that the FIFs generated by those IFSs with constant parameters usually have obvious self-similarity character, which could lead to the loss of flexibility, and might cause obvious errors in fitting and approximation of some complicated curves and non-stationary data that show less self-similarity. This complication can be rectified by defining variable or function scaling offering greater advantages. In addition, the concept of optimization of scaling parameters is much more useful as it provides better approximation than both the constant and variable scalings. This chapter discusses the fractal functions with variable scalings and optimized scalings in detail with suitable examples.

3.1 Variable Scalings

This section discusses the influence of 'variable' scaling parameters in association with self-similarity of the fractal interpolation scheme. In 2013, Wang and Yu [1] have generalized the construction of IFS with constant scaling parameters to the case

A. Gowrisankar et al., *Fractal Functions Communication with Fractional Calculus*,
https://doi.org/10.1007/978-3-031-96993-5_3

of IFS with variable scaling parameters to generate generalized FIFs. In addition, the analytical properties including the smoothness, stability and sensitivity of FIFs with variable scaling parameters are discussed. The IFS expressed in Eq. (2.3) is rewritten as

$$L_n(x) = a_n x + b_n,$$
$$F_n(x, y) = \alpha_n(x)y + q_n(x), \ \forall\, n = 1, 2, \ldots, N-1,$$

where $\alpha_n(x)$ is Lipschitz function on I satisfying

$$\|\alpha_n\|_\infty = \sup\{|\alpha_n(x)| : x \in I\} < 1, \ n = 1, 2, \ldots, N-1.$$

Then the fractal interpolation function $f : I \to \mathbb{R}$ with variable scaling factors satisfies the functional equation

$$f(L_n(x)) = \alpha_n(x)y + q_n(x), \ \forall\, x \in I_n.$$

It is to be noted that suppose all $\alpha_n(x)$ are taken to be constants then the associated FIF coincides with the FIF generated by IFS (2.3) discussed in Chap. 2. The graphs of FIFs with constant scalings display obvious self-similarity, but in some practical phenomena there might be some weaker self-similarity. In such circumstances, the choice of variable scalings helps to capture less self-similar patterns to a certain extent, thereby enlarging the flexibility and applicability of graphs of FIFs with variable scaling parameters.

Example 3.1 For the provided data set $\{(0, 0.4), (0.3, -0.2), (0.5, 0.2), (0.8, 0.4), (1, -0.6)\}$, four different set of scaling parameters are considered,

(a) constant parameters: $\alpha_1 = 0.2, \alpha_2 = 0.3, \alpha_3 = 0.2, \alpha_4 = 0.4$
(b) variable parameters: $\alpha_n(x) = \cos(2x)/3 + 0.4$, for all $n = 1, 2, 3, 4$
(c) variable parameters: $\alpha_n(x) = x/5 + 0.2$, for all $n = 1, 2, 3, 4$
(d) variable parameters: $\alpha_1(x) = \cos(2x)/3 + 0.4, \alpha_2(x) = x/5 + 0.2, \alpha_3(x) = 0.5\exp(-2x), \alpha_4(x) = 0.8\exp(x)/3$.

The graphs of linear FIF with constant parameters and variable parameters are demonstrated in Fig. 3.1. Comparing the Fig. 3.1a with Fig. 3.1b, c and d, the irregularity demonstrated in the latter three graphs is weaker than the linear fractal function in the former. This complex attribute assures that the fractal functions with variable scaling may have more applicability when the real-world phenomena is less irregular and self-similar.

Example 3.2 In this example, different quadratic fractal functions are constructed for the provided data set $\{(0, 0), (1/5, 1/4), (2/5, 2/5), (3/5, 1/2), (4/5, 3/4), (1, 0)\}$ with four different set of scaling parameters as considered below,

(a) constant parameters: $\alpha_n = 0.3$, for all $n = 1, 2, 3, 4, 5$
(b) constant parameters: $\alpha_1 = 0.5, \alpha_2 = -0.5, \alpha_3 = 0.2, \alpha_4 = -0.2, \alpha_4 = 0.3$

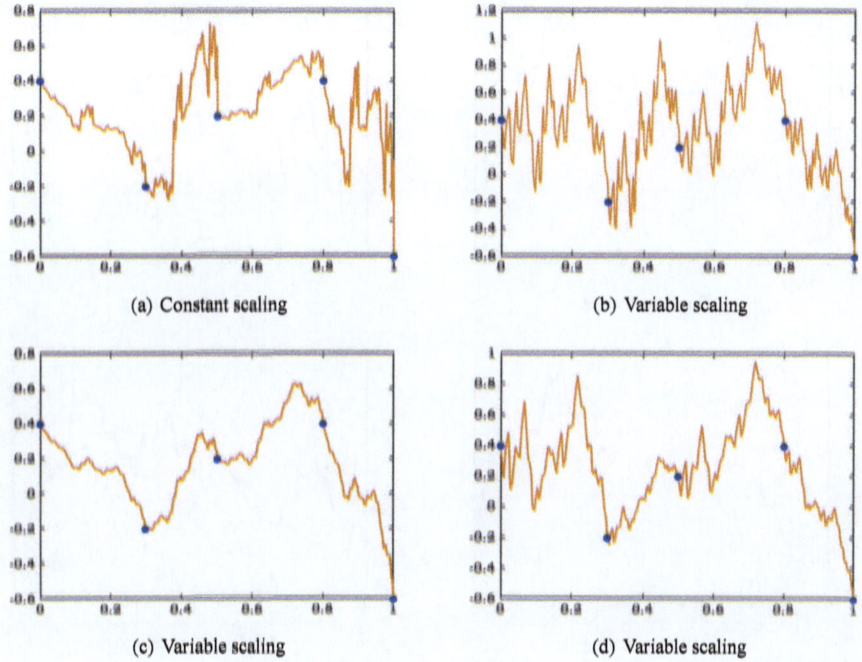

Fig. 3.1 Graphical illustration of linear fractal interpolation function with **a** $\alpha_1 = 0.2, \alpha_2 = 0.3, \alpha_3 = 0.2, \alpha_4 = 0.4$, **b** $\alpha_n(x) = \cos(2x)/3 + 0.4$, for all $n = 1, 2, 3, 4$, **c** $\alpha_n(x) = x/5 + 0.2$, for all $n = 1, 2, 3, 4$ and **d** $\alpha_1(x) = \cos(2x)/3 + 0.4$, $\alpha_2(x) = x/5 + 0.2$, $\alpha_3(x) = 0.5\exp(-2x)$, $\alpha_4(x) = 0.8\exp(x)/3$

(c) variable parameters: $\alpha_n(x) = 0.8\exp(x)/3$, for all $n = 1, 2, 3, 4, 5$
(d) variable parameters: $\alpha_1(x) = \cos(2x)/3 + 0.5, \alpha_2(x) = x/5 + 0.2, \alpha_3(x) = 0.5\exp(-2x)$, $\alpha_4(x) = 0.8\exp(x)/3$, $\alpha_5(x) = \sin(x)/2 + 0.2$.

The graphs of quadratic fractal functions with constant parameters and variable parameters are demonstrated in Fig. 3.2. Figure 3.2a and b might look similar to the Fig. 3.2c and d, respectively. But, in each subinterval, the irregularity varies, in particular the quadratic fractal function with variable scaling provides greater flexibility than the function with constant scaling.

Example 3.3 To approximate the given continuous function x^3 with the base function $b(x) = x$ using α-fractal functions, four different set of scaling parameters are considered as follows,

(a) constant parameters: $\alpha_n = 0.4$, for all $n = 1, 2, 3, 4$
(b) constant parameters: $\alpha_1 = 0.5, \alpha_2 = -0.3, \alpha_3 = 0.2, \alpha_4 = 0.6$
(c) variable parameters: $\alpha_i(x) = (x + 0.5)/2$, for all $n = 1, 2, 3, 4$
(d) variable parameters: $\alpha_1(x) = \sin(x)/3 + 0.1, \alpha_2(x) = x/2 + 0.1, \alpha_3(x) = 0.2\exp(2x)$, $\alpha_4(x) = 0.2\exp(x)/3$.

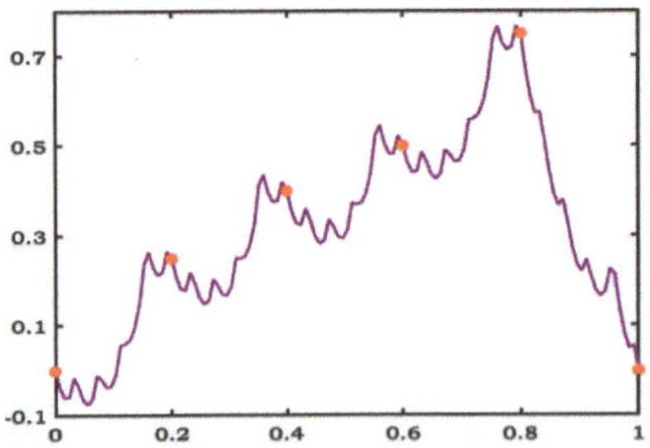

(a) Constant scaling

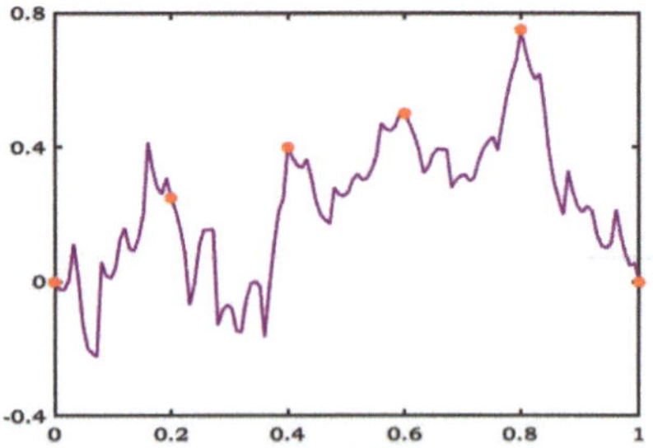

(b) Constant scaling

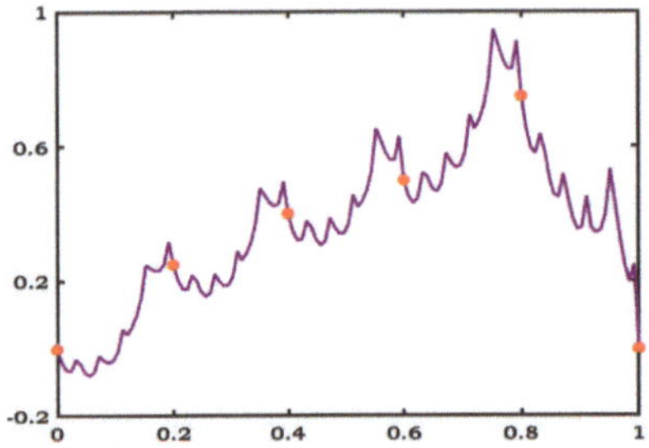

(c) Variable scaling

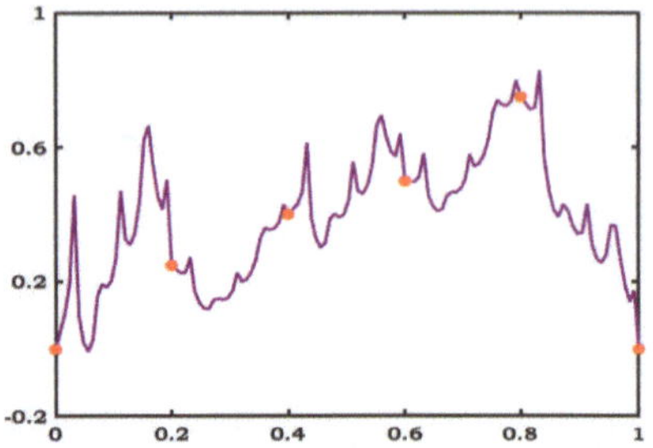

(d) Variable scaling

Fig. 3.2 Graphical illustration of quadratic fractal interpolation function with **a** $\alpha_n = 0.3$, for all $n = 1, 2, 3, 4, 5$, **b** $\alpha_1 = 0.5, \alpha_2 = -0.5, \alpha_3 = 0.2, \alpha_4 = -0.2, \alpha_4 = 0.3$, **c** $\alpha_n(x) = 0.8\exp(x)/3$, for all $n = 1, 2, 3, 4, 5$ and **d** $\alpha_1(x) = \cos(2x)/3 + 0.5, \alpha_2(x) = x/5 + 0.2, \alpha_3(x) = 0.5\exp(-2x), \alpha_4(x) = 0.8\exp(x)/3, \alpha_5(x) = \sin(x)/2 + 0.2$

The graphs of different α-fractal functions with constant parameters and variable parameters are demonstrated in Fig. 3.3. This example helps to observe the role of scaling parameter in the α-fractal function, comparing Fig. 3.3a and b, the latter gives a better approximation than the former. Similarly, when Fig. 3.3c and d are compared, choice of variable scaling varying at each interval provides a better approximation.

Appreciating the importance of variable scaling, researchers have carefully taken the concept of FIF with variable parameters to the theories studied with constant parameters, a few of which are discussed in the upcoming chapters. For additional information, see the articles [2–7].

3.2 Parameter Identification

The parameter identification problem deals with the right choice of scaling parameters associated with FIFs for the effective modeling of unknown data. Further, this appropriate choice of parameters aid to generate fractal functions preserving some characteristics inherent in the original function. In 1992, the article [8] has discussed

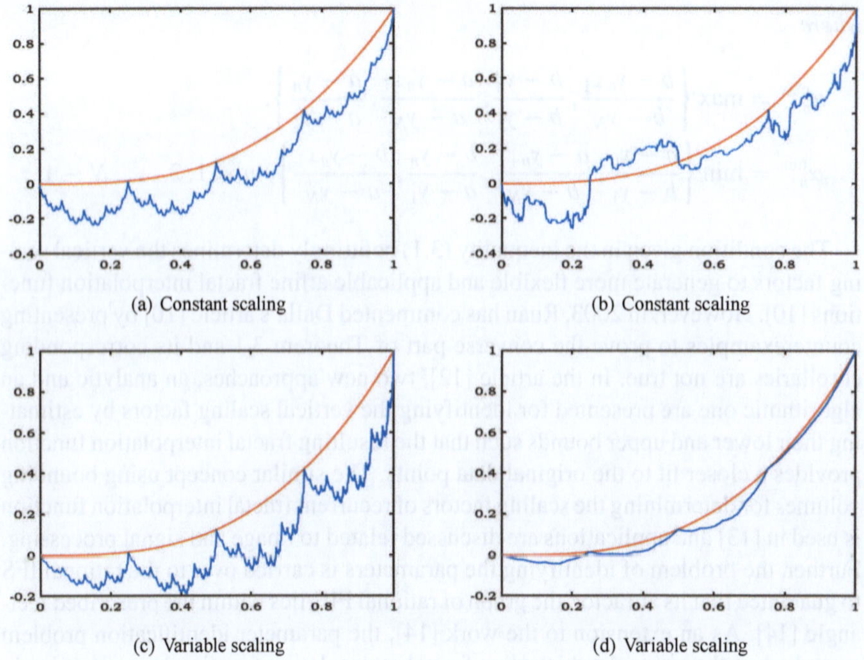

Fig. 3.3 Graphical illustration of α-fractal function with **a** $\alpha_n = 0.4$, for all $n = 1, 2, 3, 4$, **b** $\alpha_1 = 0.5$, $\alpha_2 = -0.3$, $\alpha_3 = 0.2$, $\alpha_4 = 0.6$, **c** $\alpha_i(x) = (x + 0.5)/2$, for all $n = 1, 2, 3, 4$ and **d** $\alpha_1(x) = \sin(x)/3 + 0.1$, $\alpha_2(x) = x/2 + 0.1$, $\alpha_3(x) = 0.2\exp(2x)$, $\alpha_4(x) = 0.2\exp(x)/3$

the geometrical and analytical approach for identifying right scaling parameters of FIF. In addition, it has provided an algorithm for determining the interpolation points and scaling parameters for any arbitrary data set as an extension to the work by Mazel and Hayes in [9]. Later, in 1999, Dalla [10] has developed the concept of right parameter identification to interpolating the data set within the prescribed rectangle (domain). In [11], Barnsley has defined the affine fractal interpolation in the 'bigger' compact metric space $X = I \subset \mathbb{R} \times \mathbb{R}$ and has not interpreted any particular conditions for the vertical scaling factors such that the attractor G remains within the domain X. To fill this gap, the work [10] has determined the vertical scaling factors solely depending on the provided data to handle situations if the original domain of interval is given. The following is the debut theorem concerning the parameter identification in the affine fractal interpolation function.

Theorem 3.1 *The graph of an affine fractal interpolation function remains within the prescribed rectangle $R = I \times [a, b]$ iff the vertical scalings α_n satisfy $|\alpha_n| < 1$ and*

$$\alpha_n^{\min} \leq \alpha_n \leq \alpha_n^{\max}, \tag{3.1}$$

where

$$\alpha_n^{\max} = \max\left\{\frac{b - y_{n+1}}{b - y_N}, \frac{b - y_n}{b - y_1}, \frac{a - y_{n+1}}{a - y_N}, \frac{a - y_n}{a - y_1}\right\},$$

$$\alpha_n^{\min} = \min\left\{\frac{a - y_n}{b - y_1}, \frac{a - y_{n+1}}{b - y_N}, \frac{b - y_n}{a - y_1}, \frac{b - y_{n+1}}{a - y_N}\right\}, \quad n = 1, 2, \ldots, N - 1.$$

The condition given in the inequality (3.1) befittingly determines the vertical scaling factors to generate more flexible and applicable affine fractal interpolation functions [10]. However, in 2003, Ruan has commented Dalla's article [10] by presenting counterexamples to prove the converse part of Theorem 3.1 and its corresponding corollaries are not true. In the article [12], two new approaches, an analytic and an algorithmic one are presented for identifying the vertical scaling factors by estimating their lower and upper bounds such that the resulting fractal interpolation function provides a closer fit to the original data points. The similar concept using bounding volumes for determining the scaling factors of recurrent fractal interpolation function is used in [13] and applications are discussed related to image and signal processing. Further, the problem of identifying the parameters is carried over to the rational IFS to guarantee that its attractor, the graph of rational FIF, lies within the prescribed rectangle [14]. As an extension to the work [14], the parameter identification problem is studied on the rational cubic spline fractal interpolation function by appropriately imposing additional constraints on the scaling factors and shape parameters [15].

3.3 Optimized Scalings

The main purpose of optimizing the scaling factor is to effectively apply fractal interpolation to the real-world scenarios. Here, some of the optimized scaling factors utilized in real-world problems are discussed. The scaling factor α_n in the fractal interpolation function is particularly called as 'vertical' scaling factor as it shrinks the fractal function vertically, that is, in the y direction. Hence, while optimizing the scaling factor, the variable y has to be concentrated rather than the variable x. With this clarification, let us see how the vertical scaling parameters are modified for optimized interpolation. In [16], while analyzing the turbulent flow data, the fractal interpolation function is used with the scaling factor defined by

$$|\alpha_n| = \frac{w_n(L)}{L}, \tag{3.2}$$

here, L is the length of a line segment parallel to the y-axis.

Next, we can discuss another optimized scaling factor which is most commonly used in many applications such as to study the epidemiological data [17] and to analyze the global climate dynamics [18]. For linear fractal interpolation, we have

$$F_n(x, y) = S_n^* y + c_n x + d_n, \tag{3.3}$$

where $|S_n^*| < 1$ (here the notation S_n^* is used merely to differentiate from constant scaling and variable scaling). By fixing the parameter x (horizontal axis) and incrementing the y with Δy, one can get

$$F_n(x, y + \Delta y) = S_n^*(y + \Delta y) + c_n x + d_n. \tag{3.4}$$

Subtracting the above two Eqs. (3.3) and (3.4), we get $\Delta F_n = S_n^* \Delta y_n$, which ensures that the mapping $w_n(x, y)$ defined in IFS (2.4) is contractive and we obtain,

$$S_n^* = \frac{\Delta F_n}{\Delta y_n} = \frac{F_n(x, y + \Delta y) - F_n(x, y)}{y_{n+1} - y_n}. \tag{3.5}$$

Now, considering the local information of $F_n(x, y)$ (i.e., linear map), the vertical length between $F_n(x, y + \Delta y)$ and $F_n(x, y)$ is Δy_n, since there is only change in the y axis in Eqs. (3.3) and (3.4). Next to cover the overall characteristics of the data, denominator is taken as a Euclidean metric in R^2 between $(y_{\min}, y_{\max})$ and (y_n, y_{n+1}). Thus, the Eq. (3.5) is reduced to the following equation,

$$S_n^* = \frac{y_{n+1} - y_n}{\epsilon\sqrt{(y_{\max} - y_{\min})^2 + (y_{n+1} - y_n)^2}}, \tag{3.6}$$

where $\epsilon = 1 + rand(\beta)$, β represents a random number generated by the computer in the interval (0, 1). The S_n^* is anyway an optimized scaling without the factor ϵ or $\epsilon = 1$. The purpose of choosing $\epsilon = 1 + rand(\beta)$ is to incarnate the characteristics of the fractal. In a deeper sense, ϵ is always greater than 1, so the value of S_n^* changes in the interval (0, 1) during each simulation. When ϵ is increased, S_n^* approaches 0 generating less irregular functions, and suppose ϵ is decreased, S_n^* approaches 1 generating more irregular functions. As a consequence, different values of ϵ generate different fractal interpolation functions with varying irregularity. Therefore, in addition to the optimization, the ϵ provides a comfort opportunity to enjoy the incarnations of fractal functions. The maximal value $y_{\max}$ and the minimal value $y_{\min}$ can be chosen in the interval centered by the point to be interpolated.

In the literature [19], the optimal scaling parameter selection associated with the α-fractal function is discussed, offering an outline of applications pertaining to energy minimization problems in signal processing. The article [20] evidences that the application of fractal interpolation with optimized scalings has been extended for pre-processing time series in combination with deep-learning techniques. A few real-world applications of fractal interpolation with optimized scaling are discussed in Chap. 9.

References

1. H.-Y. Wang, J.-S. Yu, Fractal interpolation functions with variable parameters and their analytical properties. J. Approx. Theory **175**, 1–18 (2013)
2. Md.N. Akhtar, M.G.P. Prasad, M.A. Navascués, Box dimension of α-fractal function with variable scaling factors in subintervals. Chaos, Solitons Fractals **103**, 440–449 (2017)
3. A. Gowrisankar, M.G.P. Prasad, Riemann-Liouville calculus on quadratic fractal interpolation function with variable scaling factors. J. Anal. **27**(2), 347–363 (2019)
4. L. Jiang, H.-J. Ruan, Box dimension of generalized affine fractal interpolation functions. J. Fractal Geom. **10**(3), 279–302 (2023)
5. T.M.C. Priyanka, A. Gowrisankar, Riemann-Liouville fractional integral of non-affine fractal interpolation function and its fractional operator. Eur. Phys. J. Spec. Top. **230**(21–22), 3789–3805 (2021)
6. T.M.C. Priyanka, A. Gowrisankar A. Agathiyan, Weyl-Marchaud fractional derivative of a vector valued fractal interpolation function with function contractivity factors. J. Anal. **31**(1), 657–689 (2023)
7. C.H. Yun, M.G. Ri, Box-counting dimension and analytic properties of hidden variable fractal interpolation functions with function contractivity factors. Chaos, Solitons Fractals **134**, 109700 (2020)
8. S.M. David, H.H. Moson, Using iterated function systems to model discrete sequences. IEEE Trans Signal Process. **40**(7), 1724–1734 (1992)
9. D.S. Mazel, M.H. Hayes, Fractal modeling of time-series data, in *Twenty-Third Asilomar Conference on Signals, Systems and Computers*, vol. 1 (IEEE Computer Society, 1989), pp. 182–183
10. L. Dalla, V. Drakopoulos, On the parameter identification problem in the plane and the polar fractal interpolation functions. J. Approx. Theory **101**(2), 289–302 (1999)
11. M.F. Barnsley, Fractal functions and interpolation. Constr. Approx. **2**, 303–329 (1986)
12. P. Manousopoulos, V. Drakopoulos, T. Theoharis, Parameter identification of 1D fractal interpolation functions using bounding volumes. J. Comput. Appl. Math. **233**(4), 1063–1082 (2009)
13. P. Manousopoulos, V. Drakopoulos, T. Theoharis, Parameter identification of 1D recurrent fractal interpolation functions with applications to imaging and signal processing. J. Math. Imaging Vis. **40**, 162–170 (2011)
14. P. Viswanathan, A.K.B. Chand, A C^1-rational cubic fractal interpolation function: convergence and associated parameter identification problem. Acta Appl. Math. **136**(1), 19–41 (2015)
15. S.K. Katiyar, A.K.B. Chand, S. Jha, Parameter identification for constrained data using a new class of rational fractal functions. Numer. Anal. Appl. **14**(3), 225–237 (2021)
16. M.A. Marvasti, W.C. Strahle, Fractal geometry analysis of turbulent data. Signal Process. **41**(2), 191–201 (1995)
17. A. Gowrisankar, T.M.C. Priyanka, S. Banerjee, Omicron: a mysterious variant of concern. Eur. Phys. J. Plus **137**(1), 1–8 (2022)
18. A. Gowrisankar, T.M.C. Priyanka, A. Saha, L. Rondoni, K. Hassan, S. Banerjee, Greenhouse gas emissions: a rapid submerge of the world. Chaos: Interdiscip. J. Nonlinear Sci. **32**(6) (2022)
19. M.A. Navascués, P. Viswanathan, Energy minimizing associate fractal functions. Revista de la Real Academia de Ciencias Exactas, Físicas y Naturales. Serie A. Matemáticas **113**(2), 1025–1039 (2019)
20. A. Băicoianu, C.G. Gavrilă, C.M. Păcurar, V.D. Păcurar, Fractal interpolation in the context of prediction accuracy optimization. Eng. Appl. Artif. Intell. **133**, 108380 (2024)

Chapter 4
Classical Integration on Fractal Functions

4.1 Introduction

In the field of fractal geometry, fractal interpolation functions have got a variety of interesting applications not only in the mathematical sciences but also in other areas of applied sciences, for instance [1–5]. The fractal interpolation function has been introduced with the aim of analyzing rough functions (or data) in the real world. Comparing with traditional interpolants, fractal interpolants seem to be more flexible. The reason behind the feasibility is that FIFs can be utilized for the best approximation of complicated mathematical structures with simple iterative procedure. As discussed in Chap. 2, Barnsley has interpreted the fractal interpolation function in two ways: (i) FIF as an attractor of the iterated function system, (ii) FIF as a fixed point of the Read-Bajraktarević operator defined on a suitable complete space of $\mathbb{R}$. Moreover, in general, fractal functions are not always differentiable, as a consequence classical differentiation cannot be applied to them. Nevertheless, the integral of fractal functions can be studied since they are continuous everywhere and bounded. The study on classical integral of linear fractal function has been investigated by Barnsley and Harrington (Ref. [6]). In [6], the authors proved that the integrals of FIFs are also FIFs demonstrating which are again attractors of special iterated function systems obeying the definition. In this chapter, the integral of various FIFs such as linear FIF, quadratic FIF, α-FIF and hidden variable FIF with variable scaling factors are investigated. Further, newly obtained FIFs with variable scaling factors are shown as the generalization of FIFs with constant scaling factors. The new interpolation data corresponding to the given data set are found out explicitly.

A. Gowrisankar et al., *Fractal Functions Communication with Fractional Calculus*,
https://doi.org/10.1007/978-3-031-96993-5_4

4.2 Integral on Linear and Quadratic Fractal Functions

In this section, firstly let us discuss the definite and indefinite integral of general fractal interpolation functions. The following lemma presents the integral of general FIF over different intervals that could be utilized for succeeding theorems.

Lemma 4.1 *Let f be a fractal interpolation function generated by the IFS (2.3). Then*

$$\int_{x_n}^{L_n(x)} f(t)\,dt = a_n \int_{x_1}^{x} f(L_n(t))\,dt, \tag{4.1}$$

$$\int_{x_n}^{x_{n+1}} f(t)\,dt = a_n \int_{x_1}^{x_N} f(L_n(t))\,dt, \tag{4.2}$$

$$\int_{x_n}^{x} f(t)\,dt = a_n \int_{x_1}^{\frac{x-b_n}{a_n}} f(L_n(t))\,dt. \tag{4.3}$$

$$\int_{L_j(x_n)}^{L_j(x_{n+1})} f(t)\,dt = a_j \int_{x_n}^{x_{n+1}} f\left(L_j(t)\right)dt. \tag{4.4}$$

Proof These equations result from a standard change of variable, considering the joint-up conditions. □

The integral of classical fractal interpolation function with constant scaling factors defined in the closed interval I is computed in the following theorem.

Theorem 4.1 *Let f be a fractal interpolation function as defined above, such that $\sum_{i=1}^{N-1} a_i\alpha_i \neq 1$. Then*

$$\int_{x_1}^{x_N} f(t)\,dt = \frac{\sum_{i=1}^{N-1} a_i \int_{x_1}^{x_N} q_i(t)\,dt}{1-\sum_{i=1}^{N-1} a_i\alpha_i}, \tag{4.5}$$

$$\int_{x_1}^{x_n} f(t)\,dt = \frac{\sum_{i=1}^{n-1} a_i\alpha_i}{1-\sum_{i=1}^{N-1} a_i\alpha_i} \sum_{j=1}^{N-1} a_j \int_{x_1}^{x_N} q_j(t)\,dt + \sum_{j=1}^{n-1} a_j \int_{x_1}^{x_N} q_j(t)\,dt, \tag{4.6}$$

$$\int_{x_n}^{x_{n+1}} f(t)\,dt = a_n\alpha_n \frac{\sum_{i=1}^{N-1} a_i \int_{x_1}^{x_N} q_i(t)\,dt}{1-\sum_{i=1}^{N-1} a_i\alpha_i} + a_n \int_{x_1}^{x_N} q_n(t)\,dt. \tag{4.7}$$

Proof By the change of variable in (4.2) and using functional equation of fractal function f,

$$\int_{x_1}^{x_N} f(t)\,dt = \sum_{i=1}^{N-1} a_i \int_{x_1}^{x_N} \alpha_i f(t) + q_i(t)\,dt$$
$$= \sum_{i=1}^{N-1} a_i\alpha_i \int_{x_1}^{x_N} f(t)\,dt + \sum_{i=1}^{N-1} a_i \int_{x_1}^{x_N} q_i(t)\,dt.$$

This last equality implies the first formula.

For the second formula, following the same argument, we have

$$\int_{x_1}^{x_n} f(t)\,dt = \sum_{i=1}^{n-1} a_i \int_{x_1}^{x_N} \alpha_i f(t) + q_i(t)\,dt$$
$$= \sum_{i=1}^{n-1} a_i\alpha_i \int_{x_1}^{x_N} f(t)\,dt + \sum_{i=1}^{n-1} a_i \int_{x_1}^{x_N} q_i(t)\,dt$$
$$= \frac{\sum_{i=1}^{n-1} a_i\alpha_i}{1-\sum_{i=1}^{N-1} a_i\alpha_i} \sum_{j=1}^{N-1} a_j \int_{x_1}^{x_N} q_j(t)\,dt + \sum_{j=1}^{n-1} a_j \int_{x_1}^{x_N} q_j(t)\,dt.$$

Similarly, for the third formula,

$$\int_{x_n}^{x_{n+1}} f(t)\,dt = a_n \int_{x_1}^{x_N} f(L_n(t))\,dt$$
$$= a_n \int_{x_1}^{x_N} \alpha_n f(t) + q_n(t)\,dt$$
$$= a_n\alpha_n \frac{\sum_{i=1}^{N-1} a_i \int_{x_1}^{x_N} q_i(t)\,dt}{1-\sum_{i=1}^{N-1} a_i\alpha_i} + a_n \int_{x_1}^{x_N} q_n(t)\,dt,$$

which concludes the proof. □

Remark 4.1 The integral of a fractal interpolation function in the interval I is a weighted average of all integrals of q_n in the same interval, multiplied by a factor that depends on the contractive factors of the problem. The average is weighted by the corresponding scaling factors of the L_n functions.

It is also observed that

$$\int_{x_N}^{x_1} f(t)\,dt = -\frac{\sum_{i=1}^{N-1} a_i \int_{x_1}^{x_N} q_i(t)\,dt}{1-\sum_{i=1}^{N-1} a_i\alpha_i},$$

Theorem 4.2 investigates the integral of classical fractal interpolation function with variable scaling factors in the closed interval I.

Theorem 4.2 *Let f be a fractal interpolation function with variable scaling factors as defined above, such that $\sum_{i=1}^{N-1} a_i\alpha_i(x_N) \neq 1$. Then*

$$\int_{x_1}^{x_N} f(t)\,dt \frac{\sum_{i=1}^{N-1} a_i \int_{x_1}^{x_N} q_i(t)\,dt - a_i \int_{x_1}^{x_N} \left(\alpha_i'(t) \int_{x_1}^{t} f(u)du\right) dt}{1 - \sum_{i=1}^{n-1} a_i \alpha_i(x_N)},$$

$$\int_{x_1}^{x_n} f(t)\,dt = \sum_{i=1}^{n-1} a_i \alpha_i(x_N) \frac{\sum_{i=1}^{N-1} a_i \left(\int_{x_1}^{x_N} q_i(t)\,dt - \int_{x_1}^{x_N} \left(\alpha_i'(t) \int_{x_1}^{t} f(u)du\right) dt\right)}{1 - \sum_{i=1}^{N-1} a_i \alpha_i(x_N)}$$
$$- \sum_{i=1}^{n-1} a_i \int_{x_1}^{x_N} \left(\alpha_i'(t) \int_{x_1}^{t} f(u)du\right) dt + \sum_{i=1}^{n-1} a_i \int_{x_1}^{x_N} q_i(t)\,dt.$$

Proof By the change of variable in (4.2) and using the function of fractal function f,

$$\begin{aligned}
\int_{x_1}^{x_N} f(t)\,dt &= \sum_{i=1}^{N-1} a_i \int_{x_1}^{x_N} (\alpha_i(t) f(t) + q_i(t))\,dt \\
&= \sum_{i=1}^{N-1} a_i \alpha_i(x_N) \int_{x_1}^{x_N} f(t)dt - a_i \int_{x_1}^{x_N} \left(\alpha_i'(t) \int_{x_1}^{t} f(u)du\right) dt + \sum_{i=1}^{N-1} a_i \int_{x_1}^{x_N} q_i(t)\,dt \\
&= \frac{\sum_{i=1}^{N-1} a_i \int_{x_1}^{x_N} q_i(t)\,dt - a_i \int_{x_1}^{x_N} \left(\alpha_i'(t) \int_{x_1}^{t} f(u)du\right) dt}{1 - \sum_{i=1}^{N-1} a_i \alpha_i(x_N)}.
\end{aligned}$$

For the second formula, and $n \in \{1, \ldots, N-1\}$ we have

$$\begin{aligned}
\int_{x_1}^{x_n} f(t)\,dt &= \sum_{i=1}^{n-1} a_i \int_{x_1}^{x_N} (\alpha_i(t) f(t) + q_i(t))\,dt \\
&= \sum_{i=1}^{n-1} a_i \left(\alpha_i(x_N) \int_{x_1}^{x_N} f(t)dt - \int_{x_1}^{x_N} \left(\alpha_i'(t) \int_{x_1}^{t} f(u)du\right) dt\right) + \sum_{i=1}^{n-1} a_i \int_{x_1}^{x_N} q_i(t)\,dt \\
&= \sum_{i=1}^{n-1} a_i \alpha_i(x_N) \frac{\sum_{i=1}^{N-1} a_i \left(\int_{x_1}^{x_N} q_i(t)\,dt - \int_{x_1}^{x_N} \left(\alpha_i'(t) \int_{x_1}^{t} f(u)du\right) dt\right)}{1 - \sum_{i=1}^{N-1} a_i \alpha_i(x_N)} \\
&\quad - \sum_{i=1}^{n-1} a_i \int_{x_1}^{x_N} \left(\alpha_i'(t) \int_{x_1}^{t} f(u)du\right) dt + \sum_{i=1}^{n-1} a_i \int_{x_1}^{x_N} q_i(t)\,dt,
\end{aligned}$$

which concludes the proof. □

Based on the fundamental theory of fractal interpolation functions described by Barnsley [4] we have the following result.

Theorem 4.3 *The IFS $\{I; L_n, n = 1, \ldots, N-1\}$ is hyperbolic and have a unique attractor. The attractor is I and, for each $x \in I$, there exists a sequence $\xi = \xi_1 \xi_2 \cdots \xi_m \cdots$ such that, $\xi_m \in \{1, \ldots, N-1\}$ and*

$$x = L_\xi(y) = L_{\xi_1 \xi_2 \cdots \xi_m \cdots}(y) := L_{\xi_1} \circ L_{\xi_2} \circ \cdots \circ L_{\xi_m} \cdots (y), \forall y \in I.$$

For some $x \in I$, this sequence is finite, with $y = x_1$, i.e.,

$$x = L_\xi(y) := L_{\xi_1} \circ L_{\xi_2} \circ \cdots \circ L_{\xi_m}(x_1),$$

where $m \in \mathbb{N}$.

With this formulation, the computation of the definite integral, for any $x \in I$, is obtained by a recursive process based on the self-similarity of the fractal functions,

$$\int_{x_1}^{x_N} f(t)\,dt = \sum_{n=1}^{N-1} \int_{x_n}^{x_{n+1}} f(t)\,dt,$$

$$\int_{x_n}^{x_{n+1}} f(t)\,dt = \sum_{j=1}^{N-1} \int_{L_j(x_n)}^{L_j(x_{n+1})} f(t)\,dt,$$

$$\int_{L_{\xi_2}(x_n)}^{L_{\xi_2}(x_{n+1})} f(t)\,dt = \sum_{\xi_1=1}^{N-1} \int_{L_{\xi_1\xi_2}(x_n)}^{L_{\xi_1\xi_2}(x_{n+1})} f(t)\,dt.$$

The general formulation needs the formula for the integral of tiny sub-subintervals.

Lemma 4.2 *Let f be a fractal interpolation function as defined above. Then, $\forall\, m \in \mathbb{N}$,*

$$\int_{L_{\xi_1\xi_2\cdots\xi_m}(x_1)}^{L_{\xi_1\xi_2\cdots\xi_m\xi_{m+1}}(x_1)} f(t)\,dt = \prod_{n=1}^{m} \left(a_{\xi_n}\alpha_{\xi_n}\right) \int_{x_1}^{x_{\xi_{m+1}}} f(t)\,dt$$
$$+ \sum_{j=1}^{m} \prod_{n=1}^{j-1} \left(a_{\xi_n}\alpha_{\xi_n}\right) a_{\xi_j} \int_{L_{\xi_{j+1}\cdots\xi_m}(x_1)}^{L_{\xi_{j+1}\cdots\xi_{m+1}}(x_1)} q_{\xi_j}(t)\,dt. \tag{4.8}$$

Proof This is done by mathematical induction.

For $n = 1$, by a change of variables of type (4.4), and the definition of f,

$$\int_{L_{\xi_1}(x_1)}^{L_{\xi_1\xi_2}(x_1)} f(t)\,dt = \int_{L_{\xi_1}(x_1)}^{L_{\xi_1}(x_{\xi_2})} f(t)\,dt$$
$$= a_{\xi_1} \int_{x_1}^{x_{\xi_2}} f\left(L_{\xi_1}(t)\right) dt$$
$$= a_{\xi_1}\alpha_{\xi_1} \int_{x_1}^{x_{\xi_2}} f(t)\,dt + a_{\xi_1} \int_{x_1}^{L_{\xi_2}(x_1)} q_{\xi_1}(t)\,dt.$$

Suppose that (4.8) is valid for some $m \in \mathbb{N}$. Then,

$$\begin{aligned}\int_{L_{\xi_1\xi_2\cdots\xi_{m+1}}(x_1)}^{L_{\xi_1\xi_2\cdots\xi_{m+1}\xi_{m+2}}(x_1)} f(t)\,dt &= \int_{L_{\xi_1}\left(L_{\xi_2\cdots\xi_{m+1}}(x_1)\right)}^{L_{\xi_1}\left(L_{\xi_2\cdots\xi_{m+1}\xi_{m+2}}(x_1)\right)} f(t)\,dt \\
&= a_{\xi_1}\int_{L_{\xi_2\cdots\xi_{m+1}}(x_1)}^{L_{\xi_2\cdots\xi_{m+1}\xi_{m+2}}(x_1)} f\left(L_{\xi_1}(t)\right)dt \\
&= a_{\xi_1}\alpha_{\xi_1}\int_{L_{\xi_2\cdots\xi_{m+1}}(x_1)}^{L_{\xi_2\cdots\xi_{m+1}\xi_{m+2}}(x_1)} f(t)\,dt + a_{\xi_1}\int_{L_{\xi_2\cdots\xi_{m+1}}(x_1)}^{L_{\xi_2\cdots\xi_{m+1}\xi_{m+2}}(x_1)} q_{\xi_1}(t)\,dt \\
&= a_{\xi_1}\alpha_{\xi_1}\prod_{n=2}^{m+1}\left(a_{\xi_n}\alpha_{\xi_n}\right)\int_{x_1}^{x_{\xi_{m+2}}} f(t)\,dt \\
&\quad + a_{\xi_1}\alpha_{\xi_1}\sum_{j=2}^{m+1}\prod_{n=2}^{j-1}\left(a_{\xi_n}\alpha_{\xi_n}\right)a_{\xi_j}\int_{L_{\xi_{j+1}\cdots\xi_{m+1}}(x_1)}^{L_{\xi_{j+1}\cdots\xi_{m+2}}(x_1)} q_{\xi_j}(t)\,dt \\
&\quad + a_{\xi_1}\int_{L_{\xi_2\cdots\xi_{m+1}}(x_1)}^{L_{\xi_2\cdots\xi_{m+1}\xi_{m+2}}(x_1)} q_{\xi_1}(t)\,dt \\
&= \prod_{n=1}^{m+1}\left(a_{\xi_m}\alpha_{\xi_m}\right)\int_{x_1}^{x_{\xi_{m+2}}} f(t)\,dt \\
&\quad + \sum_{j=1}^{m+1}\prod_{n=1}^{j-1}\left(a_{\xi_n}\alpha_{\xi_n}\right)a_{\xi_j}\int_{L_{\xi_{j+1}\cdots\xi_{m+1}}(x_1)}^{L_{\xi_{j+1}\cdots\xi_{m+2}}(x_1)} q_{\xi_j}(t)\,dt,\end{aligned}$$

which concludes the proof. □

Now the integral for any $x \in I$ is possible to compute.

Theorem 4.4 *Let f be a fractal interpolation function as defined above, such that $\sum_{i=1}^{N-1} a_i\alpha_i \neq 1$. Let $x \in I$, with corresponding sequence $\xi = \xi_1\xi_2\cdots\xi_n\cdots$ as in Theorem 4.3. Then*

$$\int_{x_1}^{x} f(t)\,dt = \lim_{m\to+\infty}\int_{x_1}^{L_{\xi_1\xi_2\cdots\xi_m}(x_1)} f(t)\,dt,$$

where

$$\begin{aligned}\int_{x_1}^{L_{\xi_1\xi_2\cdots\xi_m}(x_1)} f(t)\,dt &= \sum_{j=1}^{m}\prod_{n=1}^{j-1}\left(a_{\xi_n}\alpha_{\xi_n}\right)\int_{x_1}^{x_{\xi_j}} f(t)\,dt \\
&\quad + \sum_{i=2}^{m-1}\sum_{j=i}^{m}\prod_{n=1}^{j-i}\left(a_{\xi_n}\alpha_{\xi_n}\right)a_{\xi_{j-i+1}}\int_{L_{\xi_{j-i+1}\cdots\xi_{j-1}}(x_1)}^{L_{\xi_{j-i+1}\cdots\xi_{j-1}\xi_j}(x_1)} q_{\xi_{j-i+1}}(t)\,dt\end{aligned} \tag{4.9}$$

and $\int_{x_1}^{x_{\xi_j}} f(t)\,dt$ is given by (4.6).

Proof Formula (4.9) is proved by mathematical induction.

For $m = 1$,

$$
\begin{aligned}
\int_{x_1}^{L_{\xi_1}(x_1)} f(t)\,dt &= \sum_{j=1}^{1}\prod_{n=1}^{j-1}\left(a_{\xi_n}\alpha_{\xi_n}\right)\int_{x_1}^{x_{\xi_j}} f(t)\,dt \\
&\quad + \sum_{i=1}^{1-1}\sum_{j=i+1}^{1}\prod_{n=1}^{j-i-1}\left(a_{\xi_n}\alpha_{\xi_n}\right)a_{\xi_{j-i}}\int_{L_{\xi_{j-i+1}\cdots\xi_{j-1}}(x_1)}^{L_{\xi_{j-i+1}\cdots\xi_{j-1}\xi_j}(x_1)} q_{\xi_{j-i}}(t)\,dt \\
&= \int_{x_1}^{x_{\xi_1}} f(t)\,dt.
\end{aligned}
$$

Supposing that (4.8) is valid for some $m \in \mathbb{N}$. By definition of f, and by Lemma 4.2,

$$
\begin{aligned}
\int_{x_1}^{L_{\xi_1\xi_2\cdots\xi_{m+1}}(x_1)} f(t)\,dt &= \int_{x_1}^{L_{\xi_1\xi_2\cdots\xi_m}(x_1)} f(t)\,dt + \int_{L_{\xi_1\xi_2\cdots\xi_m}(x_1)}^{L_{\xi_1\xi_2\cdots\xi_m\xi_{m+1}}(x_1)} f(t)\,dt \\
&= \int_{x_1}^{L_{\xi_1\xi_2\cdots\xi_m}(x_1)} f(t)\,dt \\
&\quad + a_{\xi_1}\alpha_{\xi_1}\int_{L_{\xi_2\cdots\xi_m}(x_1)}^{L_{\xi_2\cdots\xi_m\xi_{m+1}}(x_1)} f(t)\,dt + a_{\xi_1}\int_{L_{\xi_2\cdots\xi_m}(x_1)}^{L_{\xi_2\cdots\xi_m\xi_{m+1}}(x_1)} q_{\xi_1}(t)\,dt.
\end{aligned}
$$

By the induction step and Lemma 4.2,

$$
\begin{aligned}
\int_{x_1}^{L_{\xi_1\xi_2\cdots\xi_{m+1}}(x_1)} f(t)\,dt &= \sum_{j=1}^{m}\prod_{n=1}^{j-1}\left(a_{\xi_n}\alpha_{\xi_n}\right)\int_{x_1}^{x_{\xi_j}} f(t)\,dt \\
&\quad + \sum_{i=2}^{m-1}\sum_{j=i}^{m}\prod_{n=1}^{j-i}\left(a_{\xi_n}\alpha_{\xi_n}\right)a_{\xi_{j-i+1}}\int_{L_{\xi_{j-i+1}\cdots\xi_{j-1}}(x_1)}^{L_{\xi_{j-i+1}\cdots\xi_{j-1}\xi_j}(x_1)} q_{\xi_{j-i+1}}(t)\,dt \\
&\quad + a_{\xi_1}\alpha_{\xi_1}\prod_{n=2}^{m}\left(a_{\xi_n}\alpha_{\xi_n}\right)\int_{x_1}^{x_{\xi_{m+1}}} f(t)\,dt \\
&\quad + a_{\xi_1}\alpha_{\xi_1}\sum_{j=2}^{m}\prod_{n=2}^{j-1}\left(a_{\xi_n}\alpha_{\xi_n}\right)a_{\xi_j}\int_{L_{\xi_{j+1}\cdots\xi_m}(x_1)}^{L_{\xi_{j+1}\cdots\xi_{m+1}}(x_1)} q_{\xi_j}(t)\,dt \\
&\quad + a_{\xi_1}\int_{L_{\xi_2\cdots\xi_m}(x_1)}^{L_{\xi_2\cdots\xi_m\xi_{m+1}}(x_1)} q_{\xi_1}(t)\,dt \\
&= \sum_{j=1}^{m+1}\prod_{n=1}^{j-1}\left(a_{\xi_n}\alpha_{\xi_n}\right)\int_{x_1}^{x_{\xi_j}} f(t)\,dt \\
&\quad + \sum_{i=2}^{m-1}\sum_{j=i}^{m}\prod_{n=1}^{j-i}\left(a_{\xi_n}\alpha_{\xi_n}\right)a_{\xi_{j-i+1}}\int_{L_{\xi_{j-i+1}\cdots\xi_{j-1}}(x_1)}^{L_{\xi_{j-i+1}\cdots\xi_{j-1}\xi_j}(x_1)} q_{\xi_{j-i+1}}(t)\,dt \\
&\quad + \sum_{j=2}^{m}\prod_{n=1}^{j-1}\left(a_{\xi_n}\alpha_{\xi_n}\right)a_{\xi_j}\int_{L_{\xi_{j+1}\cdots\xi_m}(x_1)}^{L_{\xi_{j+1}\cdots\xi_{m+1}}(x_1)} q_{\xi_j}(t)\,dt + a_{\xi_1}\int_{L_{\xi_2\cdots\xi_m}(x_1)}^{L_{\xi_2\cdots\xi_m\xi_{m+1}}(x_1)} q_{\xi_1}(t)\,dt \\
&= \sum_{j=1}^{m+1}\prod_{n=1}^{j-1}\left(a_{\xi_n}\alpha_{\xi_n}\right)\int_{x_1}^{x_{\xi_j}} f(t)\,dt \\
&\quad + \sum_{i=2}^{m}\sum_{j=i}^{m+1}\prod_{n=1}^{j-i}\left(a_{\xi_n}\alpha_{\xi_n}\right)a_{\xi_{j-i+1}}\int_{L_{\xi_{j-i+1}\cdots\xi_{j-1}}(x_1)}^{L_{\xi_{j-i+1}\cdots\xi_{j-1}\xi_j}(x_1)} q_{\xi_{j-i+1}}(t)\,dt,
\end{aligned}
$$

which concludes the proof of formula (4.9).

For a x defined by an infinity sequence of ξ_n, the integral is given by the limit of the formula of the finite sequence definition. □

Theorems 4.1 and 4.2 have discussed the definite integral of classical FIF with constant and variable scaling factors. For more details, refer [9]. The following theorem explores its indefinite integral, as a result, a new fractal interpolation function is obtained. From this end, the scaling factor $\alpha_n : [x_1, x_N] \to (0, 1)$ is taken to be a continuous function such that $\|\alpha\|_\infty < 1$ and α' exists [10].

Theorem 4.5 *Let f be a fractal interpolation function with variable scaling factors determined by the IFS $\{L_n(x), F_n(x, y) : n = 1, 2, \ldots, N-1\}$, and its corresponding integral is given by, $\hat{f}(x) = \hat{y}_1 + \int_{x_1}^{x} f(t)dt$, then $\hat{f}$ is a fractal interpolation function determined by the IFS $\{L_n(x), \hat{F}_n(x, \hat{y}) : n = 1, 2, \ldots, N-1\}$, where $\hat{F}_n(x, \hat{y}) = a_n\alpha_n(x)\hat{y} + \hat{q}_n(x)$ with $\hat{f}(x_N) = \hat{y}_N,\ \hat{f}(x_1) = 0,\ \sum_{j=1}^{N-1} a_j\alpha_j(x_N) \neq 1$ and for $j = 1, 2, \ldots, N-1$,*

$$\hat{q}_n(x) = \hat{y}_n - a_n \int_{x_1}^{x} \left(\alpha_n'(u) \int_{x_1}^{u} f(t)dt\right) du + a_n \int_{x_1}^{x} q_n(u)du,$$

$$\hat{y}_{n+1} = \sum_{j=1}^{n} \left\{a_j\alpha_j(x_N)\hat{y}_N - a_j \int_{x_1}^{x_N} \left(\alpha_j'(u) \int_{x_1}^{u} f(t)dt\right) du + a_j \int_{x_1}^{x_N} q_j(u)du\right\},$$

$$\hat{y}_N = \frac{\sum_{j=1}^{N-1} \left\{-a_j \int_{x_1}^{x_N} \left(\alpha_j'(u) \int_{x_1}^{u} f(t)dt\right) du + a_j \int_{x_1}^{x_N} q_j(u)du\right\}}{1 - \sum_{j=1}^{N-1} a_j\alpha_j(x_N)}.$$

Proof By the change of variables (4.1),

$$\begin{aligned}\hat{f}(L_n(x)) &= \hat{y}_1 + \int_{x_1}^{L_n(x)} f(t)dt \\ &= \hat{y}_1 + \int_{x_1}^{x_n} f(t)dt + \int_{x_n}^{L_n(x)} f(t)dt \\ &= \hat{y}_n + a_n \int_{x_1}^{x} f(L_n(u))du.\end{aligned}$$

Then using the functional equation of f,

$$\begin{aligned}\hat{f}(L_n(x)) &= \hat{y}_n + a_n \int_{x_1}^{x} \alpha_n(u)f(u)du + a_n \int_{x_1}^{x} q_n(u)du \\ &= \hat{y}_n + a_n\alpha_n(x) \int_{x_1}^{x} f(t)dt - a_n \int_{x_1}^{x} \left(\alpha_n'(u) \int_{x_1}^{u} f(t)dt\right) du + a_n \int_{x_1}^{x} q_n(u)du \\ &= \hat{y}_n + a_n\alpha_n(x)\hat{f}(x) - a_n \int_{x_1}^{x} \left(\alpha_n'(u) \int_{x_1}^{u} f(t)dt\right) du + a_n \int_{x_1}^{x} q_n(u)du \\ &= a_n\alpha_n(x)\hat{f}(x) + \hat{q}_n(x) \\ &= \hat{F}_n(x, \hat{f}(x)).\end{aligned}$$

For simplification, use the notation $\hat{q}_n(x) = \hat{y}_n - a_n \int_{x_1}^{x} \left(\alpha_n'(u) \int_{x_1}^{u} f(t)dt\right) du + a_n \int_{x_1}^{x} q_n(u)du$. To find a new set of interpolation data $\{(x_n, \hat{y}_n) : n = 1, 2, \ldots, N\}$, consider the following equation,

$$\hat{f}(L_n(x)) = \hat{y}_n + a_n\alpha_n(x)\hat{f}(x) - a_n \int_{x_1}^{x} \left(\alpha_n'(u) \int_{x_1}^{u} f(t)dt\right) du + a_n \int_{x_1}^{x} q_n(u)du.$$

Let $x = x_N$ and $L_n(x_N) = x_{n+1}$. Then

$$\hat{y}_{n+1} - \hat{y}_n = a_n\alpha_n(x_N)\hat{y}_N - a_n \int_{x_1}^{x_N} \left(\alpha_n'(u) \int_{x_1}^{u} f(t)dt\right) du + a_n \int_{x_1}^{x_N} q_n(u)du. \tag{4.10}$$

We know that $\hat{y}_{n+1} = \hat{y}_1 + \sum_{j=1}^{n} (\hat{y}_{j+1} - \hat{y}_j)$.

$$\hat{y}_{n+1} = \sum_{j=1}^{n} \left\{ a_j\alpha_j(x_N)\hat{y}_N - a_j \int_{x_1}^{x_N} \left(\alpha_j'(u) \int_{x_1}^{u} f(t)dt\right) du + a_j \int_{x_1}^{x_N} q_j(u)du \right\}.$$

Substituting $n = N - 1$, the end point $\hat{y}_N$ is obtained as

$$\hat{y}_N = \frac{\sum_{j=1}^{N-1} \left\{ a_j \int_{x_1}^{x_N} q_j(u)du - a_j \int_{x_1}^{x_N} \left(\alpha_j'(u) \int_{x_1}^{u} f(t)dt\right) du \right\}}{1 - \sum_{n=1}^{N-1} a_j\alpha_j(x_N)}.$$

The join-up conditions of $\hat{F}_n(x, \hat{y})$ are checked below:

$$\hat{q}_n(x_1) = \hat{y}_n - a_n\alpha_n(x_1)\hat{y}_1,$$
$$\hat{q}_n(x_N) = \hat{y}_n - a_n \int_{x_1}^{x_N} \left(\alpha_n'(u) \int_{x_1}^{u} f(t)dt\right) du + a_n \int_{x_1}^{x_N} q_n(u)du.$$

Comparing $\hat{q}_n(x_N)$ with Eq. (4.10), it is seen that

$$\hat{q}_n(x_N) = \hat{y}_{n+1} - a_n\alpha_n(x_N)\hat{y}_N,$$

thus yielding $\hat{F}_n(x_1, \hat{y}_1) = \hat{y}_n$ and $\hat{F}_n(x_N, \hat{y}_N) = \hat{y}_{n+1}$. Therefore, the function $\hat{f}$ is the fractal interpolation function whose graph is the attractor of the IFS $\{L_n(x), \hat{F}_n (x, \hat{y}) : n = 1, 2, \ldots, N - 1\}$. □

It is seen that Theorem 4.5 is the generalization of indefinite integral of classical FIFs with constant scaling factors.

Corollary 4.1 *Let f be a fractal interpolation function as described in Theorem 5, with constant scaling factors satisfying $\sum_{j=1}^{N-1} a_j\alpha_j \neq 1$. Then, for $j = 1, 2, \ldots, N - 1$,*

$$\hat{q}_n(x) = \hat{y}_n + a_n \int_{x_1}^{x} q_n(u)du,$$

$$\hat{y}_{n+1} = \sum_{j=1}^{n} \left\{ a_j \alpha_j \hat{y}_N + a_j \int_{x_1}^{x_N} q_j(u)du \right\},$$

$$\hat{y}_N = \frac{\sum_{j=1}^{N-1} a_j \int_{x_1}^{x_N} q_j(u)du}{1 - \sum_{j=1}^{N-1} a_j \alpha_j}.$$

Corollary 4.2 *If f is a linear fractal interpolation function with variable scaling factors associated with the IFS $\{(L_n(x), F_n(x, y)) : n = 1, 2, \ldots, N-1\}$, then the integral of f,*

$$\hat{f}(x) = \hat{y}_1 + \int_{x_1}^{x} f(t)dt,$$

generated by the IFS $\{(L_n(x), \hat{F}_n(x, \hat{y})) : n = 1, 2, \ldots, N-1\}$ is also a linear FIF, where $\hat{F}_n(x, \hat{y}) = a_n \alpha_n(x) \hat{f}(x) + \hat{q}_n(x)$ with $\sum_{j=1}^{N-1} a_j \alpha_j(x_N) \neq 1$, and for $j = 1, 2, \ldots, N-1$,

$$\hat{q}_n(x) = \hat{y}_n - a_n \alpha_n(x) \hat{y}_1 - a_n \int_{x_1}^{x} \left(\alpha_n'(t) \int_{x_1}^{x} f(t)dt \right) dt + \frac{a_n c_n}{2}(x^2 - x_1^2) + a_n d_n (x - x_1),$$

$$\hat{y}_{n+1} = \hat{y}_1 + \sum_{j=1}^{n} \left\{ a_j \alpha_j(x_N) \hat{y}_N - a_j \alpha_j(x_N) \hat{y}_1 - a_j \int_{x_1}^{x_N} \left(\alpha_n'(t) \int_{x_1}^{x_N} f(t)dt \right) dt \right.$$
$$\left. + \frac{a_j c_j}{2}(x_N^2 - x_1^2) + a_j d_j (x_N - x_1) \right\},$$

$$\hat{y}_N = \hat{y}_1 + \frac{\sum_{j=1}^{N-1} \left\{ - a_j \int_{x_1}^{x_N} \left(\alpha_j'(t) \int_{x_1}^{x_N} f(t)dt \right) dt + \frac{a_j c_j}{2}(x_N^2 - x_1^2) + a_j d_j (x_N - x_1) \right\}}{1 - \sum_{j=1}^{N-1} a_j \alpha_j(x_N)}.$$

Remark 4.2 If the integral of f is defined at the end point, then by

$$\hat{f}(x) = \hat{y}_N - \int_{x}^{x_N} f(t)dt,$$

here, the FIF $\hat{f}$ is generated by the IFS $\{(L_n(x), \hat{F}_n(x, \hat{y})) : n = 1, 2, \ldots, N-1\}$ where $\hat{F}_n(x, \hat{y}) = a_n \alpha_n(x) \hat{f}(x) + \hat{q}_n(x)$ with $\sum_{j=1}^{N} a_j \alpha_j(x_1) \neq 1$, and for $j = 1, 2, \ldots, N-1$,

$$\hat{q}_n(x) = \hat{y}_{n+1} - a_n\alpha_n(x)\hat{y}_N + a_n\int_x^{x_N}\left(\alpha_n'(t)\int_x^{x_N} f(t)dt\right)dt - \frac{a_nc_n}{2}(x_N^2 - x^2) - a_nd_n(x_N - x),$$

$$\hat{y}_n = \hat{y}_N - \sum_{j=n}^{N-1}\left\{a_j\alpha_j(x_1)\hat{y}_N - a_j\alpha_j(x_1)\hat{y}_1 - a_j\int_{x_1}^{x_N}\left(\alpha_j'(t)\int_{x_1}^{x_N} f(t)dt\right)dt\right.$$
$$\left. + \frac{a_jc_j}{2}(x_N^2 - x_1^2) + a_jd_j(x_N - x_1)\right\},$$

$$\hat{y}_1 = \hat{y}_N - \frac{\sum_{j=1}^{N-1}\left\{-a_j\int_{x_1}^{x_N}\left(\alpha_j'(t)\int_{x_1}^{x_N} f(t)dt\right)dt + \frac{a_jc_j}{2}(x_N^2 - x_1^2) + a_jd_j(x_N - x_1)\right\}}{1 - \sum_{j=1}^{N-1} a_j\alpha_j(x_1)}.$$

Remark 4.3 In Corollary 4.2, the integral of linear FIF with variable scaling factors is discussed. In [7], Barnsley has evaluated the integral of linear FIF with constant scaling factors. Hence, the reader can easily observe that Corollary 4.2 is the generalization of Theorem 1 in [7], i.e., the proof can be obtained if the vertical scaling factors in Corollary 4.2 are chosen as constants instead of functions,

$$\hat{f}(x) = \hat{y}_1 + \int_{x_1}^{x} f(t)dt,$$

$$\hat{q}_n(x) = \hat{y}_n - a_n\alpha_n\hat{y}_1 + a_n\int_{x_1}^{x} q_n(t)dt,$$

$$\hat{y}_{n+1} = \hat{y}_1 + \sum_{j=1}^{n}\{a_j\alpha_j(\hat{y}_N - \hat{y}_1) + a_j\int_{x_1}^{x_N} q_j(t)dt\},$$

$$\hat{y}_N = \hat{y}_1 + \frac{\sum_{j=1}^{N-1} a_j\int_{x_1}^{x_N} q_j(t)dt}{1 - \sum_{j=1}^{N-1} a_j\alpha_j}.$$

Example 4.1 Consider the interpolation data $\{(0, 0), (1/3, 1/4), (2/3, 1/4), (1, 0)\}$ with constant scaling factors $\alpha_1 = 0.5,\ \alpha_2 = -0.5,\ \alpha_3 = 0.4$. Fig. 4.1a represents the graph of linear fractal interpolation function and Fig. 4.1b illustrates the integral of f for the given data set. The graphical representation of linear FIF with variable scaling factors $\alpha_1(x) = 0.25\sin x + 0.6,\ \alpha_2(x) = 0.15\cos x + 0.7,\ \alpha_3(x) = x^2/3 + 0.1$ is demonstrated in Fig. 4.1c and its integral is presented in Fig. 4.1d. In Fig. 4.1 black dots denote the data points and the associated fractal function is represented by the curve.

The integral of quadratic fractal interpolation function is estimated as a special case of Theorem 4.5.

Corollary 4.3 *Let f be an quadratic fractal interpolation function with variable scaling factors determined by the IFS $\{L_n(x), F_n(x, y) : n = 1, 2, \dots, N-1\}$, and its integral is given by, $\hat{f}(x) = \hat{y}_1 + \int_{x_1}^{x} f(t)dt$, then $\hat{f}$ is a fractal interpolation function generated by the IFS $\{L_n(x), \hat{F}_n(x, \hat{y}) : n = 1, 2, \dots, N-1\}$, where $\hat{F}_n(x, \hat{y}) = a_n\alpha_n(x)\hat{f}(x) + \hat{q}_n(x)$ with $\hat{f}(x_1) = \hat{y}_1 = 0$, and $\sum_{j=1}^{N-1} a_j\alpha_j(x_N) \neq 1$, for $j = 1, 2, \dots, N-1$,*

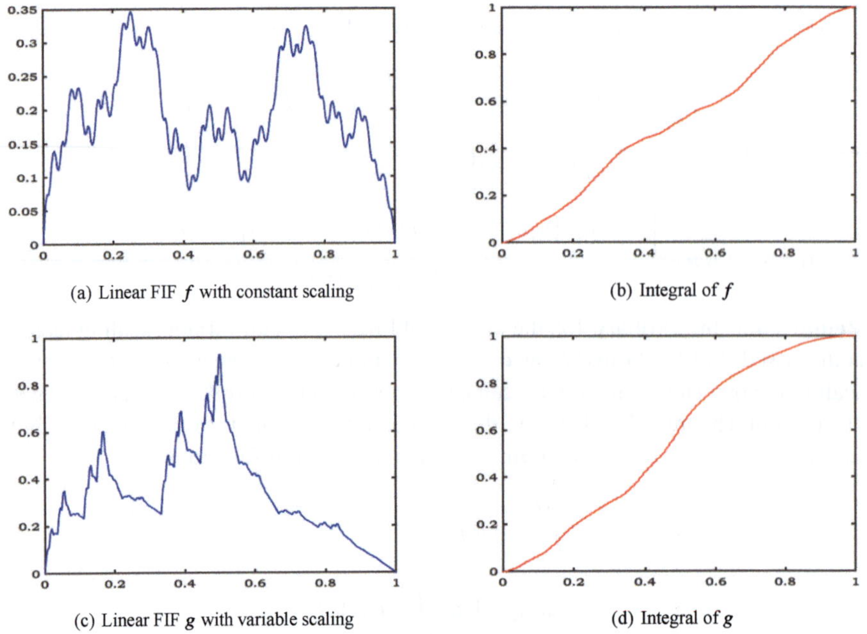

(a) Linear FIF f with constant scaling

(b) Integral of f

(c) Linear FIF g with variable scaling

(d) Integral of g

Fig. 4.1 Graphical representation of **a** linear FIF f with constant scaling factors, **b** its classical integral $\hat{f}$, **c** linear FIF g with variable scaling factors and **d** its classical integral $\hat{g}$

$$\hat{q}_n(x) = \hat{y}_n - a_n \int_{x_1}^{x} \left(\alpha_n'(u) \int_{x_1}^{u} f(t)dt\right) du + \frac{a_n}{3}(x^3 - x_1^3) + \frac{a_n c_n}{2}(x^2 - x_1^2) + a_n d_n (x - x_1),$$

$$\hat{y}_{n+1} = \sum_{j=1}^{n} \Big\{ a_j \alpha_j(x_N)\hat{y}_N - a_j \int_{x_1}^{x_N} \left(\alpha_j'(u) \int_{x_1}^{u} f(t)dt\right) du + \frac{a_j}{3}(x_N^3 - x_1^3)$$
$$+ \frac{a_j c_j}{2}(x_N^2 - x_1^2) + a_j d_j (x_N - x_1) \Big\},$$

$$\hat{y}_N = \frac{\sum_{j=1}^{N-1} \Big\{ - a_j \int_{x_1}^{x_N} \left(\alpha_j'(u) \int_{x_1}^{u} f(t)dt\right) du + \frac{a_j}{3}(x_N^3 - x_1^3) + \frac{a_j c_j}{2}(x_N^2 - x_1^2) + a_j d_j (x_N - x_1) \Big\}}{1 - \sum_{j=1}^{N-1} a_j \alpha_j(x_N)}.$$

Corollary 4.4 *Let f be a fractal interpolation function as described in Corollary 4.3, and defined at the end points as $\hat{f}(x) = \hat{y}_N - \int_x^{x_N} f(t)dt$, with $\hat{f}(x_N) = \hat{y}_N = 0$, $\sum_{j=1}^{N-1} a_j \alpha_j(x_1) \neq 1$. Then, for $j = 1, 2, \ldots, N-1$,*

$$\hat{q}_n(x) = \hat{y}_{n+1} + a_n \int_x^{x_N} \left(\alpha_n'(u) \int_u^{x_N} f(t)dt\right) du - \frac{a_n}{3}(x_N^3 - x^3) - \frac{a_n c_n}{2}(x_N^2 - x^2) - a_n d_n (x_N - x),$$

$$\hat{y}_n = -\sum_{j=n}^{N-1} \left\{ a_j \alpha_j(x_1)\hat{y}_1 + a_j \int_{x_1}^{x_N} \left(\alpha_j'(u) \int_u^{x_N} f(t)dt\right) du - \frac{a_j}{3}(x_N^3 - x_1^3) \right.$$
$$\left. - \frac{a_j c_j}{2}(x_N^2 - x_1^2) - a_j d_j (x_N - x_1) \right\},$$

$$\hat{y}_1 = -\frac{\sum_{j=1}^{N-1} \left\{ a_j \int_{x_1}^{x_N} \left(\alpha_j'(u) \int_u^{x_N} f(t)dt\right) du - \frac{a_j}{3}(x_N^3 - x_1^3) - \frac{a_j c_j}{2}(x_N^2 - x_1^2) - a_j d_j (x_N - x_1) \right\}}{1 - \sum_{j=1}^{N-1} a_j \alpha_j(x_1)}.$$

Example 4.2 In this example, quadratic fractal interpolation functions g, h and f are constructed for the given data set $J = \{(0, 0), (1/4, 1/2), (1/2, 1/3), (3/4, 1/2), (1, 1)\}$. Suppose the vertical scaling factors are chosen as constants $\alpha_1 = 0.5$, $\alpha_2 = -0.5$, $\alpha_3 = 0.4$, $\alpha_4 = -0.8$, then the quadratic FIFs g with constant scaling factors can be generated. For the same set of data J, different quadratic FIFs h with variable scaling factors and its indefinite integral $\hat{h}$ can be generated by choosing the scaling factors as variables $\alpha_1(x) = 0.25 \sin x + 0.6$, $\alpha_2(x) = 0.15 \cos x + 0.7$, $\alpha_3(x) = x^2/3 + 0.1$, $\alpha_4(x) = \frac{1}{5}\exp(x) + 0.1$, $\forall x \in J$. Figure 8.1a and b represent the graphs of quadratic FIF g with constant scalings and its indefinite integral $\hat{g}$ respectively. Similarly, the graph of quadratic FIF h with variable scalings and its indefinite integral $\hat{h}$ are illustrated in Fig. 4.2c and d. When both the graphs are compared, it can be observed that the graph of quadratic FIF offers more flexibility. Also one can note that the smoothness of integrals of fractal functions in sub-figures (c) and (d) with variable scalings is greater than the integrals with constant scalings in sub-figures (a) and (b).

Further, using (4.5), the definite integral of the quadratic fractal interpolation function f with constant scaling factors $\alpha_1 = 0.6$, $\alpha_2 = -0.6$, $\alpha_3 = 0.8$, $\alpha_4 = -0.8$ for J over the interval [0, 1] is explicitly computed as

$$\int_0^1 f(t)\,dt = \frac{\sum_{i=1}^4 a_i \int_0^1 q_i(t)\,dt}{1 - \sum_{i=1}^4 a_i \alpha_i},$$

where

$$a_1 \int_0^1 q_1(t)\,dt = -0.055,\; a_2 \int_0^1 q_2(t)\,dt = 0.136,\; a_3 \int_0^1 q_3(t)\,dt = -0.039,\; a_4 \int_0^1 q_4(t)\,dt = 0.245.$$

Therefore,

$$\int_0^1 f(t)\,dt = 0.29.$$

In Theorem 4.5, by choosing the vertical scaling factors as variable and defining the integral of quadratic FIF at the initial point, $\hat{f}$ is shown as a fractal interpolation function.

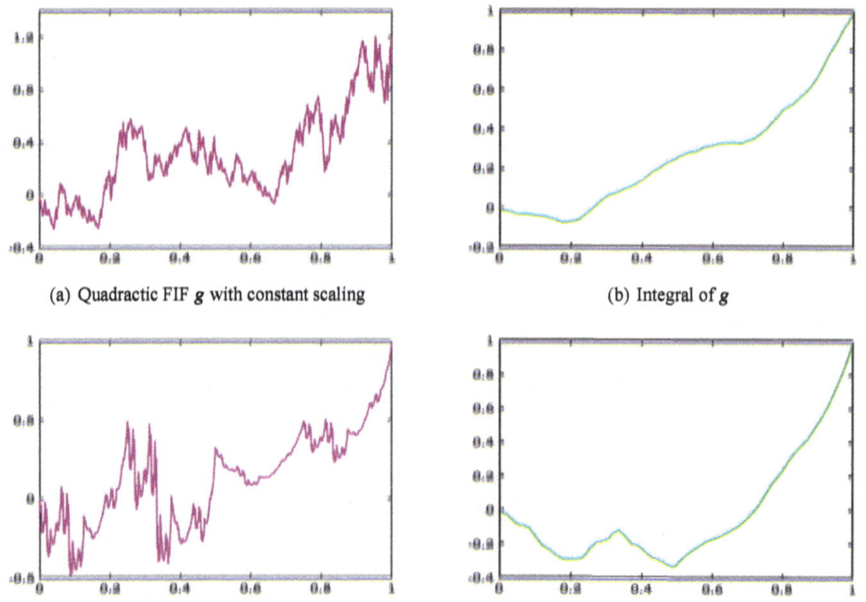

(a) Quadractic FIF g with constant scaling

(b) Integral of g

(c) Quadractic FIF h with variable scaling

(d) Integral of h

Fig. 4.2 Graphical representation of **a** quadratic FIF g with constant scaling factors, **b** its classical integral $\hat{g}$, **c** linear FIF h with variable scaling factors and **d** its classical integral $\hat{h}$

Corollary 4.5 *Let f be a fractal interpolation function as described in Theorem 5, with constant scaling factors satisfying $\sum_{j=1}^{N-1} a_j\alpha_j \neq 1$. Then, for $j = 1, 2, \ldots, N-1$,*

$$\hat{q}_n(x) = \hat{y}_n + \frac{a_n}{3}(x^3 - x_1^3) + \frac{a_n c_n}{2}(x^2 - x_1^2) + a_n d_n(x - x_1),$$

$$\hat{y}_{n+1} = \sum_{j=1}^{n} \left\{ a_j\alpha_j\hat{y}_N + \frac{a_j}{3}(x_N^3 - x_1^3) + \frac{a_j c_j}{2}(x_N^2 - x_1^2) + a_j d_j(x_N - x_1) \right\},$$

$$\hat{y}_N = \frac{\sum_{j=1}^{N-1} \left\{ \frac{a_j}{3}(x_N^3 - x_1^3) + \frac{a_j c_j}{2}(x_N^2 - x_1^2) + a_j d_j(x_N - x_1) \right\}}{1 - \sum_{j=1}^{N-1} a_j\alpha_j}.$$

Corollary 4.4 is the generalization of the integral of quadratic FIF with constant scaling factors, if its integral is predefined at the end point.

Corollary 4.6 *If f is the fractal interpolation function as described in Corollary 4.4 with constant scaling factors such that $\sum_{j=1}^{N-1} a_j\alpha_j \neq 1$. Then, for $j = 1, 2, \ldots, N-1$,*

$$\hat{q}_n(x) = \hat{y}_{n+1} - \frac{a_n}{3}(x_N^3 - x^3) - \frac{a_n c_n}{2}(x_N^2 - x^2) - a_n d_n (x_N - x),$$

$$\hat{y}_n = -\sum_{j=n}^{N-1} \left\{ - a_j \alpha_j \hat{y}_1 + \frac{a_j}{3}(x_N^3 - x_1^3) + \frac{a_j c_j}{2}(x_N^2 - x_1^2) + a_j d_j (x_N - x_1) \right\},$$

$$\hat{y}_1 = -\frac{\sum_{j=1}^{N-1} \left\{ \frac{a_j}{3}(x_N^3 - x_1^3) + \frac{a_j c_j}{2}(x_N^2 - x_1^2) + a_j d_j (x_N - x_1) \right\}}{1 - \sum_{j=1}^{N-1} a_j \alpha_j}.$$

4.3 Integral on α-Fractal Functions

In this section, α-fractal function is represented by f^α for the given continuous function f. Consider the integral of f^α as follows,

$$\hat{f}^\alpha(x) = \hat{y}_1 + \int_{x_1}^{x} f^\alpha(t) dt. \tag{4.11}$$

Equation (8.19) gives

$$\begin{aligned} \hat{f}^\alpha(L_n(x)) &= \hat{y}_1 + \int_{x_1}^{L_n(x)} f^\alpha(t) dt \\ &= \hat{y}_1 + \int_{x_1}^{x_n} f^\alpha(t) dt + \int_{x_n}^{L_n(x)} f^\alpha(t) dt, \\ &= \hat{f}^\alpha(x_n) + \int_{x_n}^{L_n(x)} f^\alpha(t) dt. \end{aligned}$$

When t is replaced by $L_n(u)$ in the second term, one can get,

$$\begin{aligned} \hat{f}^\alpha(L_n(x)) &= \hat{y}_n + a_n \int_{x_1}^{x} f^\alpha(L_n(u)) du \\ &= \hat{y}_n + a_n \int_{x_1}^{x} (\alpha_n f^\alpha(u) + q_n(u)) du \\ &= \hat{y}_n + a_n \alpha_n \int_{x_1}^{x} f^\alpha(u) du + a_n \int_{x_1}^{x} q_n(u) du \\ &= \hat{y}_n + a_n \alpha_n \hat{f}^\alpha(x) + a_n \int_{x_1}^{x} q_n(t) dt \\ &= a_n \alpha_n \hat{f}^\alpha(x) + \hat{q}_n(x) \\ &= \hat{F}_n(x, \hat{f}^\alpha(x)). \end{aligned}$$

Here, $\hat{q}_n(x) = \hat{y}_n + a_n \int_{x_1}^{x} q_n(t)dt$. Since f^α is the α-FIF, then Eq. (2.3) gives,

$$\hat{q}_n(x) = \hat{y}_n + a_n \int_{x_1}^{x} f \circ L_n(t)dt - a_n\alpha_n \int_{x_1}^{x} b(t)dt. \tag{4.12}$$

Observe that, $(F \circ L_n)'(x) = \frac{d}{dx} \int_{x_1}^{L_n(x)} f(t)dt$. By using change of variable $t = L_n(u)$, one can have

$$\begin{aligned}(F \circ L_n)'(x) &= a_n \frac{d}{dx} \int_{L_n^{-1}(x_1)}^{x} f \circ L_n(u)du \\ &= a_n f \circ L_n(x).\end{aligned}$$

Therefore, the Eq. (4.12) becomes

$$\begin{aligned}\hat{q}_n(x) &= \hat{y}_n + \int_{x_1}^{x} (F \circ L_n)'(u)du - a_n\alpha_n B(x) \\ &= \hat{y}_n + (F \circ L_n)(x) - (F \circ L_n)(x_1) - a_n\alpha_n B(x) \\ &= \hat{y}_n + (F \circ L_n)(x) - F(x_{n-1}) - a_n\alpha_n B(x) \\ &= (F \circ L_n)(x) - a_n\alpha_n B(x).\end{aligned}$$

Now, it is easy to verify that

$$\begin{aligned}\hat{q}_n(x_1) &= (F \circ L_n)(x_1) - a_n\alpha_n B(x_1) \\ &= \hat{y}_n - a_n\alpha_n \hat{y}_1, \\ \hat{q}_n(x_N) &= (F \circ L_n)(x_N) - a_n\alpha_n B(x_N) \\ &= \hat{y}_{n+1} - a_n\alpha_n \hat{y}_N,\end{aligned}$$

which clearly implies $\hat{F}_n(x_1, \hat{y}_1) = \hat{y}_n$ and $\hat{F}_n(x_N, \hat{y}_N) = \hat{y}_{n+1}$. As f^α satisfies the continuity conditions at the joints, the integral of f^α is the unique attractor of the IFS $\{L_n(x), \hat{F}_n(x, y)\} : n = 1, 2, \ldots, N-1$. To find the new set of data points of the interpolant $\hat{f}^\alpha$, consider the following equation,

$$\hat{f}^\alpha(L_n(x)) = a_n\alpha_n \hat{f}^\alpha(x) + (F \circ L_n)(x) - a_n\alpha_n B(x).$$

Take $x = x_N$ and $L_n(x_N) = x_{n+1}$, then

$$\hat{y}_{n+1} = a_n\alpha_n \hat{y}_N + F \circ L_n(x_N) - a_n\alpha_n B(x_N).$$

Substituting $n = N-1$ in the above equation, one can get the end point $\hat{y}_N$ as,

$$\hat{y}_N = \frac{F \circ L_{N-1}(x_N) - a_{N-1}\alpha_{N-1} B(x_N)}{1 - a_{N-1}\alpha_{N-1}}.$$

Thus, the integral of f^α is also a α-fractal function which passes through the data $\{(x_n, \hat{y}_n) : i = 1, 2, \ldots, N\}$. This discussion is summarized as the following theorem.

Theorem 4.6 *Let f^α be the α-FIF associated with the IFS $\{L_n(x), F_n(x, y)\} : n = 1, 2, \ldots, N-1$. Suppose $F(x) = \int_{x_1}^x f(t)dt$, $B(x) = \int_{x_1}^x b(t)dt$ with $\hat{y}_1 = F(x_1) = B(x_1) = 0$ and $F(x_N) = B(x_N)$, then the integral of f^α,*

$$\hat{f}^\alpha(x) = \hat{y}_1 + \int_{x_1}^x f^\alpha(t)dt \tag{4.13}$$

is also a α-FIF of the data $\{(x_n, \hat{y}_n) : n = 1, 2, \ldots, N\}$. Further, $\hat{f}^\alpha$ is generated by the IFS $\{L_n(x), \hat{F}_n(x, y)\} : n = 1, 2, \ldots, N-1$, where $\hat{F}_n(x, y) = a_n \alpha_n y + \hat{q}_n(x)$, for $n = 1, 2, \ldots, N-1$,

$$\begin{aligned}
\hat{q}_n(x) &= (F \circ L_n)(x) - a_n \alpha_n B(x),\\
\hat{y}_{n+1} &= a_n \alpha_n \hat{y}_N + F \circ L_n(x_N) - a_n \alpha_n B(x_N),\\
\hat{y}_N &= \frac{F \circ L_{N-1}(x_N) - a_{N-1}\alpha_{N-1} B(x_N)}{1 - a_{N-1}\alpha_{N-1}}.
\end{aligned}$$

Theorem 4.6 provides the new α-fractal interpolation function, namely $\hat{f}^\alpha(x)$, passing through the data points $\{(x_n, \hat{y}_n) : n = 1, 2, \ldots, N\}$.

The following is the precise definition of α-fractal function with variable scaling factors:

Definition 4.1 If the mappings $L_n(x)$ and $F_n(x, y)$ are defined as

$$L_n(x) = a_n x + b_n, \ F_n(x, y) = \alpha_n(x) y + q_n(x), \ \forall\, n = 1, 2, \ldots, N-1$$

with $q_n(x) = f \circ L_n(x) - \alpha_n(x) b(x)$, where b is a continuous function on I such that $b(x_1) = f(x_1)$, $b(x_N) = f(x_N)$, $b \neq f$ and the scaling factor α_n is a continuous function on I, then the fractal interpolation function satisfying the functional equation,

$$f^\alpha(x) = f(x) + \alpha_n(L_n^{-1}(x))[(f^\alpha - b) \circ L_n^{-1}(x)], \ \forall x \in I_n, \ n = 1, 2, \ldots, N-1. \tag{4.14}$$

is called the α-fractal function with variable scaling factors. Note that if $\alpha = 0$, then $f^\alpha = f$.

In the following theorem, the α-fractal function with variable scaling factors is integrated by predefining the initial condition $\hat{y}_1 = F(x_1) = B(x_1) = 0$ and a new α-fractal function is obtained along with the new interpolation data set.

Theorem 4.7 *Let f^α be a α-fractal function with variable scaling factors associated with the IFS $\{(L_n(x), F_n(x, y)) : n = 1, 2, \ldots, N-1\}$. Suppose $F(x) = \int_{x_1}^x f(t)dt$*

and $B(x)=\int_{x_1}^{x} b(t)dt$ *with* $\hat{y}_1=F(x_1)=B(x_1)=0$ *and* $F(x_N)=B(x_N)$*, then the integral of* f^{α}*,*

$$\hat{f}^{\alpha}(x)=\hat{y}_1+\int_{x_1}^{x} f^{\alpha}(t)dt$$

is also a α*-fractal function generated by the IFS* $\{(L_n(x), \hat{F}_n(x,\hat{y})) : n=1,2,\ldots, N-1\}$*, where* $\hat{F}_n(x,\hat{y})=a_n\alpha_n(x)\hat{f}^{\alpha}(x)+\hat{q}_n(x)$ *with* $a_{N-1}\alpha_{N-1}(x_N)\neq 1$*, and for* $n=1,2,\ldots,N-1$*,*

$$\hat{q}_n(x)=(F\circ L_n)(x)-a_n\alpha_n(x)B(x)-a_n\int_{x_1}^{x}\left(\alpha_n'(t)\int_{x_1}^{x} f^{\alpha}(t)dt\right)dt+a_n\int_{x_1}^{x}\alpha_n'(t)B(x)dt,$$

$$\hat{y}_{n+1}=a_n\alpha_n(x_N)\hat{y}_N-a_n\int_{x_1}^{x_N}\left(\alpha_n'(t)\int_{x_1}^{x_N} f^{\alpha}(t)dt\right)dt+(F\circ L_n)(x_N)-a_n\alpha_n(x_N)B(x_N)$$
$$+a_n\int_{x_1}^{x_N}\alpha_n'(t)B(x_N)dt,$$

$$\hat{y}_N=\left((F\circ L_{N-1})(x_N)-a_{N-1}\alpha_{N-1}(x_N)B(x_N)-a_{N-1}\int_{x_1}^{x_N}\left(\alpha_{N-1}'(t)\int_{x_1}^{x_N} f^{\alpha}(t)dt\right)dt\right.$$
$$\left.+a_{N-1}\left(\int_{x_1}^{x_N}\alpha_{N-1}'(t)B(x_N)dt\right)\right)/(1-a_{N-1}\alpha_{N-1}(x_N)).$$

Proof Consider the following equation,

$$\hat{f}^{\alpha}(L_n(x))=\hat{y}_1+\int_{x_1}^{x_n} f^{\alpha}(t)dt+\int_{x_n}^{L_n(x)} f^{\alpha}(t)dt$$
$$=\hat{y}_n+a_n\int_{x_1}^{x} f^{\alpha}(L_n(u))du.$$

Using the functional Eq. (4.14), one can get

$$\hat{f}^{\alpha}(L_n(x))=\hat{y}_n+a_n\int_{x_1}^{x}(\alpha_n(u)f^{\alpha}(u)+q_n(u))du$$
$$=\hat{y}_n+a_n\int_{x_1}^{x}\alpha_n(u)f^{\alpha}(u)du+a_n\int_{x_1}^{x}(f\circ L_n(u)-\alpha_n(u)b(u))du$$
$$=\hat{y}_n+a_n\alpha_n(x)\int_{x_1}^{x}f^{\alpha}(t)dt-a_n\int_{x_1}^{x}\left(\alpha_n'(t)\int_{x_1}^{x}f^{\alpha}(t)dt\right)dt+\int_{x_1}^{x}(F\circ L_n)'(t)dt$$
$$-a_n\int_{x_1}^{x}\alpha_n(t)b(t)dt$$
$$=\hat{y}_n+a_n\alpha_n(x)\int_{x_1}^{x}f^{\alpha}(t)dt-a_n\int_{x_1}^{x}\left(\alpha_n'(t)\int_{x_1}^{x}f^{\alpha}(t)dt\right)dt+(F\circ L_n)(x)-(F\circ L_n)(x_1)$$
$$-a_n\alpha_n(x)\int_{x_1}^{x}b(t)dt+a_n\int_{x_1}^{x}\left(\alpha_n'(t)\int_{x_1}^{x}b(t)dt\right)dt$$
$$=a_n\alpha_n(x)\hat{f}^{\alpha}(x)-a_n\int_{x_1}^{x}\left(\alpha_n'(t)\int_{x_1}^{x}f^{\alpha}(t)dt\right)dt+(F\circ L_n)(x)-a_n\alpha_n(x)B(x)$$
$$+a_n\int_{x_1}^{x}\alpha_n'(t)B(x)dt$$
$$=a_n\alpha_n(x)\hat{f}^{\alpha}(x)+\hat{q}_n(x)$$
$$=\hat{F}_n(x,\hat{f}^{\alpha}(x)).$$

Denote $\hat{q}_n(x) = (F \circ L_n)(x) - a_n\alpha_n(x)B(x) - a_n \int_{x_1}^{x} \left(\alpha_n'(t) \int_{x_1}^{x} f^\alpha(t)dt\right) dt + a_n \int_{x_1}^{x} \alpha_n'(t)B(x)dt$. To get the interpolation points $(x_n, \hat{y}_n)$, consider

$$f^\alpha(L_n(x)) = a_n\alpha_n(x)\hat{f}^\alpha(x) - a_n \int_{x_1}^{x} \left(\alpha_n'(t) \int_{x_1}^{x} f^\alpha(t)dt\right) dt + (F \circ L_n)(x) - a_n\alpha_n(x)B(x) + a_n \int_{x_1}^{x} \alpha_n'(t)B(x)dt.$$

Let $x = x_N$ and $L_n(x_N) = x_{n+1}$,

$$\begin{aligned}\hat{y}_{n+1} = a_n\alpha_n(x_N)\hat{y}_N - a_n \int_{x_1}^{x_N} \left(\alpha_n'(t) \int_{x_1}^{x_N} f^\alpha(t)dt\right) dt + (F \circ L_n)(x_N) - a_n\alpha_n(x_N)B(x_N)\\ + a_n \int_{x_1}^{x_N} \alpha_n'(t)B(x_N)dt.\end{aligned} \tag{4.15}$$

Substituting $n = N - 1$ in the above equation, one can get

$$\begin{aligned}\hat{y}_N = \Bigg((F \circ L_{N-1})(x_N) - a_{N-1}\alpha_{N-1}(x_N)B(x_N) - a_{N-1} \int_{x_1}^{x_N} \left(\alpha_{N-1}'(t) \int_{x_1}^{x_N} f^\alpha(t)dt\right) dt\\ + a_{N-1}\left(\int_{x_1}^{x_N} \alpha_{N-1}'(t)B(x_N)dt\right)\Bigg) \Big/ (1 - a_{N-1}\alpha_{N-1}(x_N)).\end{aligned}$$

Checking the continuity conditions of $\hat{q}_n(x)$,

$$\begin{aligned}&\hat{q}_n(x_1) = \hat{y}_n - a_n\alpha_n(x_1)\hat{y}_1,\\ &\hat{q}_n(x_N) = (F \circ L_n)(x_N) - a_n\alpha_n(x_N)B(x_N) - a_n \int_{x_1}^{x_N} \left(\alpha_n'(t) \int_{x_1}^{x_N} f^\alpha(t)dt\right) dt + a_n \int_{x_1}^{x_N} \alpha_n'(t)B(x_N)dt.\end{aligned}$$

Comparing $\hat{q}_n(x_N)$ with Eq. (4.15), one can get, $\hat{q}_n(x_N) = \hat{y}_{n+1} - a_n\alpha_n(x_N)\hat{y}_N$ which implies $\hat{F}_n(x_1, \hat{y}_1) = \hat{y}_n$ and $\hat{F}_n(x_N, \hat{y}_N) = \hat{y}_{n+1}$. Thus, the integral of f^α is a α-fractal function interpolating the new data set $\{(x_n, \hat{y}_n) : n = 1, 2, \ldots, N\}$. □

Remark 4.4 If the integral of f^α is defined at the end point, then

$$\hat{f}^\alpha(x) = \hat{y}_N - \int_{x}^{x_N} f^\alpha(t)dt,$$

here, the α-fractal function $\hat{f}^\alpha$ is generated by the IFS $\{(L_n(x), \hat{F}_n(x, \hat{y})) : n = 1, 2, \ldots, N-1\}$ where $\hat{F}_n(x, \hat{y}) = a_n\alpha_n(x)\hat{f}^\alpha(x) + \hat{q}_n(x)$ with $a_1\alpha_1(x_1) \neq 1$, and for $n = 1, 2, \ldots, N - 1$,

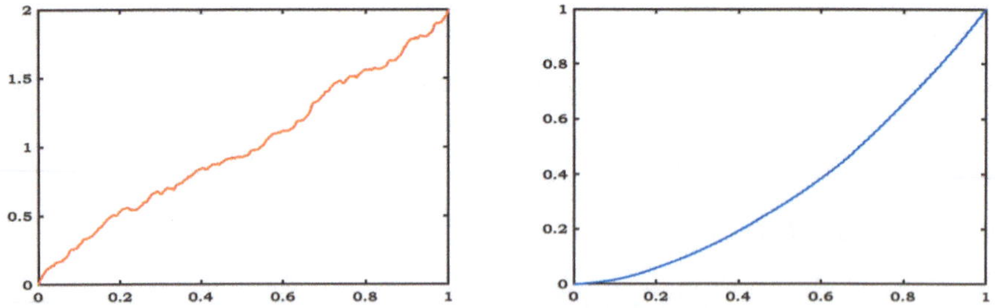

(a) f^α (b) $\hat{f}^\alpha$

Fig. 4.3 Graphs of **a** α-fractal function of $f(x) = x^2 + \sqrt{x}$ and **b** its classical integral

$$\hat{q}_n(x) = (F \circ L_n)(x) - a_n\alpha_n(x)B(x) + a_n \int_x^{x_N} \left(\alpha_n'(t) \int_x^{x_N} f^\alpha(t)dt\right) dt - a_n \int_x^{x_N} \alpha_n'(t)B(x)dt,$$

$$\hat{y}_n = a_n\alpha_n(x_1)\hat{y}_1 + a_n \int_{x_1}^{x_N} \left(\alpha_n'(t) \int_{x_1}^{x_N} f^\alpha(t)dt\right) dt + (F \circ L_n)(x_1) - a_n\alpha_n(x_1)B(x_1)$$
$$+ a_n \int_{x_1}^{x_N} \alpha_n'(t)B(x_1)dt,$$

$$\hat{y}_1 = \left((F \circ L_1)(x_1) - a_1\alpha_1(x_1)B(x_1) + a_1 \int_{x_1}^{x_N} \left(\alpha_1'(t) \int_{x_1}^{x_N} f^\alpha(t)dt\right) dt\right.$$
$$\left. - a_1 \left(\int_{x_1}^{x_N} \alpha_1'(t)B(x_1)dt\right)\right) \Big/ (1 - a_1\alpha_1(x_1)).$$

Remark 4.5 In Theorem 4.9, the integral of α-fractal function with variable scaling factors is discussed. In [8], the integral of α-fractal function with constant scaling factors has been investigated. Hence, Theorem 4.7 is the generalization of Theorem 1 in [8], i.e., if the vertical scaling factors α_n in Theorem 4.7 are chosen as constants instead of functions, then $\alpha_n'(t) = 0$ and $\alpha_{N-1}'(t) = 0$. Thus, one can get the proof of Theorem 1 in [8] as follows:

$$\hat{f}^\alpha(x) = \hat{y}_1 + \int_{x_1}^{x} f^\alpha(t)dt,$$
$$\hat{q}_n(x) = (F \circ L_n)(x) - a_n\alpha_n B(x),$$
$$\hat{y}_{n+1} = a_n\alpha_n\hat{y}_N + F \circ L_n(x_N) - a_n\alpha_n B(x_N),$$
$$\hat{y}_N = \frac{F \circ L_{N-1}(x_N) - a_{N-1}\alpha_{N-1}B(x_N)}{1 - a_{N-1}\alpha_{N-1}}.$$

Example 4.3 Consider the continuous function $f(x) = x^2 + \sqrt{x}$ and the base function $b(x) = 2x$. Let $\alpha_n(x) = -0.6,\ \forall i = 1, 2, 3$. Then, the functions f and b satisfy the condition in Theorem 4.9. The graphs of α-fractal function of f and its integral corresponding to the given data are presented in Fig. 4.3a and b respectively.

4.4 Integral on Hidden Variable Fractal Functions

The definition of hidden variable fractal interpolation function with variable scaling factors is briefly given below:

Definition 4.2 In order to define a hidden variable fractal interpolation function for the generalized data set $\{(x_n, y_n, z_n) \in I \times \mathbb{R}^2 : n = 1, 2, \ldots, N\}$, where $\{z_n : n = 1, 2, \ldots, N\}$ are real numbers, choose $L_n(x)$ as in Eq. (2.1) and

$$F_n(x, y, z) = A_n(y, z)^T + \big(p_n(x), q_n(x)\big)^T$$

where A_n are upper-triangular matrices $\begin{bmatrix} \delta_n(x) & \beta_n(x) \\ 0 & \gamma_n(x) \end{bmatrix}$ and $p_n(x)$, $q_n(x)$ are the real valued continuous functions for all $j \in J$. Note that the free variables and the constrained free variables are chosen as variable scaling factors. For the construction of HFIF, the generalized IFS on $\mathbb{R}^3$ is defined by

$$\{(L_n(x), F_n(x, y, z)) : n = 1, 2, \ldots, N-1\}. \tag{4.16}$$

Then, the IFS (4.16) has a unique attractor G^* which is the graph of the vector-valued function $\mathbf{f} = (f_1, f_2)$. Define the operator $\mathcal{T}$ on the space of continuous functions from I to $\mathbb{R}^2$ by

$$(\mathcal{T}\mathbf{g})(x) = F_n\big(L_n^{-1}(x), \mathbf{g} \circ \big(L_n^{-1}(x)\big)\big), \text{ for } x \in I_n, \ : n = 1, 2, \ldots, N-1.$$

By the Banach contraction principle, the contraction operator T' has a unique fixed point $\mathbf{f}$ which satisfies the functional equation,

$$(\mathcal{T}\mathbf{f})(x) = F_n\big(L_n^{-1}(x), \mathbf{f} \circ \big(L_n^{-1}(x)\big)\big), \text{ for } x \in I_n, \ : n = 1, 2, \ldots, N-1.$$

Then, the two components f_1 and f_2 of the vector-valued function $\mathbf{f}$ satisfies,

$$\begin{aligned} f_1\big(L_n(x)\big) &= \mathcal{T}_1 f_1(L_n(x)) = F_{1n}(x, f_1(x), f_2(x)) = \delta_n(x) f_1(x) + \beta_n(x) f_2(x) + p_n(x), \\ f_2\big(L_n(x)\big) &= \mathcal{T}_2 f_2(L_n(x)) = F_{2n}(x, f_2(x)) = \gamma_n(x) f_2(x) + q_n(x), \ x \in I. \end{aligned} \tag{4.17}$$

where $\mathcal{T}_1 f_1$ and $\mathcal{T}_2 f_2$ are the component wise operators of $\mathcal{T}$.

In Theorem 4.10, the integral of hidden variable FIF with variable scaling factors is explored and it is proved that the integral of f_1 and f_2 are again HFIF and FIF respectively.

Theorem 4.8 *Let* $\mathbf{f}(x) = (f_1, f_2)$ *be the hidden variable fractal interpolation function with variable scaling factors determined by the IFS* $\{(L_n(x), F_n(x, y, z)) : n = 1, 2, \ldots, N-1\}$. *If the integrals of* f_1 *and* f_2 *are defined by*

$$\hat{f}_1(x) = \hat{y}_1 + \int_{x_1}^{x} f_1(t)dt; \; \hat{f}_2(x) = \hat{z}_1 + \int_{x_1}^{x} f_2(t)dt$$

with $\|A'_n\| < 1$, *where* $A'_n = a_n \begin{bmatrix} \delta_n(x) & \beta_n(x) \\ 0 & \gamma_n(x) \end{bmatrix}$, *then* $\hat{f}(x) = (\hat{f}_1(x), \hat{f}_2(x))$ *is also a hidden variable FIF. Further* $\hat{f}_1(x)$ *and* $\hat{f}_2(x)$ *are generated by the IFSs* $\{(L_n(x), \hat{F}_{1n}(x, \hat{y}, \hat{z})) : n = 1, 2, \ldots, N-1\}$ *and* $\{(L_n(x), \hat{F}_{2n}(x, \hat{z})) : n = 1, 2, \ldots, N-1\}$ *respectively, where* $\hat{F}_{1n}(x, \hat{y}, \hat{z}) = a_n\delta_n(x)\hat{f}_1(x) + a_n\beta_n(x)\hat{f}_2(x) + \hat{p}_n(x)$, $\hat{F}_{2n}(x, \hat{z}) = a_n\gamma_n(x)\hat{f}_2(x) + \hat{q}_n(x)$ *with* $\sum_{j=1}^{N-1} a_j\delta_j(x_N) \neq 1$, $\sum_{j=1}^{N-1} a_j\gamma_j(x_N) \neq 1$ *and for* $n = 1, 2, \ldots, N-1$,

$$\begin{aligned}
\hat{p}_n(x) &= \hat{y}_n - a_n\delta_n(x)\hat{y}_1 - a_n\beta_n(x)\hat{z}_1 - a_n \int_{x_1}^{x} \left(\delta'_n(t) \int_{x_1}^{x} f_1(t)dt\right) dt \\
&\quad - a_n \int_{x_1}^{x} \left(\beta'_n(t) \int_{x_1}^{x} f_2(t)dt\right) dt + a_n \int_{x_1}^{x} p_n(t)dt, \\
\hat{q}_n(x) &= \hat{z}_n - a_n\gamma_n(x)\hat{z}_1 - a_n \int_{x_1}^{x} \left(\gamma'_n(t) \int_{x_1}^{x} f_2(t)dt\right) dt + a_n \int_{x_1}^{x} q_n(t)dt, \\
\hat{y}_{n+1} &= \hat{y}_1 + \sum_{j=1}^{n} \Big\{ a_j\delta_j(x_N)\hat{y}_N + a_j\beta_j(x_N)\hat{z}_N - a_j\delta_j(x_N)\hat{y}_1 - a_j\beta_n(x_N)\hat{z}_1 \\
&\quad - a_j \int_{x_1}^{x_N} \left(\delta'_j(t) \int_{x_1}^{x_N} f_1(t)dt\right) dt - a_j \int_{x_1}^{x_N} \left(\beta'_j(t) \int_{x_1}^{x_N} f_2(t)dt\right) dt + a_j \int_{x_1}^{x_N} p_j(t)dt \Big\}, \\
\hat{y}_N &= \hat{y}_1 + \left(\sum_{j=1}^{N-1} \Big\{ a_j\beta_j(x_N)\hat{z}_N - a_j\delta_j(x_N)\hat{z}_1 - a_j \int_{x_1}^{x_N} \left(\delta'_j(t) \int_{x_1}^{x_N} f_1(t)dt\right) dt \right. \\
&\quad \left. - a_j \int_{x_1}^{x_N} \left(\beta'_j(t) \int_{x_1}^{x_N} f_2(t)dt\right) dt + a_j \int_{x_1}^{x_N} p_j(t)dt \Big\} \right) \Big/ \left(1 - \sum_{j=1}^{N-1} a_j\delta_j(x_N)\right), \\
\hat{z}_{n+1} &= \hat{z}_1 + \sum_{j=1}^{n} \Big\{ a_j\gamma_j(x_N)\hat{z}_N - a_j\gamma_j(x_N)\hat{z}_1 - a_j \int_{x_1}^{x_N} \left(\gamma'_j(t) \int_{x_1}^{x_N} f_2(t)dt\right) dt + a_j \int_{x_1}^{x_N} q_j(t)dt \Big\}, \\
\hat{z}_N &= \hat{z}_1 + \frac{\sum_{j=1}^{N-1} \Big\{ - a_j \int_{x_1}^{x_N} \left(\gamma'_j(t) \int_{x_1}^{x_N} f_2(t)dt\right) dt + a_j \int_{x_1}^{x_N} q_j(t)dt \Big\}}{1 - \sum_{j=1}^{N-1} a_j\gamma_j(x_N)}.
\end{aligned}$$

Proof Consider the following equation,

$$\begin{aligned}
\hat{f}_1(L_n(x)) &= \hat{y}_1 + \int_{x_1}^{x_n} f_1(t)dt + \int_{x_n}^{L_n(x)} f_1(t)dt \\
&= \hat{y}_n + a_n \int_{x_1}^{x} f_1(L_n(u))du.
\end{aligned}$$

As f_1 satisfies $\hat{f}_1(L_n(u)) = \delta_n(x)f_1(x) + \beta_n(x)f_2(x) + p_n(x)$, then

$$\begin{aligned}
\hat{f}_1(L_n(x)) &= \hat{y}_n + a_n \int_{x_1}^{x} \delta_n(t) f_1(t)dt + a_n \int_{x_1}^{x} \beta_n(t) f_2(t)dt + a_n \int_{x_1}^{x} p_n(t)dt \\
&= \hat{y}_n + a_n\delta_n(x) \int_{x_1}^{x} f_1(t)dt - a_n \int_{x_1}^{x} \left(\delta_n'(t) \int_{x_1}^{x} f_1(t)dt\right) dt + a_n\beta_n(x) \int_{x_1}^{x} f_2(t)dt \\
&\quad - a_n \int_{x_1}^{x} \left(\beta_n'(t) \int_{x_1}^{x} f_2(t)dt\right) dt + a_n \int_{x_1}^{x} p_n(t)dt \\
&= \hat{y}_n + a_n\delta_n(x) \int_{x_1}^{x} f_1(t)dt + a_n\beta_n(x) \int_{x_1}^{x} f_2(t)dt - a_n \int_{x_1}^{x} \left(\delta_n'(t) \int_{x_1}^{x} f_1(t)dt\right) dt \\
&\quad - a_n \int_{x_1}^{x} \left(\beta_n'(t) \int_{x_1}^{x} f_2(t)dt\right) dt + a_n \int_{x_1}^{x} p_n(t)dt \\
&= \hat{y}_n + a_n\delta_n(x)\hat{f}_1(x) + a_n\beta_n(x)\hat{f}_2(x) - a_n\delta_n(x)\hat{y}_1 - a_n\beta_n(x)\hat{z}_1 \\
&\quad - a_n \int_{x_1}^{x} \left(\delta_n'(t) \int_{x_1}^{x} f_1(t)dt\right) dt - a_n \int_{x_1}^{x} \left(\beta_n'(t) \int_{x_1}^{x} f_2(t)dt\right) dt + a_n \int_{x_1}^{x} p_n(t)dt \\
&= a_n\delta_n(x)\hat{f}_1(x) + a_n\beta_n(x)\hat{f}_2(x) + \hat{p}_n(x) \\
&= \hat{F}_{1n}(x, \hat{f}_1(x), \hat{f}_2(x)).
\end{aligned}$$

Denote,

$\hat{p}_n(x) = \hat{y}_n - a_n\delta_n(x)\hat{y}_1 - a_n\beta_n(x)\hat{z}_1 - a_n \int_{x_1}^{x} \left(\delta_n'(t) \int_{x_1}^{x} f_1(t)dt\right) dt - a_n \int_{x_1}^{x} \left(\beta_n'(t) \int_{x_1}^{x} f_2(t)dt\right) dt + a_n \int_{x_1}^{x} p_n(t)dt.$

Now, consider

$$\begin{aligned}
\hat{f}_2(L_n(x)) &= \hat{z}_1 + \int_{x_1}^{x_n} f_2(t)dt + \int_{x_n}^{L_n(x)} f_2(t)dt \\
&= \hat{z}_n + a_n \int_{x_1}^{x} f_2(L_n(u))du.
\end{aligned}$$

Since f_2 satisfies the functional equation $f_2(L_n(u)) = \gamma_n f_2(u) + q_n(x)$, then

$$\begin{aligned}
\hat{f}_2(L_n(x)) &= \hat{z}_n + a_n \int_{x_1}^{x} \gamma_n(t) f_2(t)dt + a_n \int_{x_1}^{x} q_n(t)dt \\
&= \hat{z}_n + a_n\gamma_n(x) \int_{x_1}^{x} f_2(t)dt - a_n \int_{x_1}^{x} \left(\gamma_n'(t) \int_{x_1}^{x} f_2(t)dt\right) dt + a_n \int_{x_1}^{x} q_n(t)dt \\
&= \hat{z}_n + a_n\gamma_n(x)\hat{f}_2(x) - a_n\gamma_n(x)\hat{z}_1 - a_n \int_{x_1}^{x} \left(\gamma_n'(t) \int_{x_1}^{x} f_2(t)dt\right) dt + a_n \int_{x_1}^{x} q_n(t)dt \\
&= a_n\gamma_n(x)\hat{f}_2(x) + q_n(x) \\
&= \hat{F}_{2n}(x, \hat{f}_2(x)).
\end{aligned}$$

For simplification, denote $\hat{q}_n(x) = \hat{z}_n - a_n\gamma_n(x)\hat{z}_1 - a_n \int_{x_1}^{x} \left(\gamma_n'(t) \int_{x_1}^{x} f_2(t)dt\right) dt + a_n \int_{x_1}^{x} q_n(t)dt$. To find the new interpolation data $(x_n, \hat{y}_n)$, consider

$$\begin{aligned}
\hat{f}_1(L_n(x)) &= \hat{y}_n + a_n\delta_n(x)\hat{f}_1(x) + a_n\beta_n(x)\hat{f}_2(x) - a_n\delta_n(x)\hat{y}_1 - a_n\beta_n(x)\hat{z}_1 \\
&\quad - a_n \int_{x_1}^{x} \left(\delta_n'(t) \int_{x_1}^{x} f_1(t)dt\right) dt - a_n \int_{x_1}^{x} \left(\beta_n'(t) \int_{x_1}^{x} f_2(t)dt\right) dt + a_n \int_{x_1}^{x} p_n(t)dt.
\end{aligned}$$

Let $x = x_N$ and $L_n(x_N) = x_{n+1}$.

$$\hat{y}_{n+1} - \hat{y}_n = a_n\delta_n(x_N)\hat{y}_N + a_n\beta_n(x_N)\hat{z}_N - a_n\delta_n(x_N)\hat{y}_1 - a_n\beta_n(x_N)\hat{z}_1 \\ - a_n\int_{x_1}^{x_N}\left(\delta_n'(t)\int_{x_1}^{x_N} f_1(t)dt\right)dt - a_n\int_{x_1}^{x_N}\left(\beta_n'(t)\int_{x_1}^{x_N} f_2(t)dt\right)dt + a_n\int_{x_1}^{x_N} p_n(t)dt. \tag{4.18}$$

The equation $\hat{y}_{n+1} = \hat{y}_1 + \sum_{j=1}^{n}(\hat{y}_{j+1} - \hat{y}_j)$ provides,

$$\hat{y}_{n+1} = \hat{y}_1 + \sum_{j=1}^{n}\Big\{a_j\delta_j(x_N)\hat{y}_N + a_j\beta_j(x_N)\hat{z}_N - a_j\delta_j(x_N)\hat{y}_1 - a_j\beta_n(x_N)\hat{z}_1 \\ - a_j\int_{x_1}^{x_N}\left(\delta_j'(t)\int_{x_1}^{x_N} f_1(t)dt\right)dt - a_j\int_{x_1}^{x_N}\left(\beta_j'(t)\int_{x_1}^{x_N} f_2(t)dt\right)dt + a_j\int_{x_1}^{x_N} p_j(t)dt\Big\},$$

When $n = N-1$ is substituted in the above equation, one can get the end point $\hat{y}_N$ as follows,

$$\hat{y}_N = \hat{y}_1 + \Bigg(\sum_{j=1}^{N-1}\Big\{a_j\beta_j(x_N)\hat{z}_N - a_j\delta_j(x_N)\hat{z}_1 - a_j\int_{x_1}^{x_N}\left(\delta_j'(t)\int_{x_1}^{x_N} f_1(t)dt\right)dt \\ -a_j\int_{x_1}^{x_N}\left(\beta_j'(t)\int_{x_1}^{x_N} f_2(t)dt\right)dt + a_j\int_{x_1}^{x_N} p_j(t)dt\Big\}\Bigg)\Big/\left(1 - \sum_{j=1}^{N-1} a_j\delta_j(x_N)\right),$$

To get the new interpolation points $(x_n, \hat{z}_n)$ corresponding to the function f_2, consider

$$\hat{f}_1(L_n(x)) = \hat{z}_n + a_n\gamma_n(x)\hat{f}_2(x) - a_n\gamma_n(x)\hat{z}_1 - a_n\int_{x_1}^{x}\left(\gamma_n'(t)\int_{x_1}^{x} f_2(t)dt\right)dt + a_n\int_{x_1}^{x} q_n(t)dt.$$

Let $x = x_N$ and $L_n(x_N) = x_{n+1}$, then

$$\hat{z}_{n+1} - \hat{z}_n = a_n\gamma_n(x_N)\hat{z}_N - a_n\gamma_n(x_N)\hat{z}_1 - a_n\int_{x_1}^{x_N}\left(\gamma_n'(t)\int_{x_1}^{x_N} f_2(t)dt\right)dt + a_n\int_{x_1}^{x_N} q_n(t)dt. \tag{4.19}$$

The equation $\hat{z}_{n+1} = \hat{z}_1 + \sum_{j=1}^{n}(\hat{z}_{j+1} - \hat{z}_j)$ yields,

$$\hat{z}_{n+1} = \hat{z}_1 + \sum_{j=1}^{n}\Big\{a_j\gamma_j(x_N)\hat{z}_N - a_j\gamma_j(x_N)\hat{z}_1 - a_j\int_{x_1}^{x_N}\left(\gamma_j'(t)\int_{x_1}^{x_N} f_2(t)dt\right)dt + a_j\int_{x_1}^{x_N} q_j(t)dt\Big\}.$$

Substituting $n = N-1$, the end point $\hat{z}_N$ is obtained as

$$\hat{z}_N = \hat{z}_1 + \frac{\sum_{j=1}^{N-1}\Big\{-a_j\int_{x_1}^{x_N}\left(\gamma_j'(t)\int_{x_1}^{x_N} f_2(t)dt\right)dt + a_j\int_{x_1}^{x_N} q_j(t)dt\Big\}}{1 - \sum_{j=1}^{N-1} a_j\gamma_j(x_N)}.$$

The continuity conditions of $\hat{p}_n(x)$ and $\hat{q}_n(x)$ are verified below:

$$\hat{p}_n(x_1) = \hat{y}_n - a_n\delta_n(x_1)\hat{y}_1 - a_n\beta_n(x_1)\hat{z}_1,$$

$$\hat{p}_n(x_N) = \hat{y}_n - a_n\delta_n(x_N)\hat{y}_1 - a_n\beta_n(x_N)\hat{z}_1 - a_n\int_{x_1}^{x_N}\left(\delta_n'(t)\int_{x_1}^{x_N} f_1(t)dt\right)dt - a_n\int_{x_1}^{x_N}\left(\beta_n'(t)\int_{x_1}^{x_N} f_2(t)dt\right)dt + a_n\int_{x_1}^{x_N} p_n(t)dt.$$

When $\hat{p}_n(x_N)$ is compared with Eq. (4.18), one can get

$$\hat{p}_n(x_N) = \hat{y}_{n+1} - a_n\delta_n(x_N)\hat{y}_N - a_n\beta_n(x_N)\hat{z}_N.$$

Similarly,

$$\hat{q}_n(x_1) = \hat{z}_n - a_n\gamma_n(x_1)\hat{z}_1,$$

$$\hat{q}_n(x_N) = \hat{z}_n - a_n\gamma_n(x_N)\hat{z}_1 - a_n\int_{x_1}^{x_N}\left(\gamma_n'(t)\int_{x_1}^{x_N} f_2(t)dt\right)dt + a_n\int_{x_1}^{x_N} q_n(t)dt.$$

When $\hat{q}_n(x_N)$ is compared with Eq. (4.19), one can get $\hat{q}_n(x_N) = \hat{z}_{n+1} - a_n\gamma_n(x_N)\hat{z}_N$. This clearly shows that $\hat{F}_n(x, \hat{y}, \hat{z})$ satisfies the end-point conditions $\hat{F}_n(x_1, \hat{y}_1, \hat{z}_1) = (\hat{y}_n, \hat{z}_n)$ and $\hat{F}_n(x_N, \hat{y}_N, \hat{z}_N) = (\hat{y}_{n+1}, \hat{z}_{n+1})$. Hence, the function $\hat{\mathbf{f}} = (\hat{f}_1, \hat{f}_2)$ is a hidden variable FIF generated by the IFS $\{(L_n(x), \hat{F}_n(x, \hat{y}, \hat{z})) : n = 1, 2, \ldots, N-1\}$. □

Remark 4.6 If the integrals of f_1 and f_2 are defined at the end points, then the above result is defined by

$$\hat{f}_1 = \hat{y}_N - \int_x^{x_N} f_1(t)dt; \; \hat{f}_2 = \hat{z}_N - \int_x^{x_N} f_2(t)dt$$

the HFIF $\hat{f}_1$ and FIF $\hat{f}_2$ are generated by the IFSs $\{(L_n(x), \hat{F}_{1n}(x, \hat{y}, \hat{z})) : n = 1, 2, \ldots, N-1\}$ and $\{(L_n(x), \hat{F}_{2n}(x, \hat{z})) : n = 1, 2, \ldots, N-1\}$ respectively, where $\hat{F}_{1n}(x, \hat{y}, \hat{z}) = a_n\delta_n(x)\hat{f}_1(x) + a_n\beta_n(x)\hat{f}_2(x) + \hat{p}_n(x)$, $\hat{F}_{2n}(x, \hat{z}) = a_n\gamma_n(x)\hat{f}_2(x) + \hat{q}_n(x)$ with $\sum_{j=1}^{N-1} a_j\delta_j(x_1) \neq 1$, $\sum_{j=1}^{N-1} a_j\gamma_j(x_1) \neq 1$ and for $n = 1, 2, \ldots, N-1$ such that

$$\hat{p}_n(x) = \hat{y}_{n+1} - a_n\delta_n(x)\hat{y}_N - a_n\beta_n(x)\hat{z}_N + a_n\int_x^{x_N}\left(\delta_n'(t)\int_x^{x_N} f_1(t)dt\right)dt + a_n\int_x^{x_N}\left(\beta_n'(t)\int_{x_1}^{x} f_2(t)dt\right)dt - a_n\int_x^{x_N} p_n(t)dt,$$

$$\hat{q}_n(x) = \hat{z}_{n+1} - a_n\gamma_n(x)\hat{z}_N + a_n\int_x^{x_N}\left(\gamma_n'(t)\int_x^{x_N} f_2(t)dt\right)dt - a_n\int_x^{x_N} q_n(t)dt,$$

$$\hat{y}_n = \hat{y}_N - \sum_{j=n}^{N} -1\Big\{a_j\delta_j(x_1)\hat{y}_N + a_j\beta_j(x_1)\hat{z}_N - a_j\delta_j(x_1)\hat{y}_1 - a_j\beta_j(x_1)\hat{z}_1$$

$$-a_j\int_{x_1}^{x_N}\left(\delta_j'(t)\int_{x_1}^{x_N} f_1(t)dt\right)dt - a_j\int_{x_1}^{x_N}\left(\beta_j'(t)\int_{x_1}^{x_N} f_2(t)dt\right)dt + a_j\int_{x_1}^{x_N} p_j(t)dt\Big\},$$

$$\hat{y}_1 = \hat{y}_N - \left(\sum_{j=1}^{N-1}\Big\{a_j\beta_j(x_1)\hat{z}_N - a_j\delta_j(x_1)\hat{z}_1 - a_j\int_{x_1}^{x_N}\left(\delta_j'(t)\int_{x_1}^{x_N} f_1(t)dt\right)dt\right.$$

$$\left.-a_j\int_{x_1}^{x_N}\left(\beta_j'(t)\int_{x_1}^{x_N} f_2(t)dt\right)dt + a_j\int_{x_1}^{x_N} p_j(t)dt\Big\}\right)\Big/\left(1-\sum_{j=1}^{N-1} a_j\delta_j(x_1)\right),$$

$$\hat{z}_n = \hat{z}_N - \sum_{j=n}^{N-1}\Big\{a_j\gamma_j(x_1)\hat{z}_N - a_j\gamma_j(x_N)\hat{z}_1 - a_j\int_{x_1}^{x_N}\left(\gamma_j'(t)\int_{x_1}^{x_N} f_2(t)dt\right)dt + a_j\int_{x_1}^{x_N} q_j(t)dt\Big\},$$

$$\hat{z}_1 = \hat{z}_N - \frac{\sum_{j=1}^{N-1}\Big\{-a_j\int_{x_1}^{x_N}\left(\gamma_j'(t)\int_{x_1}^{x_N} f_2(t)dt\right)dt + a_j\int_{x_1}^{x_N} q_j(t)dt\Big\}}{1-\sum_{j=1}^{N-1} a_j\gamma_j(x_1)}.$$

Remark 4.7 In Theorem 4.8, the integral of hidden variable FIF with variable scaling factors is proved as a hidden variable FIF. If the vertical scaling factors in Theorem 4.8 are chosen as constants instead of functions then one can get:

$$\hat{f}_1(x) = \hat{y}_1 + \int_{x_1}^{x} f_1(t)dt;\ \hat{f}_2(x) = \hat{z}_1 + \int_{x_1}^{x} f_2(t)dt,$$

$$\hat{p}_n(x) = \hat{y}_n - a_n\delta_n\hat{y}_1 - a_n\beta_n\hat{z}_1 + a_n\int_{x_1}^{x} p_n(t)dt,$$

$$\hat{q}_n(x) = \hat{z}_n - a_n\gamma_n\hat{z}_1 + a_n\int_{x_1}^{x} q_n(t)dt,$$

$$\hat{y}_{n+1} = \hat{y}_1 + \sum_{j=1}^{n}\{a_j\delta_j(\hat{y}_N - \hat{y}_1) + a_j\beta_j(\hat{z}_N - \hat{z}_1) + a_j\int_{x_1}^{x_N} p_j(t)dt\},$$

$$\hat{y}_N = \hat{y}_1 + \frac{\sum_{j=1}^{N-1}\{a_j\beta_j(\hat{z}_N - \hat{z}_1) + a_j\int_{x_1}^{x_N} p_j(t)dt\}}{1-\sum_{j=1}^{N-1} a_j\delta_j},$$

$$\hat{z}_{n+1} = \hat{z}_1 + \sum_{j=1}^{n}\Big\{a_j\gamma_j(\hat{z}_N - \hat{z}_1) + a_j\int_{x_1}^{x_N} q_j(t)dt\Big\},$$

$$\hat{z}_N = \hat{z}_1 + \frac{\sum_{j=1}^{N-1} a_j\int_{x_1}^{x_N} q_j(t)dt}{1-\sum_{j=1}^{N-1} a_j\gamma_j}.$$

Example 4.4 Let $\{(0, 10), (30, 50), (60, 70), (100, 20)\}$ be the given set of interpolation data. By adding the hidden variables $z_1 = 10,\ z_2 = 30, z_3 = 80, z_4 = 50$, the given data set is extended to a generalized interpolation data set. Let $\delta_n = \{0.4, -0.4, 0.4\}$ be the free variables, $\beta_n = \{0.3, 0.3, 0.3\}$ be the constrained free variables with respect to hidden variables $\gamma_n = \{0.4, -0.4, 0.4\}$. The graphical representation of the non-self-affine function f_1 and its integral is presented in Fig. 4.4a and b respectively. Figure 6.3c and d illustrates the graphs of self-affine function f_2 and its integral.

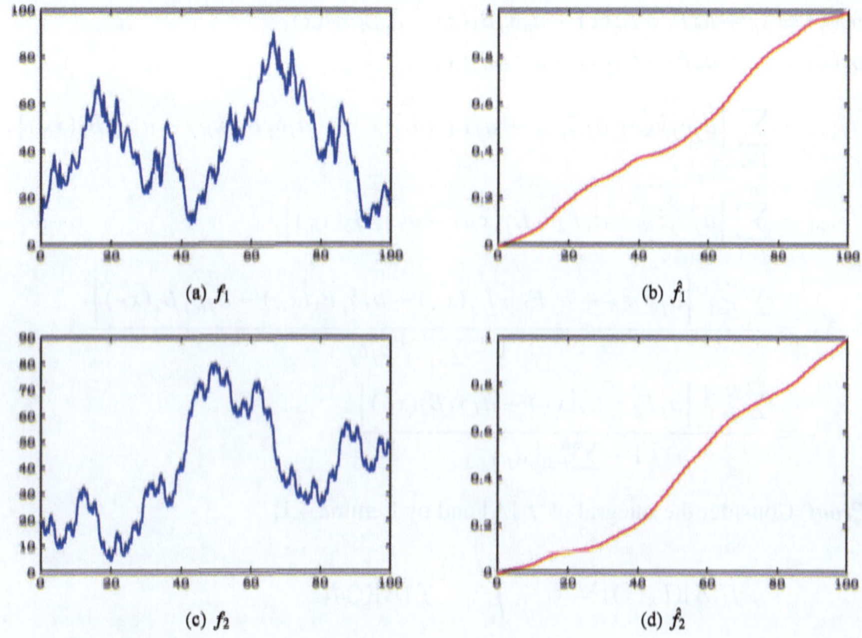

(a) f_1 (b) $\hat{f}_1$ (c) f_2 (d) $\hat{f}_2$

Fig. 4.4 Graphs of **a** non self-referential function f_1, **b** its integral $\hat{f}_1$, **c** self-referential function f_2 and its integral $\hat{f}_2$

4.5 Integral on A-Fractal Functions

In this section, the construction of A-fractal function is presented briefly. Furthermore, the indefinite integral of A-fractal function is evaluated and the necessary condition of the block matrix is provided for the integral of A-fractal function to be an A-fractal function. In Theorem 4.9, the integral of A-fractal function with constant scaling factors is investigated and it is demonstrated that its integral is again an attractor of the IFS.

Theorem 4.9 *Suppose* $f[A]$ *is the A-fractal function with constant scaling factors generated by* $\{L_n(x), F_n(x, y, z) : n = 1, 2, \ldots, N-1\}$. *If the integral of A-fractal function is componentwise defined as* $\hat{f}_1[\mathbf{A}](x) = \hat{y}_1 + \int_{x_1}^{x} f_1(t)dt$, $\hat{f}_2[\mathbf{A}](x) = \hat{z}_1 + \int_{x_1}^{x} f_2(t)dt$ *and the integrals of* $\boldsymbol{f}$ *and* $\mathbf{b}$ *are given by* $F(x) = \int_{x_1}^{x} \boldsymbol{f}(t)dt$, $B(x) = \int_{x_1}^{x} \boldsymbol{b}(t)dt$ *with* $F_1(x_1) = B_1(x_1) = \hat{f}_1[\mathbf{A}](x_1) = 0$, $F_1(x_N) = B_1(x_N) = \hat{f}_1[\mathbf{A}](x_N)$, $F_2(x_1) = B_2(x_1) = \hat{f}_1[\mathbf{A}](x_1) = 0$ *and* $F_2(x_N) = B_2(x_N) = \hat{f}_1[\mathbf{A}](x_N)$, *then* $\hat{f}[\mathbf{A}] = (\hat{f}_1[\mathbf{A}], \hat{f}_2[\mathbf{A}])$ *is also a A-fractal function. Further,* $\hat{f}_1[\mathbf{A}]$ *and* $\hat{f}_2[\mathbf{A}]$ *are generated by the IFSs* $\{L_n(x), \hat{F}_{1n}(x, \hat{y}, \hat{z}) : n = 1, 2, \ldots, N-1\}$ *and* $\{L_n(x), \hat{F}_{2n}(x, \hat{z}) : n = 1, 2, \ldots, N-1\}$ *respectively, where* $\hat{F}_{1n}(x, \hat{y}, \hat{z}) = a_n\delta_n \hat{f}_1[\mathbf{A}](x) + a_n\beta_n \hat{f}_2[\mathbf{A}](x) + \hat{p}_n(x)$ *and* $\hat{F}_{2n}(x, \hat{z}) = a_n\gamma_n \hat{f}_2[\mathbf{A}](x) + \hat{q}_n(x)$ *with* $\hat{f}_1[\mathbf{A}](x_n) = \hat{y}_n$, $\hat{f}_2[\mathbf{A}](x_n) = \hat{z}_n$, $\sum_{j=1}^{N-1} a_j\delta_j \neq 1$ *and* $\sum_{j=1}^{N-1} a_j\gamma_j \neq 1$, *for* $j = 1, 2, \ldots, N-1$,

$$\hat{p}_n(x) = \hat{y}_n + a_n F_1 \circ L_n(x) - a_n \delta_n B_1(x) - a_n \beta_n B_2(x),$$
$$q_n(x) = \hat{z}_n + a_n F_2 \circ L_n(x) - a_n \gamma_n B_2(x),$$
$$\hat{y}_{n+1} = \sum_{j=1}^{n} \Big\{a_j\delta_j\hat{y}_N + a_j\beta_j\hat{z}_N + a_j F_1 \circ L_j(x_N) - a_j\delta_j B_1(x_N) - a_j\beta_j B_2(x_N)\Big\},$$
$$\hat{z}_{n+1} = \sum_{j=1}^{n} \Big\{a_j\gamma_j\hat{z}_N + a_j F_2 \circ L_j(x_N) - a_j\gamma_j B_2(x_N)\Big\},$$
$$\hat{y}_N = \frac{\sum_{j=1}^{N-1} \Big\{a_j\beta_j\hat{z}_N + a_j F_1 \circ L_j(x_N) - a_j\delta_j B_1(x_N) - a_j\beta_j B_2(x_N)\Big\}}{1 - \sum_{j=1}^{N-1} a_j\delta_j},$$
$$\hat{z}_N = \frac{\sum_{j=1}^{N-1} \Big\{a_j F_2 \circ L_j(x_N) - a_j\gamma_j B_2(x_N)\Big\}}{1 - \sum_{j=1}^{N-1} a_j\gamma_j}.$$

Proof Consider the integral of $f_1[\mathbf{A}]$ and by Lemma 4.1,

$$\begin{aligned}
\hat{f}_1[\mathbf{A}](L_n(x)) &= \hat{y}_1 + \int_{x_1}^{L_n(x)} f_1[\mathbf{A}](t)dt \\
&= \hat{y}_1 + \int_{x_1}^{x_n} f_1[\mathbf{A}](t)dt + \int_{x_n}^{L_n(x)} f_1[\mathbf{A}](t)dt \\
&= \hat{y}_n + a_n \int_{x_1}^{x} f_1[\mathbf{A}](L_n(u))du.
\end{aligned}$$

Then, using the function equation of $f_1[\mathbf{A}]$,

$$\begin{aligned}
\hat{f}_1[\mathbf{A}](L_n(x)) &= \hat{y}_n + a_n \int_{x_1}^{x} \delta_n f_1[\mathbf{A}](t)dt + a_n \int_{x_1}^{x} \beta_n f_2[\mathbf{A}](t)dt + a_n \int_{x_1}^{x} p_n(t)dt \\
&= \hat{y}_n + a_n\delta_n(\hat{f}_1[\mathbf{A}](x) - \hat{y}_1) + a_n\beta_n(\hat{f}_2[\mathbf{A}](x) - \hat{z}_1) + a_n \int_{x_1}^{x} p_n(t)dt \\
&= \hat{y}_n + a_n\delta_n \hat{f}_1[\mathbf{A}](x) + a_n\beta_n \hat{f}_2[\mathbf{A}](x) + a_n \int_{x_1}^{x} f_1 \circ L_n(t)dt - a_n \int_{x_1}^{x} \delta_n b_1(t)dt \\
&\quad - a_n \int_{x_1}^{x} \beta_n b_2(t)dt \\
&= \hat{y}_n + a_n\delta_n \hat{f}_1[\mathbf{A}](x) + a_n\beta_n \hat{f}_2[\mathbf{A}](x) + a_n F_1 \circ L_n(x) - a_n\delta_n B_1(x) - a_n\beta_n B_2(x) \\
&= a_n\delta_n \hat{f}_1[\mathbf{A}](x) + a_n\beta_n \hat{f}_2[\mathbf{A}](x) + \hat{p}_n(x) \\
&= \hat{F}_{1n}(x, \hat{f}_1[\mathbf{A}](x), \hat{f}_2[\mathbf{A}](x)).
\end{aligned}$$

Use the notation $\hat{p}_n(x) = \hat{y}_n + a_n F_1 \circ L_n(x) - a_n\delta_n B_1(x) - a_n\beta_n B_2(x)$. Now, consider

$$\begin{aligned}\hat{f}_2[\mathbf{A}](L_n(x)) &= \hat{z}_1 + \int_{x_1}^{L_n(x)} f_2[\mathbf{A}](t)dt\\ &= \hat{z}_1 + \int_{x_1}^{x_n} f_2[\mathbf{A}](t)dt + \int_{x_n}^{L_n(x)} f_2[\mathbf{A}](t)dt\\ &= \hat{z}_n + a_n \int_{x_1}^{x} f_2[\mathbf{A}](L_n(u))du.\end{aligned}$$

Now, applying the functional equation of $f_2[\mathbf{A}]$, one can get

$$\begin{aligned}\hat{f}_2[\mathbf{A}](L_n(x)) &= \hat{z}_n + a_n \int_{x_1}^{x} \gamma_n f_2[\mathbf{A}](t)dt + a_n \int_{x_1}^{x} q_n(t)dt\\ &= \hat{z}_n + a_n\gamma_n \hat{f}_2[\mathbf{A}](x) + a_n \int_{x_1}^{x} f_2 \circ L_n(t)dt - a_n\gamma_n \int_{x_1}^{x} b_2(t)dt\\ &= \hat{z}_n + a_n\gamma_n \hat{f}_2[\mathbf{A}](x) + a_n F_2 \circ L_n(x) - a_n\gamma_n B_2(x)\\ &= a_n\gamma_n \hat{f}_2[\mathbf{A}](x) + \hat{q}_n(x)\\ &= \hat{F}_{2n}(x, \hat{f}_2[\mathbf{A}](x)).\end{aligned}$$

Denote $\hat{q}_n(x) = \hat{z}_n + a_n F_2 \circ L_n(x) - a_n\gamma_n B_2(x)$. To find the new interpolation data $(x_n, \hat{y}_n)$, consider

$$\hat{f}_1[\mathbf{A}](L_n(x)) = \hat{y}_n + a_n\delta_n \hat{f}_1[\mathbf{A}](x) + a_n\beta_n \hat{f}_2[\mathbf{A}](x) + a_n F_1 \circ L_n(x) - a_n\delta_n B_1(x) - a_n\beta_n B_2(x).$$

Let $x = x_N$ and $L_n(x_N) = x_{n+1}$. Then

$$\begin{aligned}\hat{y}_{n+1} - \hat{y}_n &= a_n\delta_n \hat{f}_1[\mathbf{A}](x_N) + a_n\beta_n \hat{f}_2[\mathbf{A}](x_N) + a_n F_1 \circ L_n(x_N) - a_n\delta_n B_1(x_N) - a_n\beta_n B_2(x_N)\\ &= a_n\delta_n \hat{y}_N + a_n\beta_n \hat{z}_N + a_n F_1 \circ L_n(x_N) - a_n\delta_n B_1(x_N) - a_n\beta_n B_2(x_N).\end{aligned}$$

The equation $\hat{y}_{n+1} = \hat{y}_1 + \sum_{j=1}^{n}(\hat{y}_{j+1} - \hat{y}_j)$ gives,

$$\hat{y}_{n+1} = \sum_{j=1}^{n} \Big\{a_j\delta_j \hat{y}_N + a_j\beta_j \hat{z}_N + a_j F_1 \circ L_j(x_N) - a_j\delta_j B_1(x_N) - a_j\beta_j B_2(x_N)\Big\}.$$

When $n = N - 1$ is substituted in the preceding equation, the end point can be obtained as

$$\hat{y}_N = \frac{\sum_{n=1}^{N-1} \Big\{a_n\beta_n \hat{z}_N + a_n F_1 \circ L_n(x_N) - a_n\delta_n B_1(x_N) - a_n\beta_n B_2(x_N)\Big\}}{1 - \sum_{n=1}^{N-1} a_n\delta_n}.$$

To get the new interpolation points $(x_n, \hat{z}_n)$, consider

$$\hat{f}_2[\mathbf{A}](L_n(x)) = \hat{z}_n + a_n\gamma_n \hat{f}_2[\mathbf{A}](x) + a_n F_2 \circ L_n(x) - a_n\gamma_n B_2(x).$$

Let $x = x_N$ and $L_n(x_N) = x_{n+1}$.

$$\hat{z}_{n+1} - \hat{z}_n = a_n\gamma_n\hat{z}_N + a_n F_2 \circ L_n(x_N) - a_n\gamma_n B_2(x_N).$$

The equation $\hat{z}_{n+1} = \hat{z}_1 + \sum_{j=1}^{n}(\hat{z}_{j+1} - \hat{z}_j)$ yields,

$$\hat{z}_{n+1} = \sum_{j=1}^{n} \left\{a_j\gamma_j\hat{z}_N + a_j F_2 \circ L_j(x_N) - a_j\gamma_j B_2(x_N)\right\}$$

for $i = 1, 2, \ldots, N-1$. Substituting $n = N-1$, the end point $\hat{z}_N$ is obtained as

$$\hat{z}_N = \frac{\sum_{j=1}^{N-1}\left\{a_j F_2 \circ L_j(x_N) - a_j\gamma_j B_2(x_N)\right\}}{1 - \sum_{j=1}^{N-1} a_j\gamma_j}.$$

As the continuous functions $\hat{p}_n(x)$ and $\hat{q}_n(x)$ satisfy the following endpoint conditions

$$\hat{p}_n(x_1) = \hat{y}_n,\ \hat{p}_n(x_N) = \hat{y}_{n+1} - a_n\delta_n\hat{y}_N - a_n\beta_n\hat{z}_N$$

and

$$\hat{q}_n(x_1) = \hat{z}_n,\ \hat{q}_n(x_N) = \hat{z}_{n+1} - a_n\gamma_n\hat{z}_N$$

which implies $\hat{F}_n(x_1, \hat{y}_1, \hat{z}_1) = (\hat{y}_n, \hat{z}_n)$ and $\hat{F}_n(x_N, \hat{y}_N, \hat{z}_N) = (\hat{y}_{n+1}, \hat{z}_{n+1})$. Therefore, the integral of the function **f**[**A**] is a A-fractal function. □

Corollary 4.7 *Let f be a fractal interpolation function as described in Theorem 4.9, and defined at the end points as $\hat{f}_1[\mathbf{A}](x) = \hat{y}_N - \int_x^{x_N} f_1(t)dt$, $\hat{f}_2[\mathbf{A}](x) = \hat{z}_N - \int_x^{x_N} f_2(t)dt$ such that the integrals of f and $\mathbf{b}$ are given by $F(x) = -\int_x^{x_N} \mathbf{f}(t)dt$, $B(x) = -\int_x^{x_N} \mathbf{b}(t)dt$ with $F_1(x_1) = B_1(x_1) = \hat{f}_1[\mathbf{A}](x_1)$, $F_1(x_N) = B_1(x_N) = \hat{f}_1[\mathbf{A}](x_N) = 0$, $F_2(x_1) = B_2(x_1) = \hat{f}_2[\mathbf{A}](x_1)$, $F_2(x_N) = B_2(x_N) = \hat{f}_2[\mathbf{A}](x_N) = 0$, $\sum_{j=1}^{N-1} a_j\delta_j \neq 1$ and $\sum_{j=1}^{N-1} a_j\gamma_j \neq 1$. Then*

$$\hat{p}_n(x) = \hat{y}_{n+1} + a_n F_1 \circ L_n(x) - a_n\delta_n B_1(x) - a_n\beta_n B_2(x),$$
$$q_n(x) = \hat{z}_{n+1} + a_n F_2 \circ L_n(x) - a_n\gamma_n B_2(x),$$
$$\hat{y}_n = \sum_{j=n}^{N-1} \left\{a_j\delta_j\hat{y}_1 + a_j\beta_j\hat{z}_1 + a_j F_1 \circ L_j(x_1) - a_j\delta_j B_1(x_1) - a_j\beta_j B_2(x_1)\right\},$$
$$\hat{y}_1 = \frac{\sum_{j=1}^{N-1}\left\{a_j\beta_j\hat{z}_1 + a_j F_1 \circ L_j(x_1) - a_j\delta_j B_1(x_1) - a_j\beta_j B_2(x_1)\right\}}{1 - \sum_{j=1}^{N-1} a_j\delta_j},$$

$$\hat{z}_n = \sum_{j=n}^{N-1} \left\{ a_j \gamma_j \hat{z}_1 + a_j F_2 \circ L_j(x_1) - a_j \gamma_j B_2(x_1) \right\},$$

$$\hat{z}_1 = \frac{\sum_{j=1}^{N-1} \left\{ a_j F_2 \circ L_j(x_1) - a_j \gamma_j B_2(x_1) \right\}}{1 - \sum_{j=1}^{N-1} a_j \gamma_j}.$$

In Theorem 4.10, by taking the vertical scaling factors as variables, the integral of A-fractal function is investigated, and the interpolation points for both the integrals of $f_1[\mathbf{A}]$ and $f_2[\mathbf{A}]$ are explicitly determined.

Theorem 4.10 *Let $f[\mathbf{A}]$ be the A-fractal function with variable scaling factors generated by $\{L_n(x), F_n(x, y, z) : n = 1, 2, \ldots, N-1\}$. Suppose the integral of A-fractal function is componentwise defined as $\hat{f}_1[\mathbf{A}](x) = \hat{y}_1 + \int_{x_1}^{x} f_1(t)dt$, $\hat{f}_2[\mathbf{A}](x) = \hat{z}_1 + \int_{x_1}^{x} f_2(t)dt$ and the integrals of f and $\mathbf{b}$ are given by $F(x) = \int_{x_1}^{x} f(t)dt$, $B(x) = \int_{x_1}^{x} \mathbf{b}(t)dt$ with $F_1(x_1) = B_1(x_1) = \hat{f}_1[\mathbf{A}](x_1) = 0$, $F_1(x_N) = B_1(x_N) = \hat{f}_1[\mathbf{A}](x_N)$, $F_2(x_1) = B_2(x_1) = \hat{f}_2[\mathbf{A}](x_1) = 0$ and $F_2(x_N) = B_2(x_N) = \hat{f}_2[\mathbf{A}](x_N)$, then $\hat{f}[\mathbf{A}] = (\hat{f}_1[\mathbf{A}], \hat{f}_2[\mathbf{A}])$ is also a A-fractal function. Further, $\hat{f}_1[\mathbf{A}]$ and $\hat{f}_2[\mathbf{A}]$ are generated by the IFSs $\{L_n(x), \hat{F}_{1n}(x, \hat{y}, \hat{z}) : n = 1, 2, \ldots, N-1\}$ and $\{L_n(x), \hat{F}_{2n}(x, \hat{z}) : n = 1, 2, \ldots, N-1\}$ respectively, where $\hat{F}_{1n}(x, \hat{y}, \hat{z}) = a_n \delta_n(x) \hat{f}_1[\mathbf{A}](x) + a_n \beta_n(x) \hat{f}_2[\mathbf{A}](x) + \hat{p}_n(x)$ and $\hat{F}_{2n}(x, \hat{z}) = a_n \gamma_n(x) \hat{f}_2[\mathbf{A}](x) + \hat{q}_n(x)$ with $\hat{f}_1[\mathbf{A}](x_n) = \hat{y}_n$, $\hat{f}_2[\mathbf{A}](x_n) = \hat{z}_n$, $\sum_{j=1}^{N-1} a_j \delta_j n(x_N) \neq 1$ and $\sum_{j=1}^{N-1} a_j \gamma_j(x_N) \neq 1$, for $j = 1, 2, \ldots, N-1$,*

$$\begin{aligned}
\hat{p}_n(x) = \hat{y}_n &- a_n \int_{x_1}^{x} \left(\delta_n'(u) \int_{x_1}^{u} f_1[\mathbf{A}](t)dt \right) du - a_n \int_{x_1}^{x} \left(\beta_n'(u) \int_{x_1}^{u} f_2[\mathbf{A}](t)dt \right) du \\
&+ a_n F_1 \circ L_n(x) - a_n \delta_n(x) B_1(x) + a_n \int_{x_1}^{x} \left(\delta_n'(u) \int_{x_1}^{u} b_1(t)dt \right) du \\
&- a_n \beta_n(x) B_2(x) + a_n \int_{x_1}^{x} \left(\beta_n'(u) \int_{x_1}^{u} b_2(t)dt \right) du,
\end{aligned}$$

$$\begin{aligned}
\hat{q}_n(x) = \hat{z}_n &- a_n \int_{x_1}^{x} \left(\gamma_n'(u) \int_{x_1}^{u} f_2[\mathbf{A}](t)dt \right) du + a_n F_2 \circ L_n(x) \\
&- a_n \gamma_n(x) B_2(x) + a_n \int_{x_1}^{x} \left(\gamma_n'(u) \int_{x_1}^{u} b_2(t)dt \right) du,
\end{aligned}$$

$$\begin{aligned}
\hat{y}_{n+1} = \sum_{j=1}^{n} \Bigg\{ & a_j \delta_j(x_N) \hat{y}_N + a_j \beta_j(x_N) \hat{z}_N - a_j \int_{x_1}^{x_N} \left(\delta_j'(u) \int_{x_1}^{u} f_1[\mathbf{A}](t)dt \right) du \\
&- a_j \int_{x_1}^{x_N} \left(\beta_j'(u) \int_{x_1}^{u} f_2[\mathbf{A}](t)dt \right) du + a_j F_1 \circ L_j(x_N) - a_j \delta_j(x_N) B_1(x_N) \\
&+ a_j \int_{x_1}^{x} \left(\delta_j'(u) \int_{x_1}^{u} b_1(t)dt \right) du - a_j \beta_j(x_N) B_2(x_N) + a_j \int_{x_1}^{x_N} \left(\beta_j'(u) \int_{x_1}^{u} b_2(t)dt \right) du \Bigg\},
\end{aligned}$$

$$\begin{aligned}
\hat{y}_N = \Bigg(\sum_{j=1}^{N-1} & a_j \beta_j(x_N) \hat{z}_N - a_j \int_{x_1}^{x_N} \left(\delta_j'(u) \int_{x_1}^{u} f_1[\mathbf{A}](t)dt \right) du - a_j \int_{x_1}^{x_N} \left(\beta_j'(u) \int_{x_1}^{u} f_2[\mathbf{A}](t)dt \right) du \\
&+ a_j F_1 \circ L_j(x_N) - a_j \delta_j(x_N) B_1(x_N) + a_j \int_{x_1}^{x_N} \left(\delta_j'(u) \int_{x_1}^{u} b_1(t)dt \right) du
\end{aligned}$$

$$-a_j\beta_j(x_N)B_2(x_N)+a_j\int_{x_1}^{x_N}\left(\beta_j'(u)\int_{x_1}^{u}b_2(t)dt\right)du\Bigg)\Bigg/\left(1-\sum_{j=1}^{N-1}a_j\delta_j(x_N)\right),$$

$$\hat{z}_{n+1}=\sum_{j=1}^{n}\Bigg\{a_j\gamma_j(x_N)\hat{z}_N-a_j\int_{x_1}^{x_N}\left(\gamma_j'(u)\int_{x_1}^{u}f_2[\mathbf{A}](t)dt\right)du$$

$$+a_jF_2\circ L_j(x_N)-a_j\gamma_j(x_N)B_2(x_N)+a_j\int_{x_1}^{x_N}\left(\gamma_j'(u)\int_{x_1}^{u}b_2(t)dt\right)du\Bigg\},$$

$$\hat{z}_N=\Bigg(\sum_{j=1}^{N-1}-a_j\int_{x_1}^{x_N}\left(\gamma_j'(u)\int_{x_1}^{u}f_2[\mathbf{A}](t)dt\right)du+a_jF_2\circ L_j(x_N)-a_j\gamma_n(x_N)B_2(x_N)$$

$$+a_j\int_{x_1}^{x_N}\left(\gamma_j'(u)\int_{x_1}^{u}b_2(t)dt\right)du\Bigg)\Bigg/\left(1-\sum_{j=1}^{N-1}a_j\gamma_j(x_N)\right).$$

Proof By the same procedure of Theorem 4.9 and employing the fixed point equation of $f_1[\mathbf{A}]$, one can get

$$\begin{aligned}
\hat{f}_1[\mathbf{A}](L_n(x)) &= \hat{y}_n+a_n\int_{x_1}^{x}\delta_n(x)f_1[\mathbf{A}](t)dt+a_n\int_{x_1}^{x}\beta_n(x)f_2[\mathbf{A}](t)dt+a_n\int_{x_1}^{x}p_n(t)dt\\
&= \hat{y}_n+a_n\delta_n(x)\int_{x_1}^{x}f_1[\mathbf{A}](t)dt+a_n\beta_n(x)\int_{x_1}^{x}f_2[\mathbf{A}](t)dt\\
&\quad -a_n\int_{x_1}^{x}\left(\delta_n'(u)\int_{x_1}^{u}f_1[\mathbf{A}](t)dt\right)du-a_n\int_{x_1}^{x}\left(\beta_n'(u)\int_{x_1}^{u}f_2[\mathbf{A}](t)dt\right)du\\
&\quad +a_n\int_{x_1}^{x}f_1\circ L_n(t)dt-a_n\int_{x_1}^{x}\delta_n(t)b_1(t)dt-a_n\int_{x_1}^{x}\beta_n(t)b_2(t)dt\\
&= \hat{y}_n+a_n\delta_n(x)\hat{f}_1[\mathbf{A}](x)-a_n\delta_n(x)\hat{y}_1+a_n\beta_n(x)\hat{f}_2[\mathbf{A}](x)-a_n\beta_n(x)\hat{z}_1\\
&\quad -a_n\int_{x_1}^{x}\left(\delta_n'(u)\int_{x_1}^{u}f_1[\mathbf{A}](t)dt\right)du-a_n\int_{x_1}^{x}\left(\beta_n'(u)\int_{x_1}^{u}f_2[\mathbf{A}](t)dt\right)du\\
&\quad +a_n\int_{x_1}^{x}f_1\circ L_n(t)dt-a_n\delta_n(x)\int_{x_1}^{x}b_1(t)dt+a_n\int_{x_1}^{x}\left(\delta_n'(u)\int_{x_1}^{u}b_1(t)dt\right)du\\
&\quad -a_n\beta_n(x)\int_{x_1}^{x}b_2(t)dt+a_n\int_{x_1}^{x}\left(\beta_n'(u)\int_{x_1}^{u}b_2(t)dt\right)du
\end{aligned}$$

$$\begin{aligned}
\hat{f}_1[\mathbf{A}](L_n(x)) &= \hat{y}_n+a_n\delta_n(x)\hat{f}_1[\mathbf{A}](x)+a_n\beta_n(x)\hat{f}_2[\mathbf{A}](x)-a_n\int_{x_1}^{x}\left(\delta_n'(u)\int_{x_1}^{u}f_1[\mathbf{A}](t)dt\right)du\\
&\quad -a_n\int_{x_1}^{x}\left(\beta_n'(u)\int_{x_1}^{u}f_2[\mathbf{A}](t)dt\right)du+a_nF_1\circ L_n(x)-a_n\delta_n(x)B_1(x)\\
&\quad +a_n\int_{x_1}^{x}\left(\delta_n'(u)\int_{x_1}^{u}b_1(t)dt\right)du-a_n\beta_n(x)B_2(x)+a_n\int_{x_1}^{x}\left(\beta_n'(u)\int_{x_1}^{u}b_2(t)dt\right)du\\
&= a_n\delta_n(x)\hat{f}_1[\mathbf{A}](x)+a_n\beta_n(x)\hat{f}_2[\mathbf{A}](x)+\hat{p}_n(x)\\
&= \hat{F}_{1n}(x,\hat{f}_1[\mathbf{A}](x),\hat{f}_2[\mathbf{A}](x)).
\end{aligned}$$

Denote,

$$\hat{p}_n(x) = \hat{y}_n - a_n \int_{x_1}^{x} \left(\delta_n'(u) \int_{x_1}^{u} f_1[\mathbf{A}](t)dt \right) du - a_n \int_{x_1}^{x} \left(\beta_n'(u) \int_{x_1}^{u} f_2[\mathbf{A}](t)dt \right) du$$
$$+ a_n F_1 \circ L_n(x) - a_n \delta_n(x) B_1(x) + a_n \int_{x_1}^{x} \left(\delta_n'(u) \int_{x_1}^{u} b_1(t)dt \right) du$$
$$- a_n \beta_n(x) B_2(x) + a_n \int_{x_1}^{x} \left(\beta_n'(u) \int_{x_1}^{u} b_2(t)dt \right) du.$$

Applying the fixed point equation of $f_2[\mathbf{A}](L_n(x))$ in the second part of Theorem 4.9, one can get

$$\begin{aligned}
\hat{f}_2[\mathbf{A}](L_n(x)) &= \hat{z}_n + a_n \int_{x_1}^{x} \gamma_n(t) f_2[\mathbf{A}](t)dt + a_n \int_{x_1}^{x} q_n(t)dt \\
&= \hat{z}_n + a_n \gamma_n(x) \int_{x_1}^{x} f_2[\mathbf{A}](t)dt - a_n \int_{x_1}^{x} \left(\gamma_n'(u) \int_{x_1}^{u} f_2[\mathbf{A}](t)dt \right) du \\
&\quad + a_n \int_{x_1}^{x} f_2 \circ L_n(t)dt - a_n \int_{x_1}^{x} \gamma_n(t) b_2(t)dt \\
&= \hat{z}_n + a_n \gamma_n(x) \hat{f}_2[\mathbf{A}](x) - a_n \gamma_n(x) \hat{z}_1 - a_n \int_{x_1}^{x} \left(\gamma_n'(u) \int_{x_1}^{u} f_2[\mathbf{A}](t)dt \right) du \\
&\quad + a_n \int_{x_1}^{x} f_2 \circ L_n(t)dt - a_n \gamma_n(x) \int_{x_1}^{x} b_2(t)dt + a_n \int_{x_1}^{x} \left(\gamma_n'(u) \int_{x_1}^{u} b_2(t)dt \right) du \\
&= \hat{z}_n + a_n \gamma_n(x) \hat{f}_2[\mathbf{A}](x) - a_n \int_{x_1}^{x} \left(\gamma_n'(u) \int_{x_1}^{u} f_2[\mathbf{A}](t)dt \right) du \\
&\quad + a_n F_2 \circ L_n(x) - a_n \gamma_n(x) B_2(x) + a_n \int_{x_1}^{x} \left(\gamma_n'(u) \int_{x_1}^{u} b_2(t)dt \right) du \\
&= a_n \gamma_n(x) \hat{f}_2[\mathbf{A}](x) + \hat{q}_n(x) \\
&= \hat{F}_{2n}(x, \hat{f}_2[\mathbf{A}](x)).
\end{aligned}$$

Denote $\hat{q}_n(x) = \hat{z}_n - a_n \int_{x_1}^{x} \left(\gamma_n'(u) \int_{x_1}^{u} f_2[\mathbf{A}](t)dt \right) du + a_n F_2 \circ L_n(x) - a_n \gamma_n(x) B_2(x) + a_n \int_{x_1}^{x} \left(\gamma_n'(u) \int_{x_1}^{u} b_2(t)dt \right) du$. For obtaining the new interpolation points $(x_n, \hat{y}_n)$, consider

$$\hat{f}_1[\mathbf{A}](L_n(x)) = \hat{y}_n + a_n \delta_n(x) \hat{f}_1[\mathbf{A}](x) + a_n \beta_n(x) \hat{f}_2[\mathbf{A}](x) - a_n \int_{x_1}^{x} \left(\delta_n'(u) \int_{x_1}^{u} f_1[\mathbf{A}](t)dt \right) du$$
$$- a_n \int_{x_1}^{x} \left(\beta_n'(u) \int_{x_1}^{u} f_2[\mathbf{A}](t)dt \right) du + a_n F_1 \circ L_n(x) - a_n \delta_n(x) B_1(x)$$
$$+ a_n \int_{x_1}^{x} \left(\delta_n'(u) \int_{x_1}^{u} b_1(t)dt \right) du - a_n \beta_n(x) B_2(x) + a_n \int_{x_1}^{x} \left(\beta_n'(u) \int_{x_1}^{u} b_2(t)dt \right) du.$$

Let $x = x_N$ and $L_n(x_N) = x_{n+1}$.

$$\hat{y}_{n+1} - \hat{y}_n = a_n \delta_n(x_N) \hat{y}_N + a_n \beta_n(x_N) \hat{z}_N - a_n \int_{x_1}^{x_N} \left(\delta_n'(u) \int_{x_1}^{u} f_1[\mathbf{A}](t)dt \right) du$$
$$- a_n \int_{x_1}^{x_N} \left(\beta_n'(u) \int_{x_1}^{u} f_2[\mathbf{A}](t)dt \right) du + a_n F_1 \circ L_n(x_N) - a_n \delta_n(x_N) B_1(x_N)$$
$$+ a_n \int_{x_1}^{x_N} \left(\delta_n'(u) \int_{x_1}^{u} b_1(t)dt \right) du - a_n \beta_n(x_N) B_2(x_N) + a_n \int_{x_1}^{x_N} \left(\beta_n'(u) \int_{x_1}^{u} b_2(t)dt \right) du.$$

The equation $\hat{y}_{n+1} = \hat{y}_1 + \sum_{j=1}^{n}(\hat{y}_{j+1} - \hat{y}_j)$ provides

$$\begin{aligned}\hat{y}_{n+1} = \sum_{j=1}^{k} \Big\{ & a_j\delta_j(x_N)\hat{y}_N + a_j\beta_j(x_N)\hat{z}_N - a_j \int_{x_1}^{x_N}\left(\delta_j'(u)\int_{x_1}^{u} f_1[\mathbf{A}](t)dt\right)du \\ & - a_j \int_{x_1}^{x_N}\left(\beta_j'(u)\int_{x_1}^{u} f_2[\mathbf{A}](t)dt\right)du + a_j F_1\circ L_j(x_N) - a_j\delta_j(x_N)B_1(x_N) \\ & + a_j \int_{x_1}^{x}\left(\delta_j'(u)\int_{x_1}^{u} b_1(t)dt\right)du - a_j\beta_j(x_N)B_2(x_N) + a_j\int_{x_1}^{x_N}\left(\beta_j'(u)\int_{x_1}^{u} b_2(t)dt\right)du\Big\}.\end{aligned}$$

Substituting $j = N-1$ in the above equation, one can get

$$\begin{aligned}\hat{y}_N = \Bigg(& \sum_{j=1}^{N-1} a_j\beta_j(x_N)\hat{z}_N - a_j\int_{x_1}^{x_N}\left(\delta_j'(u)\int_{x_1}^{u} f_1[\mathbf{A}](t)dt\right)du - a_j\int_{x_1}^{x_N}\left(\beta_j'(u)\int_{x_1}^{u} f_2[\mathbf{A}](t)dt\right)du \\ & + a_j F_1\circ L_j(x_N) - a_j\delta_j(x_N)B_1(x_N) + a_j\int_{x_1}^{x_N}\left(\delta_j'(u)\int_{x_1}^{u} b_1(t)dt\right)du \\ & - a_j\beta_j(x_N)B_2(x_N) + a_j\int_{x_1}^{x_N}\left(\beta_j'(u)\int_{x_1}^{u} b_2(t)dt\right)du\Bigg) \Big/ \left(1 - \sum_{j=1}^{N-1} a_j\delta_j(x_N)\right).\end{aligned}$$

To obtain the interpolation points $(x_n, \hat{z}_n)$, consider

$$\begin{aligned}\hat{f}_2[\mathbf{A}](L_n(x)) = {} & \hat{z}_n + a_n\gamma_n(x)\hat{f}_2[\mathbf{A}](x) - a_n\int_{x_1}^{x}\left(\gamma_n'(u)\int_{x_1}^{u} f_2[\mathbf{A}](t)dt\right)du \\ & + a_n F_2\circ L_n(x) - a_n\gamma_n(x)B_2(x) + a_n\int_{x_1}^{x}\left(\gamma_n'(u)\int_{x_1}^{u} b_2(t)dt\right)du.\end{aligned}$$

Taking $x = x_N$ and $L_n(x_N) = x_{n+1}$,

$$\begin{aligned}\hat{z}_{n+1} - \hat{z}_n = {} & a_n\gamma_n(x_N)\hat{z}_N - a_n\int_{x_1}^{x_N}\left(\gamma_n'(u)\int_{x_1}^{u} f_2[\mathbf{A}](t)dt\right)du \\ & + a_n F_2\circ L_n(x_N) - a_n\gamma_n(x_N)B_2(x_N) + a_n\int_{x_1}^{x_N}\left(\gamma_n'(u)\int_{x_1}^{u} b_2(t)dt\right)du.\end{aligned}$$

The equation $\hat{z}_{n+1} = \hat{z}_1 + \sum_{j=1}^{n}(\hat{z}_{j+1} - \hat{z}_j)$ yields

$$\begin{aligned}\hat{z}_{n+1} = \sum_{j=1}^{n}\Big\{ & a_j\gamma_j(x_N)\hat{z}_N - a_j\int_{x_1}^{x_N}\left(\gamma_j'(u)\int_{x_1}^{u} f_2[\mathbf{A}](t)dt\right)du \\ & + a_j F_2\circ L_j(x_N) - a_j\gamma_j(x_N)B_2(x_N) + a_j\int_{x_1}^{x_N}\left(\gamma_j'(u)\int_{x_1}^{u} b_2(t)dt\right)du\Big\}.\end{aligned}$$

Substituting $j = N-1$, one can obtain

$$\hat{z}_N = \left(\sum_{j=1}^{N-1} -a_j \int_{x_1}^{x_N} \left(\gamma_j'(u) \int_{x_1}^{u} f_2[\mathbf{A}](t)dt\right) du + a_j F_2 \circ L_j(x_N) - a_j \gamma_j(x_N) B_2(x_N)\right.$$
$$\left. + a_j \int_{x_1}^{x_N} \left(\gamma_j'(u) \int_{x_1}^{u} b_2(t)dt\right) du\right) \Big/ \left(1 - \sum_{j=1}^{N-1} a_j \gamma_j(x_N)\right).$$

The following continuity conditions can be easily examined,

$$\hat{p}_n(x_1) = \hat{y}_n,\ \hat{p}_n(x_N) = \hat{y}_{n+1} - a_n \delta_n(x_N)\hat{y}_N - a_n \beta_n(x_N)\hat{z}_N$$

and

$$\hat{q}_n(x_1) = \hat{z}_n,\ \hat{q}_n(x_N) = \hat{z}_{n+1} - a_n \gamma_n(x_N)\hat{z}_N,$$

which implies $\hat{F}_n(x_1, \hat{y}_1, \hat{z}_1) = (\hat{y}_n, \hat{z}_n)$ and $\hat{F}_n(x_N, \hat{y}_N, \hat{z}_N) = (\hat{y}_{n+1}, \hat{z}_{n+1})$. Therefore, the integral of the function $f[\mathbf{A}]$ with variable scaling factors is also a A-fractal function. □

Corollary 4.8 *Let f be a fractal interpolation function as described in Theorem 4.10, and defined at the end points as $\sum_{j=1}^{N-1} a_j \delta_j(x_1) \neq 1$ and $\sum_{j=1}^{N-1} a_j \gamma_j(x_1) \neq 1$. Then, for $j = 1, 2, \ldots, N-1$,*

$$\hat{p}_n(x) = \hat{y}_{n+1} + a_n \int_{x}^{x_N} \left(\delta_n'(u) \int_{u}^{x_N} f_1[\mathbf{A}](t)dt\right) du + a_n \int_{x}^{x_N} \left(\beta_n'(u) \int_{u}^{x_N} f_2[\mathbf{A}](t)dt\right) du$$
$$+ a_n F_1 \circ L_n(x) - a_n \delta_n(x) B_1(x) - a_n \int_{x}^{x_N} \left(\delta_n'(u) \int_{u}^{x_N} b_1(t)dt\right) du$$
$$- a_n \beta_n(x) B_2(x) - a_n \int_{x}^{x_N} \left(\beta_n'(u) \int_{u}^{x_N} b_2(t)dt\right) du,$$
$$\hat{q}_n(x) = \hat{z}_{n+1} + a_n \int_{x}^{x_N} \left(\gamma_n'(u) \int_{u}^{x_N} f_2[\mathbf{A}](t)dt\right) du + a_n F_2 \circ L_n(x)$$
$$- a_n \gamma_n(x) B_2(x) - a_n \int_{x}^{x_N} \left(\gamma_n'(u) \int_{u}^{x_N} b_2(t)dt\right) du,$$

$$\hat{y}_n = \sum_{j=n}^{N-1} \left\{ a_j \delta_j(x_1)\hat{y}_1 + a_j \beta_j(x_1)\hat{z}_1 + a_j \int_{x_1}^{x_N} \left(\delta_j'(u) \int_{u}^{x_N} f_1[\mathbf{A}](t)dt\right) du\right.$$
$$+ a_j \int_{x_1}^{x_N} \left(\beta_j'(u) \int_{u}^{x_N} f_2[\mathbf{A}](t)dt\right) du + a_j F_1 \circ L_j(x_1) - a_j \delta_j(x_1) B_1(x_1)$$
$$\left. - a_j \int_{x_1}^{x} \left(\delta_j'(u) \int_{u}^{x_N} b_1(t)dt\right) du - a_j \beta_j(x_1) B_2(x_1) - a_j \int_{x_1}^{x_N} \left(\beta_j'(u) \int_{u}^{x_N} b_2(t)dt\right) du\right\},$$
$$\hat{y}_1 = \left(\sum_{j=1}^{N-1} a_j \beta_j(x_1)\hat{z}_N + a_j \int_{x_1}^{x_N} \left(\delta_j'(u) \int_{u}^{x_N} f_1[\mathbf{A}](t)dt\right) du\right.$$
$$+ a_j \int_{x_1}^{x_N} \left(\beta_j'(u) \int_{u}^{x_N} f_2[\mathbf{A}](t)dt\right) du + a_j F_1 \circ L_j(x_1)$$
$$- a_j \delta_j(x_1) B_1(x_1) - a_j \int_{x_1}^{x_N} \left(\delta_j'(u) \int_{u}^{x_N} b_1(t)dt\right) du - a_j \beta_j(x_1) B_2(x_1)$$

$$-a_j \int_{x_1}^{x_N} \left(\beta_j'(u) \int_u^{x_N} b_2(t)dt \right) du \Bigg) \Bigg/ \left(1 - \sum_{j=1}^{N-1} a_j \delta_j(x_1) \right),$$

$$\hat{z}_n = \sum_{j=n}^{N-1} \Bigg\{ a_j \gamma_j(x_1) \hat{z}_N + a_j \int_{x_1}^{x_N} \left(\gamma_j'(u) \int_u^{x_N} f_2[\mathbf{A}](t)dt \right) du$$

$$+ a_j F_2 \circ L_j(x_1) - a_j \gamma_j(x_1) B_2(x_1) - a_j \int_{x_1}^{x_N} \left(\gamma_j'(u) \int_u^{x_N} b_2(t)dt \right) du \Bigg\},$$

$$\hat{z}_1 = \Bigg(\sum_{j=1}^{N-1} a_j \int_{x_1}^{x_N} \left(\gamma_j'(u) \int_u^{x_N} f_2[\mathbf{A}](t)dt \right) du + a_j F_2 \circ L_j(x_1) - a_j \gamma_j(x_1) B_2(x_1)$$

$$-a_j \int_{x_1}^{x_N} \left(\gamma_j'(u) \int_u^{x_N} b_2(t)dt \right) du \Bigg) \Bigg/ \left(1 - \sum_{j=1}^{N-1} a_j \gamma_j(x_1) \right).$$

4.6 Summary

In this chapter, the classical integral is used as a tool to generate new fractal interpolation functions which interpolate a new set of data points. The influence of scaling factors in the fractal functions is demonstrated through numerical examples by choosing them as variables and constants. The definite and indefinite integrals of a fractal function are explicitly evaluated, using the fractal property which states that any $x \in I$ is described as an image of compositions of functions. Further, the integral of non-affine fractal interpolation functions namely, the α-fractal function, hidden variable fractal function and A-fractal function are investigated with constant and variable scaling factors. This study clearly acknowledges that the integral of different types of fractal functions generate new fractal functions whenever they satisfy the endpoint conditions.

References

1. A. Băicoianu, C.G. Gavrilă, C.M. Păcurar, V.D. Păcurar, Fractal interpolation in the context of prediction accuracy optimization. Eng. Appl. Artif. Intell. **133**, 108380 (2024)
2. Y. Zhang, Q. Fan, F. Bao, Y. Liu, C. Zhang, Single-image super-resolution based on rational fractal interpolation. IEEE Trans. Image Process. **27**(8), 3782–3797 (2018)
3. Y. Jiang, S. Tang, L. Zhang, M. Xiong, Y.J. Yip. Covert voice over internet protocol communications with packet loss based on fractal interpolation. ACM Trans. Multimed. Comput., Commun., Appl. (TOMM) **12**(4), 1–20 (2016)
4. S. Raubitzek, T. Neubauer, A fractal interpolation approach to improve neural network predictions for difficult time series data. Expert Syst. Appl. **169**, 114474 (2021)
5. T.M.C. Priyanka, K. Udhayakumar, S.S. Mohanrasu, A. Gowrisankar, R. Rakkiyappan, Chaotic synchronization and fractal interpolation-based image encryption: exploring event-triggered impulsive control in variable-order fractional lur'e systems. Multimed. Tools Appl. 1–40 (2024)
6. M.F. Barnsley, A.N. Harrington, The calculus of fractal interpolation functions. J. Approx. Theory **57**(1), 14–34 (1989)

7. M.F. Barnsley, Fractal functions and interpolation. Constr. Approx. **2**, 303–329 (1986)
8. T.M.C. Priyanka, A. Gowrisankar, Riemann-Liouville fractional integral of non-affine fractal interpolation function and its fractional operator. Eur. Phys. J. Spec. Top. **230**(21–22), 3789–3805 (2021)
9. T.M.C. Priyanka, C. Serpa, A. Gowrisankar, On the classical integral of fractal functions. Fractals **31**(05), 2350057 (2023)
10. A. Agathiyan, A. Gowrisankar, T.M.C. Priyanka, Construction of new fractal interpolation functions through integration method. Results Math. **77**(3), 122 (2022)

Chapter 5
Methods of Fractional Calculus

5.1 Introduction

The subject of fractional calculus is believed to have stemmed from de L'Hop̂ital's question of "What does $\frac{d^n}{dx^n} f(x)$ mean if $n = 1/2$?" to Leibniz, when he shared the notation $\frac{d^n}{dx^n} f(x)$ to represent the nth derivative of a function f in the year 1695. Since then, many well-known mathematicians, including Euler, Abel, Liouville and Riemann, have become interested in the fractional calculus and its theory has gradually been extended to incorporate operators D^v, where v may be rational or irrational, positive or negative, real or complex. Thus, the fractional calculus is the study of differentiation and integration of any arbitrary order. The popular book of [1] on the subject of fractional calculus primarily deals with fractional differentiable equations. The book [2] describes the mathematical arguments which lead to the definitions of types of fractional integrals and fractional derivatives. The theories of differential, integral, and integro-differential equations, and special functions of mathematical physics as well as their extensions and generalizations have huge applications which include fluid flow, dynamical processes in self-similar structures, electrical networks, probability and statistics, control theory of dynamical systems, viscoelasticity, chemical physics, signal processing and so on. With this brief introduction of fractional calculus, this chapter discusses the definitions of fractional integral and fractional derivatives which are employed in the subsequent chapters to investigate the fractional calculus of fractal interpolation functions.

5.2 Riemann-Liouville Calculus

This section presents the definitions of Riemann-Liouville fractional integral and fractional derivative. The first publication in 1892 revealed G. F. Bernhard Riemann's theory of fractional integral, initially, he sought as a generalization of a Taylor series.

A. Gowrisankar et al., *Fractal Functions Communication with Fractional Calculus*,
https://doi.org/10.1007/978-3-031-96993-5_5

In 1880, A. Cayley remarked that there is a great difficulty in Riemann's theory due to the complementary function containing an infinity of arbitrary constants, in Riemann's definition of integral. After several modifications and discussions by several authors including Hargreave, Sonin and Letnikov [1–4], the following most used version is defined without the ambiguous complementary function.

Definition 5.1 Let $[a, b](-\infty < a < b < \infty)$ be a finite real sub-interval and $f : [a, b] \to \mathbb{R}$ be a continuous function. The *Riemann-Liouville fractional integral* of order v is defined by

$$I_a^v f(x) = \frac{1}{\Gamma(v)} \int_a^x (x-t)^{v-1} f(t)dt \ (x > a) \tag{5.1}$$

Here, $\Gamma(v)$ is the Gamma function defined by $\Gamma(v) = \int_0^\infty t^{v-1}e^{-t}dt$.

From the law of exponents, it is defined that $D^\mu f(x) = D^m[D^{-v} f(x)]$, where $D^{-v} = I^v$ and $v = m - \mu$, m is the smallest integer that exceeds $\mu > 0$. Then, the Riemann-Liouville fractional derivative of order v for f is defined by

$$\mathcal{D}_a^v f(x) = \frac{1}{\Gamma(n-v)} \left(\frac{d}{dx}\right)^n \int_a^x (x-t)^{n-v-1} f(t)dt \tag{5.2}$$

with $n - 1 \leq v < n$, where n is an integer and v is a real number.

Example 5.1 Consider a continuous function $f(x) = x/2$. The classical integration on $f(x)$ gives $x^2/4 + c$. The Riemann-Liouville fractional integral of $f(x)$ with $v = 0.1$ is estimated as follows:

$$\begin{aligned} I^{0.1} f(x) &= \frac{1}{\Gamma(0.1)} \int_0^x (x-t)^{0.1-1}(t/2)dt \\ &= \frac{1}{2(0.1)\Gamma(0.1)} \int_0^x (x-t)^{0.1}dt \\ &\approx 0.48x^{1.1} \end{aligned}$$

and if $v = 0.6$, then

$$\begin{aligned} I_0^{0.6} f(x) &= \frac{1}{\Gamma(0.6)} \int_0^x (x-t)^{0.6-1}(t/2)dt \\ &= \frac{1}{2(0.6)\Gamma(0.6)} \int_0^x (x-t)^{0.6}dt \\ &\approx 0.35x^{1.6}. \end{aligned}$$

The graphical illustration of $f(x) = x/2$ on the interval [0, 1], its classical integral and its Riemann-Liouville fractional integral with orders: $v = 0.1$ and $v = 0.5$ are given in Fig. 5.1a.

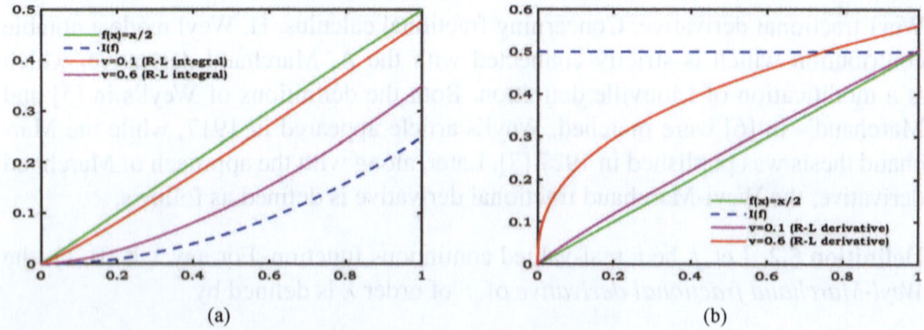

Fig. 5.1 Graphical illustration of continuous function $f(x) = x/2$ and its **a** Riemann-Liouville fractional integral with orders: $\upsilon = 0.1$ and $\upsilon = 0.5$, and **b** Riemann-Liouville fractional derivative with orders: $\upsilon = 0.1$ and **b** $\upsilon = 0.5$

Example 5.2 The ordinary derivative of $f(x) = x/2$ is $1/2$. The Riemann-Liouville fractional derivative of $f(x) = x/2$ is evaluated with order $\upsilon = 0.1$ as follows:

$$\begin{aligned}\mathcal{D}_0^{0.1} f(x) &= \frac{1}{\Gamma(1-0.1)}\left(\frac{d}{dx}\right)\int_0^x (x-t)^{-0.1}(t/2)dt\\ &= \frac{1}{2(0.9)\Gamma(0.9)}\left(\frac{d}{dx}\right)\int_0^x (x-t)^{0.9}dt\\ &\approx 0.51x^{0.9}.\end{aligned}$$

Similarly, when $\upsilon = 0.5$,

$$\begin{aligned}\mathcal{D}_0^{0.5} f(x) &= \frac{1}{\Gamma(1-0.5)}\left(\frac{d}{dx}\right)\int_0^x (x-t)^{-0.5}(t/2)dt\\ &= \frac{1}{2(0.4)\Gamma(0.4)}\left(\frac{d}{dx}\right)\int_0^x (x-t)^{0.4}dt\\ &\approx 0.56x^{0.4}.\end{aligned}$$

The graphical illustration of $f(x) = x/2$ on the interval [0, 1], its classical derivative and its Riemann-Liouville fractional derivative with orders: $\upsilon = 0.1$ and $\upsilon = 0.5$ are provided in Fig. 5.1b.

5.3 Weyl-Marchaud Fractional Derivative

The Riemann-Liouville fractional integral seemed to be inconvenient in the theory of trigonometrical series dealing with periodic functions. So, for such functions, Weyl fractional integral was suggested [4]. Then, it was natural to define the corresponding

Weyl fractional derivative. Concerning fractional calculus, H. Weyl made a notable contribution which is strictly connected with the A. Marchaud derivative, which is a modification of Liouville definition. Both the definitions of Weyl's in [5] and Marchaud's in [6] were matched. Weyl's article appeared in 1917, while the Marchaud thesis was published in 1927 [7]. Later, along with the approach of Marchaud derivative, the Weyl-Marchaud fractional derivative is defined as follows.

Definition 5.2 Let f be a real-valued continuous function. For any $\lambda \in (0, 1)$, the *Weyl-Marchaud fractional derivative* of f of order λ is defined by

$$D_0^{\lambda} f(x) = \frac{\lambda}{\Gamma(1-\lambda)} \int_0^{\infty} \frac{f(x) - f(x-y)}{y^{1+\lambda}} dy, \tag{5.3}$$

where $0 < \lambda < 1$.

When $y = x - t$ in Eq. (5.3), then the following equivalent form can be obtained

$$D_0^{\lambda} f(x) = \frac{\lambda}{\Gamma(1-\lambda)} \int_{-\infty}^{x} \frac{f(x) - f(t)}{(x-t)^{1+\lambda}} dt.$$

Example 5.3 Consider a continuous function $f(x) = x^2$. The classical integration on $f(x)$ gives $x^3/3 + c$. The Weyl-Marchaud fractional derivative of $f(x)$ with $v = 0.2$ is estimated as follows:

$$\begin{aligned} D_0^{0.2} f(x) &= \frac{0.2}{\Gamma(1-0.2)} \int_0^{x} \frac{(x^2 - t^2)}{(x-t)^{1+0.2}} dt \\ &= \frac{0.2}{0.8\Gamma(0.9)} \left\{ x^{1.8} + \int_0^{x} (x-t)^{0.8} dt \right\} \\ &\approx 0.35x^{1.8}. \end{aligned}$$

Similarly, when $\lambda = 0.5$,

$$\begin{aligned} D_0^{0.5} f(x) &= \frac{0.5}{\Gamma(1-0.5)} \int_0^{x} \frac{(x^2 - t^2)}{(x-t)^{1+0.5}} dt \\ &\approx 0.94x^{1.5}. \end{aligned}$$

The graphs of $f(x) = x^2$ on the interval [0, 1], its classical derivative and its Weyl-Marchaud fractional derivative with orders: $v = 0.2$ and $v = 0.5$ are respectively illustrated in Fig. 5.2a and b.

5.4 Caputo-Fractional Derivative

The Riemann-Liouville type of fractional differentiation has played a significant role in the applications of pure mathematics. However, practical problems like in mathematical modeling, required the definitions which allow for physically interpretable initial conditions containing f and f', and so on. In 1969, M. Caputo proposed a certain solution to this conflict by giving the following definition [8].

Definition 5.3 Let f be a continuous function on $[a, b]$. Then *the Caputo-fractional derivative* of f is defined in Eq. (5.4) by

$$^{C}D_a^{\lambda} f(x) = \frac{1}{\Gamma(n-\lambda)} \int_a^x (x-t)^{n-\lambda-1} f^{(n)}(t)dt \tag{5.4}$$

with $n - 1 \leq \lambda < n$, where n is an integer and λ is a real number..

In particular, when $n = 1$ and $0 < \lambda < 1$, then

$$^{C}D_a^{\lambda} f(x) = \frac{1}{\Gamma(1-\lambda)} \int_a^x (x-t)^{-\lambda} f'(t)dt.$$

For $\upsilon \to n$, the Caputo derivative coincides with the classical n^{th} derivative of the function f. The main advantage is that the initial conditions for Caputo-fractional derivatives take the form which is similar to the integer-order differential equations.

Example 5.4 The Caputo-fractional derivative of $f(x) = x^2$ with $\upsilon = 0.2$ is estimated as follows:

$$\begin{aligned} ^{C}D_0^{0.2} f(x) &= \frac{1}{\Gamma(1-0.2)} \int_0^x (x-t)^{-0.2}(2t)dt \\ &= \frac{2}{0.8\Gamma(0.8)} \int_0^x (x-t)^{0.8}dt \\ &\approx 1.19x^{1.8}. \end{aligned}$$

Similarly, when $\lambda = 0.5$,

$$\begin{aligned} ^{C}D_0^{0.5} f(x) &= \frac{1}{\Gamma(1-0.5)} \int_0^x (x-t)^{-0.5}(2t)dt \\ &\approx 1.51x^{1.5}. \end{aligned}$$

The graphs of $f(x) = x^2$ on the interval $[0, 1]$ and its Caputo-fractional derivative with orders: $\upsilon = 0.2$ and $\upsilon = 0.5$ are illustrated in Fig. 5.2a and b, respectively.

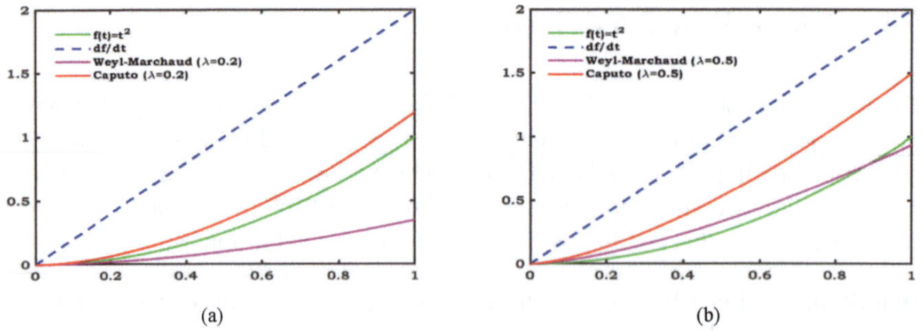

Fig. 5.2 Graphical illustration of continuous function $f(x) = x^2$ and its Weyl-Marchaud and Caputo-fractional derivatives with order: **a** $\lambda = 0.2$ and **b** $\lambda = 0.5$

References

1. S.G. Samko, A.A. Kilbas, O.I. Marichev, *Fractional Integrals and Derivatives*, vol. 1 (Gordon and breach science publishers, Yverdon Yverdon-les-Bains, Switzerland, 1993)
2. A.A. Kilbas, H.M. Srivastava, J.J Trujillo, *Theory and Applications of Fractional Differential Equations*, vol. 204 (Elsevier, 2006)
3. K. Diethelm, *The Analysis of Fractional Differential Equations: An Application-Oriented Exposition Using Differential Operators of Caputo Type*, vol. 2004 (Springer, Berlin, 2010)
4. K.S. Miller, B. Ross, An introduction to the fractional calculus and fractional differential equations (1993)
5. W.E.Y.L. Hermann, Bemerkungen zum begriff des differentialquotienten gebrochener ordnung. Vierteljahrsschrift der naturforschenden Gesellschaft in Zürich **62**, 296 (1917)
6. André Marchaud, Sur les dérivées et sur les différences des fonctions de variables réelles. J. de mathématiques pures et appliquées **6**, 337–425 (1927)
7. Fausto Ferrari, Weyl and Marchaud derivatives: a forgotten history. Mathematics **6**(1), 6 (2018)
8. I. Podlubny, *Fractional Differential Equations: An Introduction to Fractional Derivatives, Fractional Differential Equations, to Methods of Their Solution and Some of Their Applications*, vol. 198 (Elsevier, 1998)

Chapter 6
Fractional Integral Order and Fractal Scalings

6.1 Introduction

In the evolution of fractal interpolation theory, many researchers have constructed various types of non-affine fractal functions to amplify the flexibility in approximation, the α-fractal function is one among the classic example of non-affine fractal functions. In general, the fractal properties such as fractal dimension, smoothness of fractal function are much influenced by the free parameter called the vertical scaling factor. Most of the research works on the FIF have been studied with constant scaling factors which obviously provide self-similar character. Since there exists nonuniform data (or functions) showing less self-similarity, for this type of functions, variable scaling factors could be the better choice instead of constant scaling factors. This chapter discusses the fractional integral of non-affine fractal functions with variable scaling parameters. Additionally, a fractional operator developed in relation with α-fractal function is presented and some analytical properties such as linearity, boundedness of the fractional operator are discussed with respect to the order of Riemann-Liouville fractional integral.

In this chapter, the α-fractal function, denoted by f^α, is precisely recalled with numerical illustration. Let $\Delta = \{x_1, x_1, \ldots, x_N : x_n < x_{n+1}\}$ be a partition of I and $f \in C(I)$ be a continuous function such that $f(x_n) = y_n$, for all $n = 1, 2, \ldots, N$. Consider, in particular

$$q_n(x) = f \circ L_n(x) - \alpha_n b(x), \ n = 1, 2, \ldots, N-1, \tag{6.1}$$

where $b : I \to \mathbb{R}$ is a continuous function which obeys the conditions $b(x_1) = f(x_1)$ and $b(x_N) = f(x_N)$, and $b \neq f$. Note that, here α_n is the constant scaling factor. Substituting $q_n(x)$ provided in (6.1) in the functional equation (2.5), it follows that

$$f^\alpha(x) = f(x) + \alpha_n[(f^\alpha - b) \circ L_n^{-1}(x)], \ \forall x \in I_n, \ n = 1, 2, \ldots, N-1. \tag{6.2}$$

A. Gowrisankar et al., *Fractal Functions Communication with Fractional Calculus*,
https://doi.org/10.1007/978-3-031-96993-5_6

When $\alpha = 0$, clearly $f^{\alpha} = f$. In [1] Barnsley has proposed the fact that the function defined in (6.2) is the generalization of the continuous function f. Further, the smoothness of the α-fractal function f^{α} is depending on the values of the scale vector α, the position and number of data points, and the choice of the base function $b(x)$. In particular, the following choices of the base function $b(x)$ and properties of the corresponding bounded linear fractal operator were investigated (see for more details, see Chapter 2.

(a) $b(x) = (f \circ c)(x)$, where c is a increasing, real valued continuous function on I such that $c(x_1) = x_1, c(x_N) = x_N$ and c is not the identity function.
(b) In general, $b(x) = Lf(x)$, where L is a bounded linear operator with respect to the supremum norm, $\|f\|_{\infty} = \max\{|f(x)| : x \in I\}$, on $C(I)$ such that $Lf(x_1) = f(x_1), Lf(x_N) = f(x_N)$.

Example 6.1 Consider the function $f(x) = x^{1/2}$ on the closed interval $[0, 1]$ with the partition $D = \{0, 0.2, 0.4, 0.6, 0.8, 1\}$. Then, the α-fractal function f^{α} corresponding to f with constant scaling factors $\alpha_1 = 0.6, \alpha_2 = 0.5, \alpha_3 = 0.6, \alpha_4 = 0.4, \alpha_4 = 0.5$ is graphically presented in Fig. 6.1a, b for the following two different choices of the base function $b(x)$ respectively:

(a) $b(x) = (f \circ c)(x)$ where $c(x) = e^{\mu x}/e^{\mu}$, $\mu = 3$,
(b) $b(x) = v(x)f(x)$, where $v(x) = \sin(\pi x/2) + \cos(\pi x/2)$.

Note that all the base functions provided herein are taken in such a way that they coincide with the germ function f at the initial and end points. Now comparing the two figures, the reader can clearly notice the changes in all the subintervals of the α-fractal function due to changes in the base function. The α-fractal function in the second graph shows high irregularity than the first one.

Suppose $\alpha_n \in C(I)$ satisfying $\|\alpha_n\|_{\infty} < 1$. Then (6.1) gives,

$$q_n(x) = f \circ L_n(x) - \alpha_n(x)b(x), \; n = 1, 2, \ldots, N-1. \tag{6.3}$$

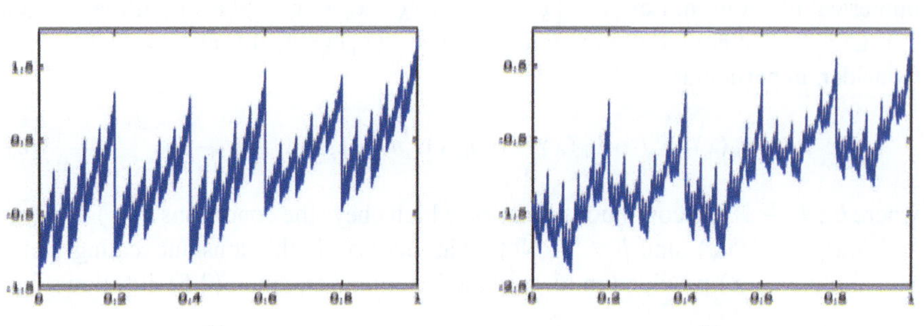

(a) (b)

Fig. 6.1 α-fractal function corresponding to the function $f(x) = x^{1/2}$ on $[0, 1]$ with the constant scaling factors $\alpha_1 = 0.6$, $\alpha_2 = 0.5, \alpha_3 = 0.6, \alpha_4 = 0.4, \alpha_5 = 0.5$: **a** $b(x) = (f \circ c)(x)$, where $c(x) = (e^{3x} - 1)/(e^3 - 1)$ and **b** $b(x) = v(x)f(x)$, where $v(x) = \sin(\pi x/2) + \cos(\pi x/2)$

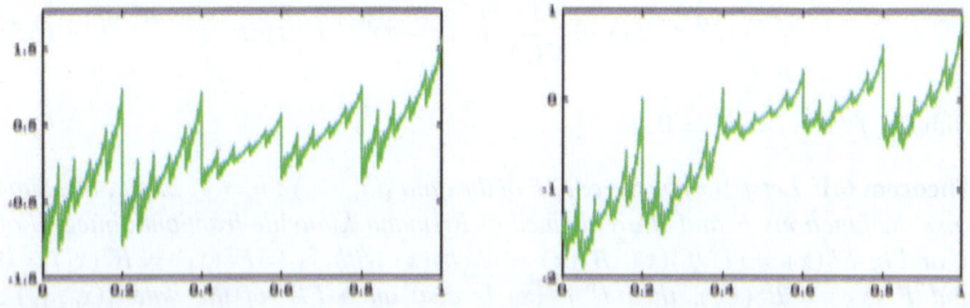

(a) (b)

Fig. 6.2 α-fractal function corresponding to the function $f(x) = \sqrt{x}$ on $[0, 1]$ with the variable scaling factors $\alpha_1(x) = 0.6 + \sin(x/3)$, $\alpha_2(x) = 0.5 + (x/2)$, $\alpha_3(x) = 0.2e^{2x}$, $\alpha_4(x) = 0.5e^x/2$, $\alpha_5(x) = 0.4 + \sin(x/2)$: **a** $b(x) = (f \circ c)(x)$, where $c(x) = (e^{3x} - 1)/(e^3 - 1)$ and **b** $b(x) = v(x)f(x)$, where $v(x) = \sin(\pi x/2) + \cos(\pi x/2)$

Hence the functional equation (6.2) becomes,

$$f^\alpha(x) = f(x) + \alpha_n(L_n^{-1}(x))[(f^\alpha - b) \circ L_n^{-1}(x)], \ \forall x \in I_n, \ n = 1, 2, \ldots, N-1. \tag{6.4}$$

Here, the base function $b(x)$ satisfies $b(x_1) = f(x_1)$ and $b(x_N) = f(x_N)$ and discrete scaling factor is replaced by variable scaling factor $\alpha_n \in C(I)$.

Example 6.2 Let $f(x) = \sqrt{x}$ be the given continuous function with the scaling functions $\alpha_1(x) = 0.6 + \sin(x/3)$, $\alpha_2(x) = 0.5 + (x/2)$, $\alpha_3(x) = 0.2e^{2x}$, $\alpha_4(x) = 0.5e^x/2\alpha_5(x) = 0.4 + \sin(x/2)$. The graphical illustrations of an α-fractal interpolation function corresponding to the given $f(x)$ are depicted in Fig. 6.2a, b for the following two different choices of the base function $b(x)$ respectively:

(a) $b(x) = (f \circ c)(x)$ where $c(x) = (e^{\mu x} - 1)/(e^\mu - 1)$, $\mu = 3$,
(b) $b(x) = v(x)f(x)$, where $v(x) = \sin(\pi x/2) + \cos(\pi x/2)$.

Both the base functions have been chosen as continuous functions and clearly they obey the stated underlying endpoint conditions. The graphs of f^α in Fig. 6.2a, b look much less dense than that of Fig. 8.1a, b, respectively and the shape of the graphs are also completely different. The only reason behind this fact is the selection of variable scaling factors.

6.2 Fractional Integral of Non-affine Fractal Functions

This section briefly describes the definition of Riemann-Liouville fractional integral of α-fractal function of order v. It mainly focus on exploring the Riemann-Liouville fractional integral of α-FIF with both constant as well as variable scaling factors.

Let f be an α-fractal function satisfying Eq. (6.2), then the Riemann-Liouville fractional integral of f^α of order v is defined by

$$(I_{x_1}^{v} f^{\alpha})(x) = \frac{1}{\Gamma(v)} \int_{x_1}^{x} (x-t)^{v-1} f^{\alpha}(t)dt \tag{6.5}$$

with $(I_{x_1}^{v} f^{\alpha})(x_1) = \hat{y}_1 = 0$.

Theorem 6.1 *Let $f^{\alpha}(x)$ be the α-FIF of the data $\{(x_n, y_n) : n = 1, 2, \ldots, N\}$. Suppose the functions F and B are defined as Riemann-Liouville fractional integral of f and b, $F^{v}(x) = (I_{x_1}^{v} f)(x)$, $B^{v}(x) = (I_{x_1}^{v} b)(x)$ with $\hat{y}_1 = F^{v}(x_1) = B^{v}(x_1) = 0$ and $F^{v}(x_N) = B^{v}(x_N)$, then $I_{x_1}^{v} f^{\alpha}(x)$ is also an α-FIF of the data $\{(x_n, \hat{y}_n) : n = 1, 2, \ldots, N\}$. Further, $I_{x_1}^{v} f^{\alpha}(x)$ is generated by the IFS $\{L_n(x), \hat{F}_n(x, y) : n = 1, 2, \ldots, N-1\}$, where $\hat{F}_n(x, y) = a_n^{v}\alpha_n y + \hat{q}_n(x)$ and $\sum_{j=1}^{N-1} a_j^{v}\alpha_j \neq 1$, for $j = 1, 2, \ldots, N-1$,*

$$\begin{aligned}
\hat{q}_n(x) &= \hat{y}_n + f_{n,v}^{\alpha}(x) + a_n^{v} F^{v} \circ L_n(x) - a_n^{v}\alpha_n B^{v}(x),\\
f_{n,v}^{\alpha}(x) &= \frac{1}{\Gamma(v)} \int_{x_1}^{x_n} \left((L_n(x) - t)^{v-1} - (x_n - t)^{v-1}\right) f^{\alpha}(t)dt,\\
F^{v} \circ L_n(x) &= \frac{1}{\Gamma(v)} \int_{x_1}^{x} (x-u)^{v-1} f \circ L_n(u)du,\\
B^{v}(x) &= \frac{1}{\Gamma(v)} \int_{x_1}^{x} (x-u)^{v-1} b(u)du,\\
\hat{y}_{n+1} &= \sum_{j=1}^{n} \left(f_{j,v}^{\alpha}(x_N) + a_j^{v}\alpha_j \hat{y}_N + a_j^{v} F^{v} \circ L_j(x_N) - a_j^{v}\alpha_j B^{v}(x_N)\right),\\
\hat{y}_N &= \frac{\sum_{j=1}^{N-1} \left(f_{j,v}^{\alpha}(x_N) + a_j^{v} F^{v} \circ L_j(x_N) - a_j^{v}\alpha_j B^{v}(x_N)\right)}{1 - \sum_{j=1}^{N-1} a_j^{v}\alpha_j}.
\end{aligned}$$

Proof The IFS corresponding to the function f^{α} with constant scaling factor α_n is given by,

$$L_n(x) = a_n x + b_n, \; F_n(x, y) = \alpha_n y + f \circ L_n(x) - \alpha_n b(x), \; n = 1, 2, \ldots, N-1. \tag{6.6}$$

The Riemann-Liouville integral of f^{α} of order v is expressed by

$$\begin{aligned}
I_{x_1}^{v} f^{\alpha}(L_n(x)) &= \frac{1}{\Gamma(v)} \int_{x_1}^{L_n(x)} (L_n(x) - t)^{v-1} f^{\alpha}(t)dt\\
&= \frac{1}{\Gamma(v)} \int_{x_1}^{x_n} (L_n(x) - t)^{v-1} f^{\alpha}(t)dt + \frac{1}{\Gamma(v)} \int_{x_n}^{L_n(x)} (L_n(x) - t)^{v-1} f^{\alpha}(t)dt\\
&= \frac{1}{\Gamma(v)} \int_{x_1}^{x_n} (x_n - t)^{v-1} f^{\alpha}(t)dt + \frac{1}{\Gamma(v)} \int_{x_1}^{x_n} \left((L_n(x) - t)^{v-1} - (x_n - t)^{v-1}\right) f^{\alpha}(t)dt\\
&\quad + \frac{1}{\Gamma(v)} \int_{x_n}^{L_n(x)} (L_n(x) - t)^{v-1} f^{\alpha}(t)dt.
\end{aligned}$$

Here, take

$$\hat{y}_n = \frac{1}{\Gamma(v)} \int_{x_1}^{x_n} (x_n - t)^{v-1} f^{\alpha}(t)dt$$

$$f_{n,v}^{\alpha}(x) = \frac{1}{\Gamma(v)} \int_{x_1}^{x_n} \left((L_n(x) - t)^{v-1} - (x_n - t)^{v-1}\right) f^{\alpha}(t)dt.$$

By applying the change of variable in third integral as $t = L_n(u)$, the above equation becomes,

$$I_{x_1}^{v} f^{\alpha}(L_n(x)) = \hat{y}_n + f_{n,v}^{\alpha}(x) + \frac{a_n}{\Gamma(v)} \int_{x_1}^{x} (L_n(x) - L_n(u))^{v-1} f^{\alpha}(L_n(u))du.$$

Being f^{α} as an α-FIF, the functional equation (6.2) gives,

$$\begin{aligned}
&= \hat{y}_n + f_{n,v}^{\alpha}(x) + \frac{a_n}{\Gamma(v)} \int_{x_1}^{x} a_n^{v-1}(x-u)^{v-1} \left(\alpha_n f^{\alpha}(u) + f \circ L_n(u) - \alpha_n b(u)\right) du \\
&= \hat{y}_n + f_{n,v}^{\alpha}(x) + \frac{a_n^v \alpha_n}{\Gamma(v)} \int_{x_1}^{x} (x-u)^{v-1} f^{\alpha}(u)du + \frac{a_n^v}{\Gamma(v)} \int_{x_1}^{x} (x-u)^{v-1} f \circ L_n(u)du \\
&\quad - \frac{a_n^v \alpha_n}{\Gamma(v)} \int_{x_1}^{x} (x-u)^{v-1} b(u)du \\
&= \hat{y}_n + f_{n,v}^{\alpha}(x) + a_n^v \alpha_n I_{x_1}^{v} f^{\alpha}(x) + a_n^v I_{x_1}^{v} f \circ L_n(x) - a_n^v \alpha_n I_{x_1}^{v} b(x) \\
&= \hat{y}_n + f_{n,v}^{\alpha}(x) + a_n^v \alpha_n I_{x_1}^{v} f^{\alpha}(x) + a_n^v F^v \circ L_n(x) - a_n^v \alpha_n B^v(x) \\
&= a_n^v \alpha_n I_{x_1}^{v} f^{\alpha}(x) + \hat{q}_n(x) \\
&= \hat{F}_n(x, I_{x_1}^{v} f^{\alpha}(x)).
\end{aligned}$$

Denote $\hat{q}_n(x) = \hat{y}_n + f_{n,v}^{\alpha}(x) + a_n^v F^v \circ L_n(x) - a_n^v \alpha_n B^v(x)$. Obviously $|a_n^v \alpha_n| < 1$. In order to find the new interpolation points of the interpolant $I_{x_1}^{v} f^{\alpha}$, consider the following equation

$$I_{x_1}^{v} f^{\alpha}(L_n(x)) = \hat{y}_n + f_{n,v}^{\alpha}(x) + a_n^v \alpha_n I_{x_1}^{v} f^{\alpha}(x) + a_n^v F^v \circ L_n(x) - a_n^v \alpha_n B^v(x). \tag{6.7}$$

Taking $x = x_N$ and $L_n(x_N) = x_{n+1}$, the Eq. (6.7) yields,

$$\hat{y}_{n+1} = \hat{y}_n + f_{n,v}^{\alpha}(x_N) + a_n^v \alpha_n I_{x_1}^{v} f^{\alpha}(x_N) + a_n^v F^v \circ L_n(x_N) - a_n^v \alpha_n B^v(x_N). \tag{6.8}$$

The equation $\hat{y}_{n+1} = \hat{y}_1 + \sum_{j=1}^{n} \left(\hat{y}_{j+1} - \hat{y}_j\right)$ gives,

$$\hat{y}_{n+1} = \sum_{j=1}^{n} \left(f_{j,v}^{\alpha}(x_N) + a_j^v \alpha_j \hat{y}_N + a_j^v F^v \circ L_j(x_N) - a_j^v \alpha_j B^v(x_N)\right),$$

for each $n = 1, 2, \ldots, N$ and substituting $i = N$, $\hat{y}_N$ is obtained as follows,

$$\hat{y}_N = \frac{\sum_{j=1}^{N-1}\left(f_{j,v}^{\alpha}(x_N) + a_j^v F^v \circ L_j(x_N) - a_j^v \alpha_j B^v(x_N)\right)}{1 - \sum_{j=1}^{N-1} a_j^v \alpha_j}.$$

The function $I_{x_1}^v f^\alpha(x)$ to be an α-FIF, $\hat{q}_n(x)$ needs to satisfy the continuity conditions. Note that

$$\begin{aligned}
\hat{q}_n(x_1) &= \hat{y}_n + f_{n,v}^{\alpha}(x_1) + a_n^v F^v \circ L_n(x_1) - a_n^v \alpha_n B^v(x_1) \\
&= \hat{y}_n - a_n^v \alpha_n \hat{y}_1, \\
\hat{q}_n(x_N) &= \hat{y}_{n+1} + f_{n,v}^{\alpha}(x_N) + a_n^v F^v \circ L_n(x_N) - a_n^v \alpha_n B^v(x_N)
\end{aligned}$$

Comparing the above equation with $\hat{y}_n$ in Eq. (6.8),

$$\hat{q}_n(x_N) = \hat{y}_n - a_n^v \alpha_n \hat{y}_N.$$

which clearly implies $\hat{F}_n(x_1, \hat{y}_1) = \hat{y}_n$ and $\hat{F}_n(x_N, \hat{y}_N) = \hat{y}_{n+1}$. Therefore the IFS $\{L_n(x), \hat{F}_n(x, y) : n = 1, 2, \ldots, N-1\}$ admits a unique attractor $I_{x_1}^v \hat{f}^\alpha$ which is also an α-FIF.

Thus, in Theorem 6.1, it is showed that the Riemann-Liouville integral of an α-FIF with the predefined initial condition $\hat{y}_1 = 0$ is also an α-FIF which interpolate the new data set $\{(x_n, \hat{y}_n) : n = 1, 2, \ldots, N\}$. The following theorem investigates the Riemann-Liouville integral of an α-FIF by predefining the end point $\hat{y}_N = 0$. Further it provides the new data set $\{(x_n, \hat{y}_n) : n = 1, 2, \ldots, N\}$ in explicit form. □

Let f be an α-fractal function, then the Riemann-Liouville fractional integral of f^α of order v is defined by

$$(I_{x_N}^v f^\alpha)(x) = -\frac{1}{\Gamma(v)} \int_x^{x_N} (x-t)^{v-1} f^\alpha(t)dt \tag{6.9}$$

with $(I_{x_N}^v f^\alpha)(x_N) = \hat{y}_N = 0$.

Theorem 6.2 *Let $f^\alpha(x)$ be the α-FIF of the data $\{(x_n, y_n) : n = 1, 2, \ldots, N\}$. Suppose the functions F and B are defined as Riemann-Liouville fractional integral of f and b, $F^v(x) = (I_{x_N}^v f)(x)$, $B^v(x) = (I_{x_N}^v b)(x)$ with $F^v(x_1) = B^v(x_1)$ and $\hat{y}_N = F^v(x_N) = B^v(x_N) = 0$, then $I_{x_N}^v f^\alpha(x)$ is also an α-FIF of the data $\{(x_n, \hat{y}_n) : n = 1, 2, \ldots, N\}$. Further, $I_{x_N}^v f^\alpha(x)$ is generated by the IFS $\{L_n(x), \hat{F}_n(x, y) : n = 1, 2, \ldots, N-1\}$, where $\hat{F}_n(x, y) = a_n^v \alpha_n y + \hat{q}_n(x)$ and $\sum_{j=1}^{N-1} a_j^v \alpha_j \neq 1$, for $n = 1, 2, \ldots, N-1$,*

$$\begin{aligned}
\hat{q}_n(x) &= \hat{y}_{n+1} - f_{n,v}^{\alpha}(x) + a_n^v F^v \circ L_n(x) - a_n^v \alpha_n B^v(x), \\
f_{n,v}^{\alpha}(x) &= \frac{1}{\Gamma(v)} \int_{x_{n+1}}^{x_N} \left((L_n(x) - t)^{v-1} - (x_{n+1} - t)^{v-1}\right) f^\alpha(t)dt, \\
F^v \circ L_n(x) &= -\frac{1}{\Gamma(v)} \int_x^{x_N} (x-u)^{v-1} f \circ L_n(u)du,
\end{aligned}$$

$$B^{\upsilon}(x) = -\frac{1}{\Gamma(\upsilon)} \int_x^{x_N} (x-u)^{\upsilon-1} b(u)du,$$

$$\hat{y}_n = -\sum_{j=n}^{N-1} \left(f_{j,\upsilon}^{\alpha}(x_1) - a_j^{\upsilon}\alpha_j \hat{y}_1 - a_j^{\upsilon} F^{\upsilon} \circ L_j(x_1) + a_j^{\upsilon}\alpha_j B^{\upsilon}(x_1) \right),$$

$$\hat{y}_1 = -\frac{\sum_{j=1}^{N-1} \left(f_{j,\upsilon}^{\alpha}(x_1) - a_j^{\upsilon} F^{\upsilon} \circ L_j(x_1) + a_j^{\upsilon}\alpha_j B^{\upsilon}(x_1) \right)}{1 - \sum_{j=1}^{N-1} a_j^{\upsilon}\alpha_j}.$$

Proof The function f^{α} is the α-FIF with constant scaling factor α_n and (6.2). Then, the Riemann-Liouville fractional integral of f^{α} of order υ is given by the definition (6.9),

$$\begin{aligned} I_{x_N}^{\upsilon} f^{\alpha}(L_n(x)) &= -\frac{1}{\Gamma(\upsilon)} \int_{L_n(x)}^{x_N} (L_n(x)-t)^{\upsilon-1} f^{\alpha}(t)dt \\ &= -\frac{1}{\Gamma(\upsilon)} \int_{L_n(x)}^{x_{n+1}} (L_n(x)-t)^{\upsilon-1} f^{\alpha}(t)dt - \frac{1}{\Gamma(\upsilon)} \int_{x_{n+1}}^{x_N} (L_n(x)-t)^{\upsilon-1} f^{\alpha}(t)dt \\ &= -\frac{1}{\Gamma(\upsilon)} \int_{x_{n+1}}^{x_N} (x_{n+1}-t)^{\upsilon-1} f^{\alpha}(t)dt - \frac{1}{\Gamma(\upsilon)} \int_{x_{n+1}}^{x_N} \left((L_n(x)-t)^{\upsilon-1} - (x_{n+1}-t)^{\upsilon-1}\right) f^{\alpha}(t)dt \\ &\quad - \frac{1}{\Gamma(\upsilon)} \int_{L_n(x)}^{x_{n+1}} (L_n(x)-t)^{\upsilon-1} f^{\alpha}(t)dt. \end{aligned}$$

Here, take

$$\hat{y}_{n+1} = -\frac{1}{\Gamma(\upsilon)} \int_{x_{n+1}}^{x_N} (x_{n+1}-t)^{\upsilon-1} f^{\alpha}(t)dt,$$

$$f_{n,\upsilon}^{\alpha}(x) = \frac{1}{\Gamma(\upsilon)} \int_{x_{n+1}}^{x_N} \left((L_n(x)-t)^{\upsilon-1} - (x_{n+1}-t)^{\upsilon-1}\right) f^{\alpha}(t)dt.$$

By using change of variable in the third integral as $t = L_n(u)$, the limits change as $L_n(x) \to x$, $x_{n+1} \Rightarrow x_N$. Thus,

$$I_{x_N}^{\upsilon} f^{\alpha}(L_n(x)) = \hat{y}_{n+1} - f_{n,\upsilon}^{\alpha}(x) - \frac{a_n}{\Gamma(\upsilon)} \int_x^{x_N} (L_n(x) - L_n(u))^{\upsilon-1} f^{\alpha}(L_n(u))du.$$

As f^{α} obeys the functional equation (6.2), then the above equation becomes,

$$\begin{aligned} I_{x_N}^{\upsilon} f^{\alpha}(L_n(x)) &= \hat{y}_{n+1} - f_{n,\upsilon}^{\alpha}(x) - \frac{a_n}{\Gamma(\upsilon)} \int_x^{x_N} a_n^{\upsilon-1}(x-u)^{\upsilon-1} \left(\alpha_n f^{\alpha}(u) + f \circ L_n(u) - \alpha_n b(u)\right) du \\ &= \hat{y}_{n+1} - f_{n,\upsilon}^{\alpha}(x) - \frac{a_n^{\upsilon}\alpha_n}{\Gamma(\upsilon)} \int_x^{x_N} (x-u)^{\upsilon-1} f^{\alpha}(u)du - \frac{a_n^{\upsilon}}{\Gamma(\upsilon)} \int_x^{x_N} (x-u)^{\upsilon-1} f \circ L_n(u)du \\ &\quad + \frac{a_n^{\upsilon}\alpha_n}{\Gamma(\upsilon)} \int_x^{x_N} (x-u)^{\upsilon-1} b(u)du \\ &= \hat{y}_{n+1} - f_{n,\upsilon}^{\alpha}(x) + a_n^{\upsilon}\alpha_n I_{x_N}^{\upsilon} f^{\alpha}(x) + a_n^{\upsilon} I_{x_N}^{\upsilon} f \circ L_n(x) - a_n^{\upsilon}\alpha_n I_{x_N}^{\upsilon} b(x) \end{aligned}$$

$$= \hat{y}_{n+1} - f^{\alpha}_{n,v}(x) + a^{v}_{n}\alpha_n I^{v}_{x_N} f^{\alpha}(x) + a^{v}_{n} F^{v} \circ L_n(x) - a^{v}_{n}\alpha_n B^{v}(x)$$
$$= a^{v}_{n}\alpha_n I^{v}_{x_N} f^{\alpha}(x) + \hat{q}_n(x)$$
$$= \hat{F}_n(x, I^{v}_{x_N} f^{\alpha}(x)).$$

Here, denote $\hat{q}_n(x) = \hat{y}_{n+1} - f^{\alpha}_{n,v}(x) + a^{v}_{n} F^{v} \circ L_n(x) - a^{v}_{n}\alpha_n B^{v}(x)$. To obtain the ordinates $\hat{y}_n$ of the interpolant $I^{v}_{x_N} f^{\alpha}$, consider the following equation

$$I^{v}_{x_N} f^{\alpha}(L_n(x)) = \hat{y}_{n+1} - f^{\alpha}_{n,v}(x) + a^{v}_{n}\alpha_n I^{v}_{x_N} f^{\alpha}(x) + a^{v}_{n} F^{v} \circ L_n(x) - a^{v}_{n}\alpha_n B^{v}(x). \tag{6.10}$$

Now take $x = x_1$ and $L_n(x_1) = x_n$, the Eq. (6.10) yields,

$$\hat{y}_n = \hat{y}_{n+1} - f^{\alpha}_{n,v}(x_1) + a^{v}_{n}\alpha_n I^{v}_{x_N} f^{\alpha}(x_1) + a^{v}_{n} F^{v} \circ L_n(x_1) - a^{v}_{n}\alpha_n B^{v}(x_1). \tag{6.11}$$

We know that $\hat{y}_n = \hat{y}_N - \sum_{j=n}^{N-1} \left(\hat{y}_{j+1} - \hat{y}_j\right)$. Thus,

$$\hat{y}_n = -\sum_{j=n}^{N-1} \left(f^{\alpha}_{j,v}(x_1) - a^{v}_{j}\alpha_j \hat{y}_1 - a^{v}_{j} F^{v} \circ L_j(x_1) + a^{v}_{j}\alpha_j B^{v}(x_1)\right),$$

for each $n = 1, 2, \dots, N$ and when $n = 1$, the initial point $\hat{y}_1$ is estimated as follows,

$$\hat{y}_1 = -\frac{\sum_{j=1}^{N-1} \left(f^{\alpha}_{j,v}(x_1) - a^{v}_{j} F^{v} \circ L_j(x_1) + a^{v}_{j}\alpha_j B^{v}(x_1)\right)}{1 - \sum_{j=1}^{N-1} a^{v}_{j}\alpha_j}.$$

It is showed below that the function $\hat{q}_n(x)$ meets the continuity conditions:

$$\hat{q}_n(x_1) = \hat{y}_{n+1} - f^{\alpha}_{n,v}(x_1) + a^{v}_{n} F^{v} \circ L_n(x_1) - a^{v}_{n}\alpha_n B^{v}(x_1)$$

Comparing the above equation with $\hat{y}_n$ in Eq. (6.11),

$$\hat{q}_n(x_1) = \hat{y}_n - a^{v}_{n}\alpha_n \hat{y}_1,$$
$$\hat{q}_n(x_N) = \hat{y}_{n+1} - f^{\alpha}_{n,v}(x_N) + a^{v}_{n} F^{v} \circ L_n(x_N) - a^{v}_{n}\alpha_n B^{v}(x_N)$$
$$= \hat{y}_{n+1} - a^{v}_{n}\alpha_n \hat{y}_N,$$

which in turn implies that the function $I^{v}_{x_N} f^{\alpha}(x)$ fulfills the endpoint conditions $\hat{F}_n(x_1, \hat{y}_1) = \hat{y}_n$, $\hat{F}_n(x_N, \hat{y}_N) = \hat{y}_{n+1}$. Also, it is easy to check that $|a^{v}_{n}\alpha_n| < 1$. Hence, it is clear that the Riemann-Liouville fractional integral of an α-FIF generated by the IFS $\{L_n(x), \hat{F}_n(x, y) : n = 1, 2, \dots, N-1\}$ with the predefined initial condition $\hat{y}_N = 0$ is also an α-FIF interpolating the new data set $\{(x_n, \hat{y}_n) : n = 1, 2, \dots, N\}$. □

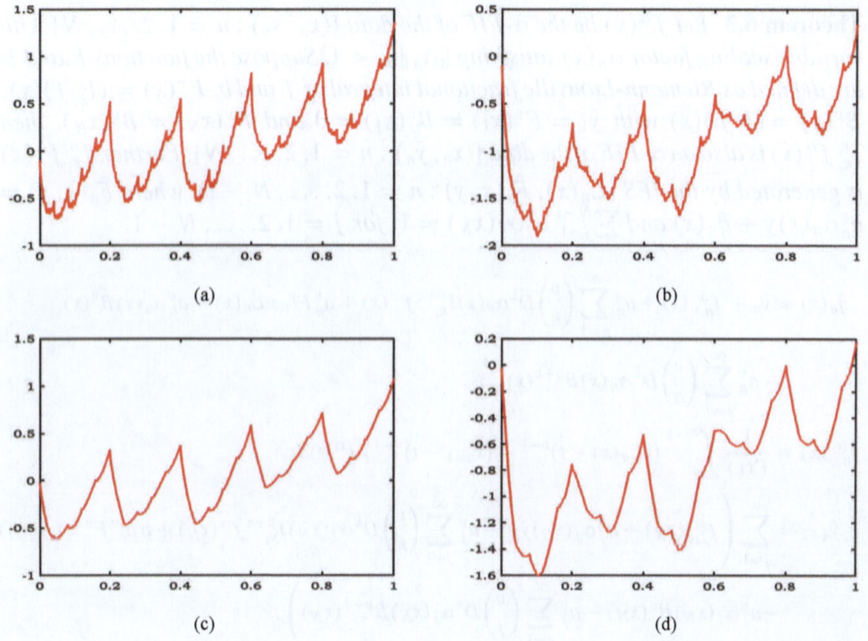

Fig. 6.3 Riemann-Liouville fractional integral of α-fractal function corresponding to the function $f(x) = \sqrt{x}$ on [0, 1] with the scale vector $\alpha_1 = 0.6$, $\alpha_2 = 0.5$, $\alpha_3 = 0.6$, $\alpha_4 = 0.4$, $\alpha_5 = 0.5$: **a** $b(x) = (f \circ c)(x)$, where $c(x) = (e^{3x} - 1)/(e^3 - 1)$, with fractional order $\upsilon = 0.5$, **b** $b(x) = \upsilon(x)f(x)$, where $\upsilon(x) = \sin(\pi x/2) + \cos(\pi x/2)$, with fractional order $\upsilon = 0.5$, **c** $b(x) = (f \circ c)(x)$, where $c(x) = (e^{3x} - 1)/(e^3 - 1)$, with fractional order $\upsilon = 0.8$ and **d** $b(x) = \upsilon(x)f(x)$, where $\upsilon(x) = \sin(\pi x/2) + \cos(\pi x/2)$, with fractional order $\upsilon = 0.8$

Example 6.3 Take the interpolation points as prescribed in Example 6.1 corresponding to the same function $f(x) = \sqrt{x}$ with same set of constant scaling factors $\alpha_1 = 0.6$, $\alpha_2 = 0.5$, $\alpha_3 = 0.6$, $\alpha_4 = 0.4$, $\alpha_5 = 0.5$. Figure 6.3a, b represent the graphs of Riemann-Liouville fractional integral of α-fractal interpolation function of order $\upsilon = 0.5$, and Fig. 6.3c, d represent the graphs of order $\upsilon = 0.8$ by changing the base function $b(x)$ respectively as follows:

(a) $b(x) = (f \circ c)(x)$ where $c(x) = (e^{\mu x} - 1)/(e^{\mu} - 1)$, $\mu = 3$,
(b) $b(x) = \upsilon(x)f(x)$, where $\upsilon(x) = \sin(\pi x/2) + \cos(\pi x/2)$.

The graphs of Riemann-Liouville fractional integral of α-FIF of order $\upsilon = 0.5$ are non-smooth when compared with the graphs with order $\upsilon = 0.8$. Moreover, there is a major change in the shapes of Riemann-Liouville fractional integral of α-FIF due to the fractional-order than the original function $f^{\alpha}(x)$ in Fig. 8.1.

In Theorems 6.1 and 6.2, the Riemann-Liouville fractional integral of α-FIFs when $\hat{y}_1 = 0$ and $\hat{y}_N = 0$ with constant scaling factors are discussed respectively. This observation motivates to explore the Riemann-Liouville fractional integral of α-FIF with variable scaling factors in the following theorems.

Theorem 6.3 *Let $f^{\alpha}(x)$ be the α-FIF of the data $\{(x_n, y_n) : n = 1, 2, \ldots, N\}$ with variable scaling factor $\alpha_n(x)$ satisfying $\|\alpha_n\|_\infty < 1$. Suppose the functions F and B are defined as Riemann-Liouville fractional integral of f and b, $F^{v}(x) = (I^{v}_{x_1} f)(x)$, $B^{v}(x) = (I^{v}_{x_1} b)(x)$ with $\hat{y}_1 = F^{v}(x_1) = B^{v}(x_1) = 0$ and $F^{v}(x_N) = B^{v}(x_N)$, then $I^{v}_{x_1} f^{\alpha}(x)$ is also an α-FIF of the data $\{(x_n, \hat{y}_n) : n = 1, 2, \ldots, N\}$. Further, $I^{v}_{x_1} f^{\alpha}(x)$ is generated by the IFS $\{L_n(x), \hat{F}_n(x, y) : n = 1, 2, \ldots, N-1\}$, where $\hat{F}_n(x, y) = a^{v}_n \alpha_n(x) y + \hat{q}_n(x)$ and $\sum_{j=1}^{N-1} a^{v}_j \alpha_j(x_N) \neq 1$, for $j = 1, 2, \ldots, N-1$,*

$$\hat{q}_n(x) = \hat{y}_n + f^{\alpha}_{n,v}(x) + a^{v}_n \sum_{n=1}^{\infty} \binom{v}{n} D^k \alpha_n(x) I^{v+k}_{x_1} f^{\alpha}(x) + a^{v}_n F^{v} \circ L_n(x) - a^{v}_n \alpha_n(x) B^{v}(x)$$
$$- a^{v}_n \sum_{k=1}^{\infty} \binom{v}{n} D^k \alpha_n(x) B^{v+k}(x),$$
$$f^{\alpha}_{n,v}(x) = \frac{1}{\Gamma(v)} \int_{x_1}^{x_{n-1}} \left((L_n(x) - t)^{v-1} - (x_{n-1} - t)^{v-1}\right) f^{\alpha}(t) dt,$$
$$\hat{y}_{n+1} = \sum_{j=1}^{n} \Bigg(f^{\alpha}_{j,v}(x_N) + a^{v}_j \alpha_j(x_N) \hat{y}_N + a^{v}_j \sum_{j=1}^{\infty} \binom{v}{n} D^k \alpha_j(x_N) I^{v+n}_{x_1} f^{\alpha}(x_N) + a_j n^{v} F^{v} \circ L_j(x_N)$$
$$- a^{v}_j \alpha_j(x_N) B^{v}(x_N) - a^{v}_j \sum_{n=1}^{\infty} \binom{v}{j} D^k \alpha_j(x_N) B^{v+j}(x_N) \Bigg),$$
$$\hat{y}_N = \sum_{j=1}^{N-1} \Bigg(f^{\alpha}_{j,v}(x_N) + a^{v}_j \sum_{n=1}^{\infty} \binom{v}{j} D^k \alpha_j(x_N) I^{v+n}_{x_1} f^{\alpha}(x_N) + a^{v}_u F^{v} \circ L_u(x_N)$$
$$- a^{v}_j \alpha_j(x_N) B^{v}(x_N) - a^{v}_j \sum_{n=1}^{\infty} \binom{v}{j} D^k \alpha_j(x_N) B^{v+j}(x_N) \Bigg) \Big/ \left(1 - \sum_{j=1}^{N-1} a^{v}_j \alpha_j(x_N)\right).$$

Proof Provided that f^{α} is a α-FIF with variable scaling factor $\alpha_n \in C(I)$, then with the similar argument of Theorem 6.1, one can have

$$I^{v}_{x_1} f^{\alpha}(L_n(x)) = \hat{y}_n + f^{\alpha}_{n,v}(x) + \frac{a_n}{\Gamma(v)} \int_{x_1}^{x} (L_n(x) - L_n(u))^{v-1} f^{\alpha}(L_n(u)) du.$$

Since f^{α} is an α-FIF associated with the functional equation (6.4), then the above equation becomes,

$$I^{v}_{x_1} f^{\alpha}(L_n(x)) = \hat{y}_n + f^{\alpha}_{n,v}(x) + \frac{a_n}{\Gamma(v)} \int_{x_1}^{x} a_n^{v-1} (x - u)^{v-1} \left(\alpha_n(u) f^{\alpha}(u) + f \circ L_n(u) - \alpha_n(u) b(u)\right) du$$
$$= \hat{y}_n + f^{\alpha}_{n,v}(x) + \frac{a^{v}_n}{\Gamma(v)} \int_{x_1}^{x} (x - u)^{v-1} \alpha_n(u) f^{\alpha}(u) du + \frac{a^{v}_n}{\Gamma(v)} \int_{x_1}^{x} (x - u)^{v-1} f \circ L_n(u) du$$
$$- \frac{a^{v}_n}{\Gamma(v)} \int_{x_1}^{x} (x - u)^{v-1} \alpha_n(u) b(u) du$$
$$= \hat{y}_n + f^{\alpha}_{n,v}(x) + a^{v}_n I^{v}_{x_1} \alpha_n(x) f^{\alpha}(x) + a^{v}_n I^{v}_{x_1} f \circ L_n(x) - a^{v}_n I^{v}_{x_1} \alpha_n(x) b(x).$$

By the Leibniz rule for fractional integrals, it follows that,

$$
\begin{aligned}
I_{x_1}^{v} f^{\alpha}(L_n(x)) &= \hat{y}_n + f_{n,v}^{\alpha}(x) + a_n^{v} \sum_{k=0}^{\infty} \binom{v}{n} D^k \alpha_n(x) I_{x_1}^{v+k} f^{\alpha}(x) + a_n^{v} I_{x_1}^{v} f \circ L_n(x) - a_n^{v} \sum_{k=0}^{\infty} \binom{v}{n} D^k \alpha_n(x) I_{x_1}^{v+k} b(x) \\
&= \hat{y}_n + f_{n,v}^{\alpha}(x) + a_n^{v} \alpha_n(x) I_{x_1}^{v} f^{\alpha}(x) + a_n^{v} \sum_{k=1}^{\infty} \binom{v}{n} D^k \alpha_n(x) I_{x_1}^{v+k} f^{\alpha}(x) + a_n^{v} I_{x_1}^{v} f \circ L_n(x) \\
&\quad - a_n^{v} \alpha_n(x) I_{x_1}^{v} b(x) - a_n^{v} \sum_{k=1}^{\infty} \binom{v}{n} D^k \alpha_n(x) I_{x_1}^{v+k} b(x) \\
&= \hat{y}_n + f_{n,v}^{\alpha}(x) + a_n^{v} \alpha_n(x) I_{x_1}^{v} f^{\alpha}(x) + a_n^{v} \sum_{k=1}^{\infty} \binom{v}{n} D^k \alpha_n(x) I_{x_1}^{v+k} f^{\alpha}(x) + a_n^{v} F^{v} \circ L_n(x) \\
&\quad - a_n^{v} \alpha_n(x) B^{v}(x) - a_n^{v} \sum_{k=1}^{\infty} \binom{v}{n} D^k \alpha_n(x) B^{v+k}(x) \\
&= a_n^{v} \alpha_n(x) I_{x_1}^{v} f^{\alpha}(x) + \hat{q}_n(x) \\
&= \hat{F}_n(x, I_{x_1}^{v} f^{\alpha}(x)).
\end{aligned}
$$

Using the notations,

$$
\begin{aligned}
\hat{q}_n(x) &= \hat{y}_n + f_{n,v}^{\alpha}(x) + a_n^{v} \sum_{k=1}^{\infty} \binom{v}{n} D^k \alpha_n(x) I_{x_1}^{v+k} f^{\alpha}(x) + a_n^{v} F^{v} \circ L_n(x) - a_n^{v} \alpha_n(x) B^{v}(x) \\
&\quad - a_n^{v} \sum_{k=1}^{\infty} \binom{v}{n} D^k \alpha_n(x) B^{v+k}(x), \\
f_{n,v}^{\alpha}(x) &= \frac{1}{\Gamma(v)} \int_{x_1}^{x_n} \left((L_n(x) - t)^{v-1} - (x_n - t)^{v-1} \right) f^{\alpha}(t) dt.
\end{aligned}
$$

Now, the following equation is considered to get the new interpolation points of the interpolant $I_{x_1}^{v} f^{\alpha}$,

$$
\begin{aligned}
I_{x_1}^{v} f^{\alpha}(L_n(x)) &= \hat{y}_n + f_{n,v}^{\alpha}(x) + a_n^{v} \alpha_n(x) I_{x_1}^{v} f^{\alpha}(x) + a_n^{v} \sum_{k=1}^{\infty} \binom{v}{n} D^k \alpha_n(x) I_{x_1}^{v+k} f^{\alpha}(x) + a_n^{v} F^{v} \circ L_n(x) \\
&\quad - a_n^{v} \alpha_n(x) B^{v}(x) - a_n^{v} \sum_{k=1}^{\infty} \binom{v}{n} D^k \alpha_n(x) B^{v+k}(x).
\end{aligned} \tag{6.12}
$$

By taking $x = x_N$ and $L_n(x_N) = x_{n+1}$ in Eq. (6.12) one can get,

$$
\begin{aligned}
\hat{y}_{n+1} &= \hat{y}_n + f_{n,v}^{\alpha}(x_N) + a_n^{v} \alpha_n(x_N) I_{x_1}^{v} f^{\alpha}(x_N) + a_n^{v} \sum_{k=1}^{\infty} \binom{v}{n} D^k \alpha_n(x_N) I_{x_1}^{v+k} f^{\alpha}(x_N) + a_n^{v} F^{v} \circ L_n(x_N) \\
&\quad - a_n^{v} \alpha_n(x_N) B^{v}(x_N) - a_n^{v} \sum_{k=1}^{\infty} \binom{v}{n} D^k \alpha_n(x_N) B^{v+k}(x_N).
\end{aligned} \tag{6.13}
$$

For each $n = 1, 2, \ldots, N-1$, one can have

$$
\hat{y}_{n+1} = \hat{y}_1 + \sum_{j=1}^{n} \left(\hat{y}_{n+1} - \hat{y}_n \right),
$$

Then from Eq. (6.13), it follows that,

$$\hat{y}_{n+1} = \sum_{j=1}^{n}\left(f_{j,v}^{\alpha}(x_N) + a_j^v \alpha_j(x_N)\hat{y}_N + a_j^v \sum_{j=1}^{\infty}\binom{v}{n} D^k \alpha_j(x_N) I_{x_1}^{v+n} f^{\alpha}(x_N) + a_j n^v F^v \circ L_j(x_N) \right.$$
$$\left. - a_j^v \alpha_j(x_N) B^v(x_N) - a_j^v \sum_{n=1}^{\infty}\binom{v}{j} D^k \alpha_j(x_N) B^{v+j}(x_N)\right).$$

When $n = N - 1$ in the above equation, the end point $\hat{y}_N$ is obtained as,

$$\hat{y}_N = \sum_{j=1}^{N-1}\left(f_{j,v}^{\alpha}(x_N) + a_j^v \sum_{n=1}^{\infty}\binom{v}{j} D^k \alpha_j(x_N) I_{x_1}^{v+n} f^{\alpha}(x_N) + a_u^v F^v \circ L_u(x_N) \right.$$
$$\left. - a_j^v \alpha_j(x_N) B^v(x_N) - a_j^v \sum_{n=1}^{\infty}\binom{v}{j} D^k \alpha_j(x_N) B^{v+j}(x_N)\right) \Big/ \left(1 - \sum_{j=1}^{N-1} a_j^v \alpha_j(x_N)\right).$$

It is clear that $\|a_n^v \alpha_n(x)\| < 1$. Further, it is necessary to check the continuity conditions of $\hat{q}_n(x)$,

$$\hat{q}_n(x_1) = \hat{y}_n + f_{n,v}^{\alpha}(x_1) + a_n^v \sum_{k=1}^{\infty}\binom{v}{n} D^k \alpha_n(x_1) I_{x_1}^{v+k} f^{\alpha}(x_1) + a_n^v F^v \circ L_n(x_1) - a_n^v \alpha_n(x_1) B^v(x_1)$$
$$- a_n^v \sum_{k=1}^{\infty}\binom{v}{n} D^k \alpha_n(x_1) B^{v+k}(x_1)$$
$$= \hat{y}_n - a_n^v \alpha_n(x_1)\hat{y}_1,$$
$$\hat{q}_n(x_N) = \hat{y}_n + f_{n,v}^{\alpha}(x_N) + a_n^v \sum_{k=1}^{\infty}\binom{v}{n} D^k \alpha_n(x_N) I_{x_1}^{v+k} f^{\alpha}(x_N) + a_n^v F^v \circ L_n(x_N) - a_n^v \alpha_n(x_N) B^v(x_N)$$
$$- a_n^v \sum_{k=1}^{\infty}\binom{v}{n} D^k \alpha_n(x_N) B^{v+k}(x_N)$$

When the above equation is compared with $\hat{y}_{n+1}$ in Eq. (6.13),

$$\hat{q}_n(x_N) = \hat{y}_{n+1} - a_n^v \alpha_n(x_N)\hat{y}_N.$$

which clearly shows that the function $I_{x_1}^v f^{\alpha}(x)$ meets the endpoint conditions $\hat{F}_n(x_1, \hat{y}_1) = \hat{y}_n$, $\hat{F}_n(x_N, \hat{y}_N) = \hat{y}_{n+1}$, hence the IFS $\{L_n(x), \hat{F}_n(x, y) : n = 1, 2, \ldots, N-1\}$ determines an α-FIF $I_{x_1}^v f^{\alpha}(x)$ which interpolates the new data set $\{(x_n, \hat{y}_n) : n = 1, 2, \ldots, N\}$. □

Thus, in Theorem 6.3 it is shown that the Riemann-Liouville fractional integral of an α-FIF predefined at the initial point of the data set $\{(x_n, \hat{y}_n) : n = 1, 2, \ldots, N\}$ is also an α-FIF. The following theorem illustrates the Riemann-Liouville fractional integral of an α-FIF by predefining the integral at the end point of the data set $\{(x_n, \hat{y}_n) : n = 1, 2, \ldots, N\}$.

Theorem 6.4 *Let $f^{\alpha}(x)$ be the α-FIF of the data $\{(x_n, y_n) : n = 1, 2, \ldots, N\}$ with variable scaling factor $\alpha_n(x)$ satisfying $\|\alpha_n\|_{\infty} < 1$. Suppose the functions F and B*

are defined as Riemann-Liouville fractional integral of f and b, $F^v(x) = (I^v_{x_N} f)(x)$, $B^v(x) = (I^v_{x_N} b)(x)$ with $F^v(x_1) = B^v(x_1)$ and $\hat{y}_N = F^v(x_N) = B^v(x_N) = 0$, then $I^v_{x_N} f^\alpha(x)$ is also an α-FIF of the data $\{(x_n, \hat{y}_n) : n = 1, 2, \ldots, N\}$. Further, $I^v_{x_N} f^\alpha(x)$ is generated by the IFS $\{L_n(x), \hat{F}_n(x, y) : n = 1, 2, \ldots, N-1\}$, where $\hat{F}_n(x, y) = a_n^v \alpha_n(x) y + \hat{q}_n(x)$ and $\sum_{j=1}^{N-1} a_j^v \alpha_j(x_1) \neq 1$, for $j = 1, 2, \ldots, N-1$,

$$\hat{q}_n(x) = \hat{y}_{n+1} - f^\alpha_{n,v}(x) + a_n^v \sum_{k=1}^{\infty} \binom{v}{n} D^k \alpha_n(x) I^{v+k}_{x_N} f^\alpha(x) + a_n^v F^v \circ L_n(x) - a_n^v \alpha_n(x) B^v(x)$$
$$- a_n^v \sum_{k=1}^{\infty} \binom{v}{n} D^k \alpha_n(x) B^{v+k}(x),$$
$$f^\alpha_{n,v}(x) = \frac{1}{\Gamma(v)} \int_{x_{n+1}}^{x_N} \left((L_n(x) - t)^{v-1} - (x_{n+1} - t)^{v-1} \right) f^\alpha(t) dt,$$
$$\hat{y}_n = - \sum_{j=n}^{N-1} \left(f^\alpha_{j,v}(x_1) - a_j^v \alpha_j(x_1) \hat{y}_1 - a_j^v \sum_{k=1}^{\infty} \binom{v}{j} D^k \alpha_j(x_1) I^{v+k}_{x_N} f^\alpha(x_1) - a_j^v F^v \circ L_j(x_1) \right.$$
$$\left. + a_j^v \alpha_j(x_1) B^v(x_1) + a_j^v \sum_{k=1}^{\infty} \binom{v}{j} D^k \alpha_j(x_1) B^{v+k}(x_1) \right),$$
$$\hat{y}_1 = - \sum_{j=1}^{N-1} \left(f^\alpha_{j,v}(x_1) - a_j^v \sum_{k=1}^{\infty} \binom{v}{j} D^k \alpha_j(x_1) I^{v+k}_{x_N} f^\alpha(x_1) - a_j^v F^v \circ L_j(x_1) \right.$$
$$\left. + a_j^v \alpha_j(x_1) B^v(x_1) + a_j^v \sum_{k=1}^{\infty} \binom{v}{j} D^k \alpha_j(x_1) B^{v+k}(x_1) \right) \Big/ \left(1 - \sum_{j=1}^{N} a_j^v \alpha_j(x_1) \right).$$

Proof Given that f^α is an α-FIF with variable scaling factor $\alpha_n \in C(I)$, then using the same argument of Theorem 6.2, one can get

$$I^v_{x_N} f^\alpha(L_n(x)) = \hat{y}_{n+1} - f^\alpha_{n,v}(x) - \frac{a_n}{\Gamma(v)} \int_x^{x_N} (L_n(x) - L_n(u))^{v-1} f^\alpha(L_n(u)) du.$$

By applying the functional equation $f^\alpha(L_n(x)) = \alpha_n(x) f^\alpha(x) + f \circ L_n(x) - \alpha_n(x) b(x)$, the above equation becomes,

$$= \hat{y}_{n+1} - f^\alpha_{n,v}(x) - \frac{a_n}{\Gamma(v)} \int_x^{x_N} a_n^{v-1} (x - u)^{v-1} \left(\alpha_n(u) f^\alpha(u) + f \circ L_n(u) - \alpha_n(u) b(u) \right) du$$
$$= \hat{y}_{n+1} - f^\alpha_{n,v}(x) - \frac{a_n^v}{\Gamma(v)} \int_x^{x_N} (x - u)^{v-1} \alpha_n(u) f^\alpha(u) du - \frac{a_n^v}{\Gamma(v)} \int_x^{x_N} (x - u)^{v-1} f \circ L_n(u) du$$
$$+ \frac{a_n^v}{\Gamma(v)} \int_x^{x_N} (x - u)^{v-1} \alpha_n(u) b(u) du$$
$$= \hat{y}_{n+1} - f^\alpha_{n,v}(x) + a_n^v I^v_{x_N} \alpha_n(x) f^\alpha(x) + a_n^v I^v_{x_N} f \circ L_n(x) - a_n^v I^v_{x_N} \alpha_n(x) b(x).$$

The following equation is obtained by using Leibniz rule for fractional integrals,

$$I^v_{x_N} f^\alpha(L_n(x)) = \hat{y}_{n+1} - f^\alpha_{n,v}(x) + a_n^v \sum_{k=0}^{\infty} \binom{v}{n} D^k \alpha_n(x) I^{v+k}_{x_N} f^\alpha(x) + a_n^v I^v_{x_N} f \circ L_n(x)$$
$$- a_n^v \sum_{k=0}^{\infty} \binom{v}{n} D^k \alpha_n(x) I^{v+k}_{x_N} b(x)$$

$$= \hat{y}_{n+1} - f^{\alpha}_{n,v}(x) + a^v_n \alpha_n(x) I^v_{x_N} f^{\alpha}(x) + a^v_n \sum_{k=1}^{\infty} \binom{v}{n} D^k \alpha_n(x) I^{v+k}_{x_N} f^{\alpha}(x) + a^v_n I^v_{x_N} f \circ L_n(x)$$

$$- a^v_n \alpha_n(x) I^v_{x_N} b(x) - a^v_n \sum_{k=1}^{\infty} \binom{v}{n} D^k \alpha_n(x) I^{v+k}_{x_N} b(x)$$

$$= \hat{y}_{n+1} - f^{\alpha}_{n,v}(x) + a^v_n \alpha_n(x) I^v_{x_N} f^{\alpha}(x) + a^v_n \sum_{k=1}^{\infty} \binom{v}{n} D^k \alpha_n(x) I^{v+k}_{x_N} f^{\alpha}(x) + a^v_n F^v \circ L_n(x)$$

$$- a^v_n \alpha_n(x) B^v(x) - a^v_n \sum_{k=1}^{\infty} \binom{v}{n} D^k \alpha_n(x) B^{v+k}(x)$$

$$= a^v_n \alpha_n(x) I^v_{x_N} f^{\alpha}(x) + \hat{q}_n(x)$$

$$= \hat{F}_n(x, I^v_{x_N} f^{\alpha}(x)).$$

Here, denote

$$\hat{q}_n(x) = \hat{y}_{n+1} - f^{\alpha}_{n,v}(x) + a^v_n \sum_{k=1}^{\infty} \binom{v}{n} D^k \alpha_n(x) I^{v+k}_{x_N} f^{\alpha}(x) + a^v_n F^v \circ L_n(x) - a^v_n \alpha_n(x) B^v(x)$$

$$- a^v_n \sum_{k=1}^{\infty} \binom{v}{n} D^k \alpha_n(x) B^{v+k}(x),$$

$$f^{\alpha}_{n,v}(x) = \frac{1}{\Gamma(v)} \int_{x_{n+1}}^{x_N} \left((L_n(x) - t)^{v-1} - (x_{n+1} - t)^{v-1}\right) f^{\alpha}(t) dt.$$

Clearly $\|a^v_n \alpha_n(x)\| < 1$. Now, consider the following equation to get the ordinates $\hat{y}_n$ of the new interpolation data points,

$$I^v_{x_N} f^{\alpha}(L_n(x)) = \hat{y}_{n+1} - f^{\alpha}_{n,v}(x) + a^v_n \alpha_n(x) I^v_{x_N} f^{\alpha}(x) + a^v_n \sum_{k=1}^{\infty} \binom{v}{n} D^k \alpha_n(x) I^{v+k}_{x_N} f^{\alpha}(x) + a^v_n F^v \circ L_n(x)$$
$$- a^v_n \alpha_n(x) B^v(x) - a^v_n \sum_{k=1}^{\infty} \binom{v}{n} D^k \alpha_n(x) B^{v+k}(x). \tag{6.14}$$

Take $x = x_1$ and $L_n(x_1) = x_n$, then the Eq. (6.14) yields,

$$\hat{y}_n = \hat{y}_{n+1} - f^{\alpha}_{n,v}(x_1) + a^v_n \alpha_n(x_1) I^v_{x_N} f^{\alpha}(x_1) + a^v_n \sum_{k=1}^{\infty} \binom{v}{n} D^k \alpha_n(x_1) I^{v+k}_{x_N} f^{\alpha}(x_1) + a^v_n F^v \circ L_n(x_1)$$
$$- a^v_n \alpha_n(x_1) B^v(x_1) - a^v_n \sum_{k=1}^{\infty} \binom{v}{n} D^k \alpha_n(x_1) B^{v+k}(x_1). \tag{6.15}$$

Using the equation $\hat{y}_n = \hat{y}_N - \sum_{j=n}^{N-1} \left(\hat{y}_{j+1} - \hat{y}_j\right)$ for each $n = 1, 2, \ldots, N-1$, one can obtain

$$\hat{y}_n = - \sum_{j=n}^{N-1} \left(f^{\alpha}_{j,v}(x_1) - a^v_j \alpha_j(x_1) \hat{y}_1 - a^v_j \sum_{k=1}^{\infty} \binom{v}{j} D^k \alpha_j(x_1) I^{v+k}_{x_N} f^{\alpha}(x_1) - a^v_j F^v \circ L_j(x_1) \right.$$

$$+a_j^{\upsilon}\alpha_j(x_1)B^{\upsilon}(x_1)+a_j^{\upsilon}\sum_{k=1}^{\infty}\binom{\upsilon}{j}D^k\alpha_j(x_1)B^{\upsilon+k}(x_1)\Bigg).$$

If $n=1$ in the above equation, the initial point $\hat{y}_1$ is estimated as follows,

$$\hat{y}_1=-\sum_{j=1}^{N-1}\Bigg(f_{j,\upsilon}^{\alpha}(x_1)-a_j^{\upsilon}\sum_{k=1}^{\infty}\binom{\upsilon}{j}D^k\alpha_j(x_1)I_{xN}^{\upsilon+k}f^{\alpha}(x_1)-a_j^{\upsilon}F^{\upsilon}\circ L_j(x_1)$$
$$+a_j^{\upsilon}\alpha_j(x_1)B^{\upsilon}(x_1)+a_j^{\upsilon}\sum_{k=1}^{\infty}\binom{\upsilon}{j}D^k\alpha_j(x_1)B^{\upsilon+k}(x_1)\Bigg)\Bigg/\Bigg(1-\sum_{j=1}^{N}a_j^{\upsilon}\alpha_j(x_1)\Bigg).$$

Further, the continuity conditions of $\hat{q}_n(x)$ is verified below:

$$\hat{q}_n(x_1)=\hat{y}_{n+1}-f_{n,\upsilon}^{\alpha}(x_1)+a_n^{\upsilon}\sum_{k=1}^{\infty}\binom{\upsilon}{n}D^k\alpha_n(x_1)I_{xN}^{\upsilon+k}f^{\alpha}(x_1)+a_n^{\upsilon}F^{\upsilon}\circ L_n(x_1)-a_n^{\upsilon}\alpha_n(x_1)B^{\upsilon}(x_1)$$
$$-a_n^{\upsilon}\sum_{k=1}^{\infty}\binom{\upsilon}{n}D^k\alpha_n(x_1)B^{\upsilon+k}(x_1)$$

When the above equation is compared with $\hat{y}_n$ in Eq. (6.15), one can see that,

$$\hat{q}_n(x_1)=\hat{y}_n-a_n^{\upsilon}\alpha_n(x_1)\hat{y}_1,$$
$$\hat{q}_n(x_N)=\hat{y}_{n+1}-f_{n,\upsilon}^{\alpha}(x_N)+a_n^{\upsilon}\sum_{k=1}^{\infty}\binom{\upsilon}{n}D^k\alpha_n(x_N)I_{xN}^{\upsilon+k}f^{\alpha}(x_N)+a_n^{\upsilon}F^{\upsilon}\circ L_n(x_N)-a_n^{\upsilon}\alpha_n(x_N)B^{\upsilon}(x_N)$$
$$-a_n^{\upsilon}\sum_{k=1}^{\infty}\binom{\upsilon}{n}D^k\alpha_n(x_N)B^{\upsilon+k}(x_N)$$
$$=\hat{y}_{n+1}-a_n^{\upsilon}\alpha_n(x_N)\hat{y}_N,$$

this shows that the function $I_{xN}^{\upsilon}f^{\alpha}(x)$ satisfies the endpoint conditions $\hat{F}_n(x_1,\hat{y}_1)=\hat{y}_n$, $\hat{F}_n(x_N,\hat{y}_N)=\hat{y}_{n+1}$, therefore the Riemann-Liouville fractional integral of an α-FIF with the predefined initial condition $\hat{y}_N=0$ is also an α-FIF interpolating the new data set $\{(x_n,\hat{y}_n):n=1,2,\ldots,N\}$. □

Example 6.4 The Riemann-Liouville fractional integral of an α-fractal interpolation function is generated by considering the data set as in Example 6.2 for the same function $f(x)=\sqrt{x}$ with the same set of variable scaling factors $\alpha_1(x)=0.6+\sin(x/3)$, $\alpha_2(x)=0.5+(x/2)$, $\alpha_3(x)=0.2e^{2x}$, $\alpha_4(x)=0.5e^{x}/2$, $\alpha_5(x)=0.4+\sin(x/2)$. Figure 6.4 represent the pictorial representations of $I^{\upsilon}f^{\alpha}(x)$ by changing the base function $b(x)$ as follows:

(a) $b(x)=(f\circ c)(x)$ where $c(x)=(e^{\mu x}-1)/(e^{\mu}-1)$, $\mu=1$,
(b) $b(x)=v(x)f(x)$, where $v(x)=\sin(\pi x/2)+\cos(\pi x/2)$.

It is observed that Fig. 6.4a–d show pleasing changes in the graphs of the Riemann-Liouville fractional integral of α-FIF than the graphs of the functions $f^{\alpha}(x)$ in Fig. 6.2. Since the scaling factors are chosen as functions instead of constants, the comparison can also be made with the figures in Example 6.3.

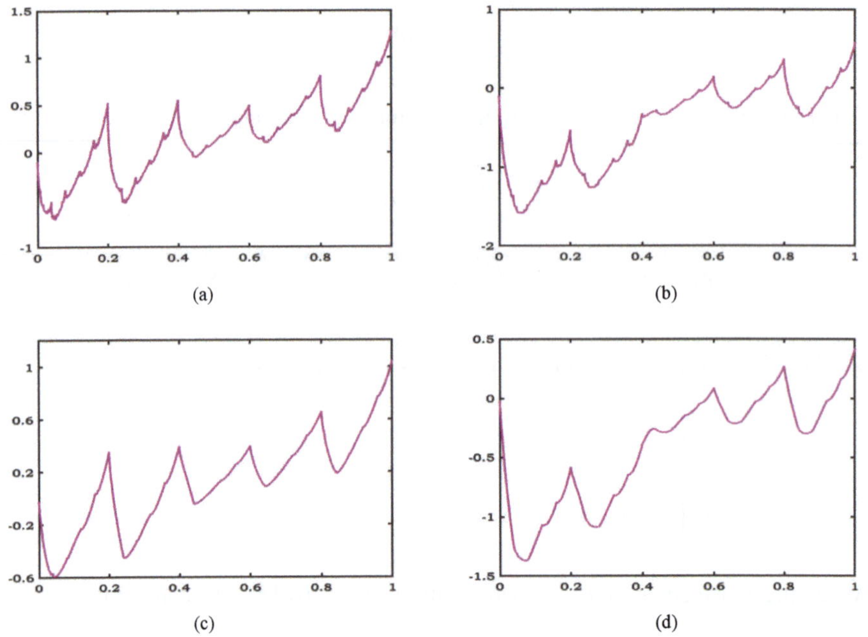

Fig. 6.4 Riemann-Liouville fractional integral of α-fractal function corresponding to the function $f(x) = \sqrt{x}$ on $[0, 1]$ with the variable scaling $\alpha_1(x) = 0.6 + \sin(x/3)$, $\alpha_2(x) = 0.5 + (x/2)$, $\alpha_3(x) = 0.2e^{2x}$, $\alpha_4(x) = 0.5e^x/2$, $\alpha_5(x) = 0.4 + \sin(x/2)$: **a** $b(x) = (f \circ c)(x)$, where $c(x) = (e^{3x} - 1)/(e^3 - 1)$, with fractional order $v = 0.5$, **b** $b(x) = v(x)f(x)$, where $v(x) = \sin(\pi x/2) + \cos(\pi x/2)$, with fractional order $v = 0.5$, **c** $b(x) = (f \circ c)(x)$, where $c(x) = (e^{3x} - 1)/(e^3 - 1)$, with fractional order $v = 0.8$ and **d** $b(x) = v(x)f(x)$, where $v(x) = \sin(\pi x/2) + \cos(\pi x/2)$, with fractional order $v = 0.8$

6.3 Fractional Operator of α-Fractal Function

In this section, the fractional operator of α-fractal function is defined in terms of Riemann-Liouville integral of f^α of order v. Further, some analytical properties of this operator are discussed.

Definition 6.1 Let $f \in C(I)$ and D be the partition of I. The α-fractal function of f corresponding to α, D and the base function $b(x)$ is denoted as f^α. Here, we consider the case $b(x) = (f \circ c)(x)$, where $c : I \to I$ is a increasing, continuous function such that $c(x_1) = x_1$, $c(x_N) = x_N$ and c is not the identity function. If the Riemann-Liouville integral of f^α of order v is symbolized as $I^v_{x_1} f^\alpha$ (refer Theorem 6.1), then the fractional operator of α-fractal function $\mathcal{F}^v : C(I) \to C(I)$ is defined by

$$\mathcal{F}^v(f(x)) = a_n^v I^v_{x_1} f(x) + a_n^v \alpha_n (I^v_{x_1} f^\alpha - I^v_{x_1} f \circ c) \circ L_n^{-1}(x) + \hat{y}_{n-1} + f^\alpha_{n,v}(L_n^{-1}(x)), \tag{6.16}$$

for all $x \in I_n \subset I$.

Let us discuss some analytical properties of this operator.

Theorem 6.5 *$\mathcal{F}^v$ is a bounded linear operator. Further,*

$$\|\mathcal{F}^v(f)\|_\infty \le 2K\|f\|_\infty \left(\frac{1+|\alpha|_\infty^2}{1-|\alpha|_\infty}\right),$$

where $K = \frac{(x_N - x_1)^v}{v\Gamma(v)}$.

Proof For given $f, g \in C(I)$ and $r, s \in \mathbb{R}$,

$$\mathcal{F}^v(rf) = a_n^v I_{x_1}^v rf(x) + a_n^v \alpha_n (I_{x_1}^v f^\alpha - I_{x_1}^v rf \circ c) \circ L_n^{-1}(x) + \hat{y}_n + f_{n,v}^\alpha (L_n^{-1}(x)),$$
$$\mathcal{F}^v(sg) = a_n^v I_{x_1}^v sg(x) + a_n^v \alpha_n (I_{x_1}^v g^\alpha - I_{x_1}^v sg \circ c) \circ L_n^{-1}(x) + \hat{y}_n' + g_{n,v}^\alpha (L_n^{-1}(x)).$$

It gives that $\mathcal{F}^v(rf + sg) = r\mathcal{F}^v(f) + s\mathcal{F}^v(g)$. Thus, $\mathcal{F}^v$ is linear operator. The uniform norm of the operator $\mathcal{F}^v f$ is given by

$$\begin{aligned}\|\mathcal{F}^v(f)\|_\infty &= \|a_n^v I_{x_1}^v f(x) + a_n^v \alpha_n (I_{x_1}^v f^\alpha - I_{x_1}^v f \circ c) \circ L_n^{-1}(x) + \hat{y}_n + f_{n,v}^\alpha (L_n^{-1}(x))\|_\infty \\ &\le |a^v|_\infty \|I_{x_1}^v f(x)\|_\infty + |a^v|_\infty |\alpha|_\infty \|(I_{x_1}^v f^\alpha - I_{x_1}^v f \circ c) \circ L_n^{-1}\|_\infty + \|\hat{y}_n + f_{n,v}^\alpha (L_n^{-1})\|_\infty.\end{aligned}$$

Here, $|a^v|_\infty = \max_{1\le n \le N-1}\{a_n^v\}$ and $|\alpha|_\infty = \max_{1\le n\le N-1}\{\alpha_n\}$. The Riemann-Liouville integral $I_{x_1}^v f(x)$ is bounded by $K\|f(x)\|_\infty$, where $K = \frac{(x_N - x_1)^v}{v\Gamma(v)}$ (Refer Lemma 2.1 in [2]). Also, $\|\hat{y}_n + f_{n,v}^\alpha (L_n^{-1})\|_\infty \le \frac{(x_n - x_1)^v}{v\Gamma(v)} \|f^\alpha\|_\infty$. Hence, the above inequality becomes

$$\|\mathcal{F}^v(f)\|_\infty \le |a^v|_\infty K\|f\|_\infty + |a^v|_\infty |\alpha|_\infty K\|(f^\alpha - f\circ c)\circ L_n^{-1}\|_\infty + \frac{(x_n - x_1)^v}{v\Gamma(v)}\|f^\alpha\|_\infty.$$

It is known that $\|f^\alpha\|_\infty \le \frac{1+|\alpha|_\infty}{1-|\alpha|_\infty}\|f\|_\infty$. Further, $|a^v|_\infty \le 1$ and $\frac{(x_n - x_1)^v}{v\Gamma(v)}\|f^\alpha\|_\infty \le K\|f^\alpha\|_\infty$. So,

$$\begin{aligned}\|\mathcal{F}^v(f)\|_\infty &\le K\|f\|_\infty + |\alpha|_\infty K\|(f^\alpha - f\circ c)\|_\infty + K\|f^\alpha\|_\infty \\ &= K\|f\|_\infty + K\|f^\alpha\|_\infty(1+|\alpha|_\infty) - K|\alpha|_\infty\|f\circ c\|_\infty \\ &\le K\|f\|_\infty + K\|f\|_\infty\left(\frac{1+|\alpha|_\infty}{1-|\alpha|_\infty}\right)(1+|\alpha|_\infty) - K|\alpha|_\infty\|f\circ c\|_\infty \\ &= K\|f\|_\infty\left(1 + \frac{(1+|\alpha|_\infty)^2}{1-|\alpha|_\infty}\right) - K|\alpha|_\infty\|f\circ c\|_\infty \\ &= K\left(\|f\|_\infty\left(\frac{2+|\alpha|_\infty+|\alpha|_\infty^2}{1-|\alpha|_\infty}\right) - |\alpha|_\infty\|f\circ c\|_\infty\right).\end{aligned}$$

By the definition of c in Eq. (6.16), $c(x) \in I$ for all $x \in I$, and thus $\|f\circ c\|_\infty \le \|f\|_\infty$. Hence,

$$\|\mathcal{F}^{v}(f)\|_{\infty} \leq 2K\|f\|_{\infty}\left(\frac{1+|\alpha|_{\infty}^{2}}{1-|\alpha|_{\infty}}\right).$$

Thus, $\mathcal{F}^{v}$ is bounded by $2K\|f\|_{\infty}\left(\frac{1+|\alpha|_{\infty}^{2}}{1-|\alpha|_{\infty}}\right)$ where $K = \frac{(x_N - x_1)^{v}}{v\Gamma(v)}$. □

Remark 6.1 The operator $\mathcal{F}^{v}$ is defined by the notion of Theorem 6.1, hence $\mathcal{F}^{v}$ depends on the fractional order v. Thus, for every fractional order $v > 0$, $\mathcal{F}^{v}(f) = I_{x_1}^{v} f^{\alpha}$ provides a α-fractal function (in particular continuous function).

In view of the considerations of the above remark, the following result seems rather natural. Consider the collection of operators

$$\{\mathcal{F}^{v} : C(I) \to C(I); v > 0\}.$$

The composition operation of the Riemann-Liouville fractional integral with respect to the fractional orders $r, s > 0$ for the function on $C([a, b])$ is defined by

$$I_a^{r}(I_a^{s} f) = I_a^{r+s} f. \tag{6.17}$$

Additionally, if $f \in C([a, b])$ and $r, s > 0$, then $I_a^{r+s} f \in C([a, b])$ (See [3]). The algebraic properties of fractional operator $\mathcal{F}^{v}$ is exhibited in the following theorem.

Theorem 6.6 *The collection of operators* $\{\mathcal{F}^{v} : C(I) \to C(I) : v > 0\}$ *forms a semigroup with respect to composition operation defined in Eq. (6.17). Further,* $\mathcal{F}^{v_1}\mathcal{F}^{v_2} f = \mathcal{F}^{v_2}\mathcal{F}^{v_1} f$.

Proof Let $v_1, v_2 > 0$ and fix $f \in C(I)$. Then,

$$\begin{aligned}
\mathcal{F}^{v_1}\mathcal{F}^{v_2}(f) &= \frac{a_n^{v_1+v_2}}{\Gamma(v_1)\Gamma(v_2)} \int_{x_1}^{x} (x-t)^{v_1-1} \int_{x_1}^{t} (t-u)^{v_2-1} f(u)\,du\,dt + \hat{y}_{n,v_1+v_2} \\
&+ \frac{a_n^{v_1+v_2}\alpha_n}{\Gamma(v_1)\Gamma(v_2)} \int_{x_1}^{x} (x-t)^{v_1-1} \int_{x_1}^{t} (t-u)^{v_2-1} (f^{\alpha}-b)\circ L_n^{-1}(u)\,du\,dt + \hat{f}^{\alpha}_{n,v_1+v_2}(L_n^{-1}(x)) \\
&= \frac{a_n^{v_1+v_2}}{\Gamma(v_1)\Gamma(v_2)} \int_{x_1}^{x}\int_{u}^{x} (x-t)^{v_1-1}(t-u)^{v_2-1} f(u)\,dt\,du + \hat{y}_{n,v_1+v_2} \\
&+ \frac{a_n^{v_1+v_2}\alpha_n}{\Gamma(v_1)\Gamma(v_2)} \int_{x_1}^{x}\int_{u}^{x} (x-t)^{v_1-1}(t-u)^{v_2-1} (f^{\alpha}-b)\circ L_n^{-1}(u)\,dt\,du + \hat{f}^{\alpha}_{n,v_1+v_2}(L_n^{-1}(x)) \\
&= \frac{a_n^{v_1+v_2}}{\Gamma(v_1)\Gamma(v_2)} \int_{x_1}^{x} f(u) \int_{u}^{x} (x-t)^{v_1-1}(t-u)^{v_2-1}\,dt\,du + \hat{y}_{n,v_1+v_2} \\
&+ \frac{a_n^{v_1+v_2}\alpha_n}{\Gamma(v_1)\Gamma(v_2)} \int_{x_1}^{x} (f^{\alpha}-b)\circ L_n^{-1}(u) \int_{u}^{x} (x-t)^{v_1-1}(t-u)^{v_2-1}\,dt\,du + \hat{f}^{\alpha}_{n,v_1+v_2}(L_n^{-1}(x)).
\end{aligned}$$

Substituting $t = u + v(x-u)$ in above equation gives

$$\begin{aligned}
\mathcal{F}^{v_1}\mathcal{F}^{v_2}(f) &= \frac{a_n^{v_1+v_2}}{\Gamma(v_1)\Gamma(v_2)} \int_{x_1}^{x} f(u) \int_{1}^{1} \{(x-u)(1-v)\}^{v_1-1}\{v(x-u)\}^{v_2-1}(x-u)\,dv\,du \\
&+ \frac{a_n^{v_1+v_2}\alpha_n}{\Gamma(v_1)\Gamma(v_2)} \int_{x_1}^{x} (f^{\alpha}-b)\circ L_n^{-1}(u) \int_{1}^{1} \{(x-u)(1-v)\}^{v_1-1}\{v(x-u)\}^{v_2-1}(x-u)\,dv\,du
\end{aligned}$$

$$
\begin{aligned}
&+\hat{y}_{n,v_1+v_2}+\hat{f}^{\alpha}_{n,v_1+v_2}(L_n^{-1}(x))\\
&=a\frac{a_n^{v_1+v_2}}{\Gamma(v_1)\Gamma(v_2)}\int_{x_1}^{x} f(u)(x-u)^{v_1+v_2-1}\int_1^1(1-v)^{v_1-1}v^{v_2-1}dvdu+\hat{f}^{\alpha}_{n,v_1+v_2}(L_n^{-1}(x))\\
&+\frac{a_n^{v_1+v_2}\alpha_n}{\Gamma(v_1)\Gamma(v_2)}\int_{x_1}^{x}(f^{\alpha}-b)\circ L_n^{-1}(u)(x-u)^{v_1+v_2-1}\int_1^1(1-v)^{v_1-1}v^{v_2-1}dvdu+\hat{y}_{n,v_1+v_2}\\
&=\frac{a_n^{v_1+v_2}}{\Gamma(v_1+v_2)}\int_{x_1}^{x} f(u)(x-u)^{v_1+v_2-1}du\\
&+\frac{a_n^{v_1+v_2}\alpha_n}{\Gamma(v_1+v_2)}\int_{x_1}^{x}(f^{\alpha}-b)\circ L_n^{-1}(u)(x-u)^{v_1+v_2-1}du+\hat{y}_{n,v_1+v_2}+\hat{f}^{\alpha}_{n,v_1+v_2}(L_n^{-1}(x))\\
&=a_n^{v_1+v_2}I_{x_1}^{v_1+v_2}f(x)+a_n^{v_1+v_2}\alpha_n I_{x_1}^{v_1+v_2}(f^{\alpha}-b)\circ L_n^{-1}(x)+\hat{y}_{n,v_1+v_2}+\hat{f}^{\alpha}_{n,v_1+v_2}(L_n^{-1}(x))\\
&=\mathcal{F}^{v_1+v_2}(f).
\end{aligned}
$$

It gives that $\mathcal{F}^{v_1+v_2}(f)\in C(I)$. Thus, the collection of operators $\{\mathcal{F}^{v}:C(I)\to C(I);\, v>0\}$ forms a semigroup. Further, $\mathcal{F}^{v_1}\mathcal{F}^{v_2}(f)=\mathcal{F}^{v_1+v_2}(f)=\mathcal{F}^{v_2+v_1}(f)=\mathcal{F}^{v_2}\mathcal{F}^{v_1}(f)$ [4]. □

It is observed that the fractional operator of a α-fractal function is a continuous bounded linear operator, and its collection exhibits the semigroup property associated with composition operation of the Riemann-Liouville fractional integral with respect to the fractional orders $r, s > 0$.

6.4 Summary

In this chapter, the Riemann-Liouville fractional integral of α-fractal function f^{α} with constant scaling factor is investigated by predefining the fractional integral at the initial point of the function f. Similarly, by predefining the fractional integral at the terminal points of the function f, the Riemann-Liouville fractional integral of α-fractal function f^{α} with variable scaling factor is discussed. In addition, the fractional operator of a α-fractal function is defined in terms of the Riemann-Liouville fractional integral and its bounds, linearity property are explored.

References

1. M.F. Barnsley, Fractal functions and interpolation. Constr. Approx. **2**, 303–329 (1986)
2. A.A. Kilbas, H.M. Srivastava, J.J Trujillo, *Theory and Applications of Fractional Differential Equations*, vol. 204 (Elsevier, 2006)
3. K. Diethelm, *The Analysis of Fractional Differential Equations: An Application-oriented Exposition Using Differential Operators of Caputo Type*, vol. 2004 (Springer, Berlin, 2010)
4. T.M.C. Priyanka, A. Gowrisankar, Riemann-Liouville fractional integral of non-affine fractal interpolation function and its fractional operator. Eur. Phys. J. Spec. Top. **230**(21), 3789–3805 (2021)

Chapter 7
Fractional Derivative Order and Fractal Scalings

7.1 Introduction

The first intriguing element for integrating fractal functions and fractional order is the fractal dimension. Fractal sets, fractal functions, and fractal surfaces have non-integer (fractional) fractal dimension, in general. This characteristic has opened up the potential of bridging the two distinct domains of fractal geometry and fractional calculus, see [1–4]. Later, motivated by the merits of fractional calculus, the trend of applying fractional derivative and fractional integral to fractal interpolation functions has gained popularity. One more non-avoidable reason behind this study is the non-differentiable nature of fractal interpolation functions despite they are continuous everywhere. For detailed study on the fractional derivative and fractional integral of fractal functions, interested readers may refer the literature [5–10]. Among the various fractional derivative methods, this chapter concentrates on the Weyl-Marchaud fractional derivative. This particular choice is purely based on its definition, it requires just a continuous function for differentiating with fractional order $\lambda \in (0, 1)$. Whereas other fractional derivative methods like Caputo fractional derivative requires a differentiable function, also Riemann fractional derivative does not match with the classical derivative when applied on a constant function. The Weyl-Marchaud fractional derivative omits the demand for differentiable function and coincides with classical derivative for a constant function.

With these advantages of Weyl-Marchaud fractional derivative, this chapter covers the study of Weyl-Marchaud fractional derivative for types of fractal interpolation functions, where the scaling factors are chosen as constants. It highlights the choice of fractional order while applying the fractional derivative to fractal functions. Pertinent examples and counter examples are discussed for the particular fractional order choice. In a broader sense, this chapter answers the question, is the fractional derivative of fractal function is again an attractor for a new iterated function system? Further, the relation between the fractional order of Weyl-Marchaud derivative and the scaling factors associated with the fractal functions is explicitly presented.

A. Gowrisankar et al., *Fractal Functions Communication with Fractional Calculus*,
https://doi.org/10.1007/978-3-031-96993-5_7

7.2 Fractional Derivative of Fractal Functions

This section begins with a fundamental theorem discussing the fractional derivative of general fractal interpolation function. Then, based on the choice of continuous function q_n as affine or non-affine, the fractional derivative of affine or non-affine fractal functions are discussed.

Consider the fractal interpolation function f associated with the iterated function system $\{(L_n(x), F_n(x, y)) : n = 1, 2, \ldots, N-1\}$, where the functions L_n and F_n are recalled from Chap. 1. That is,

$$\begin{aligned} L_n(x) &= a_n x + b_n, \\ F_n(x, y) &= \alpha_n y + q_n(x), \ \alpha_n \in (0, 1), \ n = 1, 2, \ldots, N-1. \end{aligned} \tag{7.1}$$

In the definition of Weyl-Marchaud derivative of a continuous function, we apply our fractal interpolation function, precisely in the definition, the continuous function is replaced with fractal function f obeying the functional equation

$$f(L_n(x)) = \alpha_n f(x) + q_n(x), \ \alpha_n \in (0, 1), \ n = 1, 2, \ldots, N-1. \tag{7.2}$$

With this application, the Weyl-Marchaud derivative of f is derived and verified whether the resultant function, say $D^\lambda f$ also behaves like a fractal interpolation function. For this verification, the following needs to be satisfied:

(i) the resultant function should preserve the functional equation representation,
(ii) the new scaling factor should be in the range $(0, 1)$,
(iii) the new F_n, say $\tilde{F}_n$ should obey the end point condition.

If the function $D^\lambda f$ satisfies the afore-listed conditions, then the Weyl-Marchaud fractional derivative of f is also a fractal interpolation function. The following theorem presents the Weyl-Marchaud derivative of a general fractal interpolation function f with no specific choice of q_n.

Theorem 7.1 *Consider the fractal interpolation function f associated with the IFS $\{(L_n(x), F_n(x, y)) : n = 1, 2, \ldots, N-1\}$ defined in Eq. (7.1). If the scaling factors obey $|\alpha_n| < |a_n^\lambda|$, for all $n = 1, 2 \ldots, N-1$, then the function $(D_{x_1}^\lambda f)(x)$ is also a FIF. The new IFS $\{(L_n(x), \tilde{F}_n(x, y)) : n = 1, 2, \ldots, N-1\}$ determines the function $(D_{x_1}^\lambda f)(x)$, where $\tilde{F}_n(x, y) = a_n^{-\lambda}\alpha_n y + \tilde{q}_n(x)$ and $\sum_{j=1}^{N-1} a_j^{-\lambda}\alpha_j \neq 1$, for $j = 1, 2, \ldots, N-1$. Further, the fractal function $(D_{x_1}^\lambda f)(x)$ passes through a new set of data $\{(x_n, \tilde{y}_n) : n = 1, 2, \ldots, N\}$.*

Proof The Weyl-Marchaud fractional derivative of a fractal interpolation function f is defined by,

$$(D_{x_1}^\lambda f)(L_n(x)) = \frac{\lambda}{\Gamma(1-\lambda)} \int_{x_1}^{L_n(x)} \frac{f(L_n(x)) - f(t)}{(L_n(x) - t)^{1+\lambda}} dt$$

Substitution of the functional equation given in Eq. (7.2) yields

$$\begin{aligned}(D_{x_1}^{\lambda} f)(L_n(x)) &= \tilde{y}_n + f_{n,\lambda}(x) + \frac{a_n\lambda}{\Gamma(1-\lambda)} \int_{x_1}^{x} \frac{\alpha_n f(x) + q_n(x) - \alpha_n f(u) - q_n(u)}{(a_n x + b_n - a_n u - b_n)^{1+\lambda}} du \\ &= a_n^{-\lambda}\alpha_n (D_{x_1}^{\lambda} f)(x) + \tilde{q}_n(x) \\ &= \tilde{F}_n(x, (D_{x_1}^{\lambda} f)(x)).\end{aligned}$$

The notations used are described by

$$\begin{aligned}f_{n,\lambda}(x) &= \frac{\lambda}{\Gamma(1-\lambda)} \int_{x_1}^{x_n} \left(\frac{f(L_n(x)) - f(t)}{(L_n(x) - t)^{1+\lambda}} - \frac{f(x_n) - f(t)}{(x_n - t)^{1+\lambda}} \right) dt \\ \tilde{q}_n(x) &= \tilde{y}_n - f_{n,\lambda}(x) + a_n^{-\lambda}(D_{x_N}^{\lambda} q_n)(x).\end{aligned}$$

Hence, the Weyl-Marchaud fractional derivative of a linear FIF determined by the IFS $\{(L_n(x), \tilde{F}_n(x, y)) : n = 1, 2, \ldots, N-1\}$ preserves the functional equation representation and the scaling factors are constrained to be $|\alpha_n| < |a_n^{\lambda}|$ such that $|\alpha_n a_n^{-\lambda}| < 1$. The endpoint conditions of $\tilde{F}_n$ are verified as below.

$$\begin{aligned}\tilde{F}_n(x_1, D_{x_1}^{\lambda} f)(x_1) &= \tilde{y}_n, \\ \tilde{F}_n(x_N, D_{x_1}^{\lambda} f)(x_N) &= a_n^{-\lambda}\alpha_n (D_{x_1}^{\lambda} f)(x_N) + \tilde{q}_n(x_N), \\ &= (D_{x_1}^{\lambda} f)(L_n(x_N)), \\ &= \tilde{y}_{n+1}.\end{aligned}$$

The explicit formula of new interpolation points of $(D_{x_1}^{\lambda} f)(x)$ are given by

$$\tilde{y}_{n+1} = \sum_{j=1}^{n} \left(f_{j,\lambda}(x_N) + a_j^{-\lambda}\alpha_j \tilde{y}_N + a_j^{-\lambda}(D_{x_N}^{\lambda} q_j)(x_N) \right). \tag{7.3}$$

The end point $\tilde{y}_N$ of $D_{x_1}^{\lambda} f$ can be obtained by substituting $n = N-1$ in Eq. (7.3) as follows

$$\tilde{y}_N = \frac{\sum_{j=1}^{N-1} \left(f_{j,\lambda}(x_N) - a_j^{-\lambda}(D_{x_1}^{\lambda} q_j)(x_N) \right)}{1 - \sum_{j=1}^{N-1} a_j^{-\lambda}\alpha_j}.$$

The Weyl-Marchaud fractional derivative of a fractal function generated by the IFS $\{(L_n(x), \tilde{F}_n(x, y)) : n = 1, 2, \ldots, N-1\}$ is also a fractal interpolation function interpolating the data set $\{(x_n, \tilde{y}_n) : n = 1, 2, \ldots, N\}$. □

The above theorem is proved by predefining the fractional derivative at the initial point x_1, following the same procedure one can show that the Weyl-Marchaud fractional derivative of a fractal interpolation function f predefined at x_N is also a fractal function. The result is precisely given in the below corollary.

Corollary 7.1 *Consider the fractal interpolation function f associated with the IFS $\{(L_n(x), F_n(x, y)) : n = 1, 2, \ldots, N-1\}$ defined in Eq. (7.1). If the scaling factors obey $|\alpha_n| < |a_n^\lambda|$, for all $n = 1, 2 \ldots, N-1$, then the function $(D_{x_N}^\lambda f)(x)$ is also a FIF. The new IFS $\{(L_n(x), \tilde{F}_n(x, y)) : n = 1, 2, \ldots, N-1\}$ determines the function $(D_{x_N}^\lambda f)(x)$, where $\tilde{F}_n(x, y) = a_n^{-\lambda}\alpha_n y + \tilde{q}_n(x)$ and $\sum_{j=1}^{N-1} a_j^{-\lambda}\alpha_j \neq 1$, for $j = 1, 2, \ldots, N-1$. Further, the fractal function $(D_{x_N}^\lambda f)(x)$ passes through a new set of data $\{(x_n, \tilde{y}_n) : n = 1, 2, \ldots, N\}$.*

Remark 7.1 In the definition of Weyl-Marchaud fractional derivative, the fractional order λ lies in the interval $(0, 1)$. However, Theorem 7.1 elucidate that the range of λ is not same while applying it on the fractal function. The condition $|\alpha_n| < |a_n^\lambda|$ is crucial, it recasts the range of fractional order based on the predefined scaling factors. In addition, it is indirectly depending on the given data points (see the formula of a_n defined in Chap. 2). Now, it is essential to answer the following two questions (i) Why this condition is stated? (ii) What happens if it fails? The definition of fractal function clearly tells that the scaling factors α_n are predefined in the interval $(0, 1)$. While investigating the fractional derivative of fractal function, new set of scalings are obtained as $\alpha_n a_n^{-\lambda}$. Hence, to make the resultant function $D^\lambda f$ again a fractal function, the condition $|\alpha_n| < |a_n^\lambda|$ is stated which guarantees that the new scalings $|\alpha_n a_n^{-\lambda}|$ are also less than 1. Answering the second question, the Weyl-Marchaud fractional derivative of a fractal interpolation function cannot be a fractal function if the fractional order λ disobeys the constraint $|\alpha_n| < |a_n^\lambda|$. For example, if $|\alpha_n| = 0.5$, $\forall\, n$ and if $\lambda = 0.42$ is the largest fractional order satisfying $|a_n^\lambda| < 0.5$, then no fractional order $\lambda > 0.42$ can generate new fractal functions.

Now, with Theorem 7.1 serving as the foundation, the Weyl-Marchaud fractional derivative of different fractal interpolation functions are discussed for different choice of continuous function q_n.

7.2.1 Fractional Derivative with q_n as a Polynomial of Degree 1

In Theorem 7.1, if the continuous function q_n involved in the fractal function f is taken as a linear polynomial,

$$q_n(x) = c_n x + d_n,\ n = 1, 2, \ldots, N-1,$$

where,

$$c_n = \frac{(y_{n+1} - y_n) - \alpha_n(y_N - y_1)}{x_N - x_1},$$

$$d_n = \frac{(x_N y_n - x_1 y_{n+1}) - \alpha_n (x_N y_1 - x_1 y_N)}{x_N - x_1},$$

then the Weyl-Marchaud fractional derivative of a linear fractal interpolation function can be investigated. As q_n has no effect on the fractional order constraint $|\alpha_n| < |a_n^{\lambda}|$, it remains the same. However, the new IFS $\{(L_n(x), \tilde{F}_n(x, y)) : n = 1, 2, \ldots, N-1\}$, where $\tilde{F}_n(x, y) = a_n^{-\lambda}\alpha_n y + \tilde{q}_n(x)$ and the new data set $\{(x_n, \tilde{y}_n) : n = 1, 2, \ldots, N\}$ change with respect to the linear fractal function. The result is summarized as below theorem.

Theorem 7.2 *Let f be the linear fractal interpolation function with $q_n(x) = c_n x + d_n$. Suppose the scaling factors obey $|\alpha_n| < |a_n^{\lambda}|$, for all $n = 1, 2, \ldots, N-1$, the fractional derivative of f, say $(D_{x_1}^{\lambda} f)(x)$ is also a FIF associated with $\{(x_n, \tilde{y}_n) : n = 1, 2, \ldots, N\}$. Further, the new IFS $\{(L_n(x), \tilde{F}_n(x, y)) : n = 1, 2, \ldots, N-1\}$ determines $(D_{x_1}^{\lambda} f)(x)$, where $\tilde{F}_n(x, y) = a_n^{-\lambda}\alpha_n y + \tilde{q}_n(x)$ for $n = 1, 2, \ldots, N-1$,*

$$\tilde{q}_n(x) = \tilde{y}_n + f_{n,\lambda}(x) + \frac{a_n^{-\lambda} c_n \lambda}{(1-\lambda)\Gamma(1-\lambda)}(x - x_1)^{1-\lambda},$$

$$\tilde{y}_{n+1} = \sum_{j=1}^{n}\left(f_{j,\lambda}(x_N) + a_j^{-\lambda}\alpha_j \tilde{y}_N + \frac{a_j^{-\lambda} c_j \lambda}{(1-\lambda)\Gamma(1-\lambda)}(x_N - x_1)^{1-\lambda}\right), \quad n = 1, 2, \ldots, N-1,$$

here, $\tilde{y}_{n+1}$, $n = 1, 2, \ldots, N-1$ is the new interpolation data associated with the fractional derivative of linear fractal function.

Remark 7.2 The above theorem discusses the fractional derivative for a linear fractal function by predefining $(D_{x_1}^{\lambda} f)(x_1) = 0$. Similarly, defining the fractional derivative at the end point x_N, $(D_{x_N}^{\lambda} f)(x_N) = 0$, the resultant function $D_{x_N}^{\lambda} f$ can be verified for a fractal function with the inclusion of necessary fractional order constraint $|\alpha_n| < |a_n^{\lambda}|$.

7.2.2 Fractional Derivative with q_n as a Polynomial of Degree 2

Let us consider the quadratic polynomial q_n defined by

$$q_n(x) = x^2 + c_n x + d_n, \; n = 1, 2, \ldots, N-1, \tag{7.4}$$

where the parameters c_n and d_n are estimated as

$$c_n = \frac{y_{n+1} - y_n - \alpha_n (y_N - y_1) - (x_N^2 - x_1^2)}{x_N - x_1},$$

$$d_n = -\frac{x_N (y_n - x_1^2 - \alpha_n y_1) + x_1(\alpha_n y_N + x_N^2 - y_{n+1})}{x_N - x_1}.$$

Then the associated quadratic fractal function g obeys the functional equation, $g(L_n(x)) = \alpha_n g(x) + x^2 + c_n x + d_n$, $\alpha_n \in (0, 1)$, $n = 1, 2, \ldots, N-1$. The Weyl-Marchaud fractional derivative of a quadratic fractal function g can be explored by replacing q_n in Theorem 7.1 with Eq. (7.4). Its summarization is presented in the below theorem.

Theorem 7.3 *If g is the quadratic fractal interpolation function with $q_n(x) = x^2 + c_n x + d_n$. Suppose the scaling factors obey $|\alpha_n| < |a_n^\lambda|$, for all $n = 1, 2, \ldots, N-1$, the fractional derivative of g, say $(D_{x_1}^\lambda g)(x)$ is also a FIF associated with $\{(x_n, \tilde{y}_n) : n = 1, 2, \ldots, N\}$. Further, the new IFS $\{(L_n(x), \tilde{F}_n(x, y)) : n = 1, 2, \ldots, N-1\}$ determines $(D_{x_1}^\lambda g)(x)$, where $\tilde{F}_n(x, y) = a_n^{-\lambda}\alpha_n y + \tilde{q}_n(x)$ for $n = 1, 2, \ldots, N-1$,*

$$\tilde{q}_n(x) = \tilde{y}_n + f_{n,\lambda}(x) + \frac{a_n^{-\lambda}\lambda}{\Gamma(1-\lambda)}\int_{x_1}^{x}\frac{x^2-u^2}{(x-u)^{1+\lambda}}du + \frac{a_n^{-\lambda}c_n\lambda}{(1-\lambda)\Gamma(1-\lambda)}(x-x_1)^{1-\lambda},$$

$$\tilde{y}_{n+1} = \sum_{j=1}^{n}\left(f_{j,\lambda}(x_N) + a_j^{-\lambda}\alpha_n\tilde{y}_N + \frac{a_j^{-\lambda}\lambda}{\Gamma(1-\lambda)}\int_{x_1}^{x_N}\frac{x_N^2-u^2}{(x_N-u)^{1+\lambda}}du + \frac{a_j^{-\lambda}c_n\lambda}{(1-\lambda)\Gamma(1-\lambda)}(x_N-x_1)^{1-\lambda}\right),$$

here, $\tilde{y}_{n+1}$, $n = 1, 2, \ldots, N-1$ is the new interpolation data associated with the fractional derivative of linear fractal function.

Remark 7.3 In Theorem 7.3, the fractional derivative of a quadratic fractal function is discussed by predefining $(D_{x_1}^\lambda g)(x_1) = 0$. Similarly, defining the fractional derivative at the end point x_N, $(D_{x_N}^\lambda g)(x_N) = 0$, the resultant function $D_{x_N}^\lambda g$ can be verified for a fractal function with the inclusion of necessary fractional order constraint $|\alpha_n| < |a_n^\lambda|$.

7.2.3 Special Cases

Case (i)
The following special case of q_n generates a special type of fractal interpolation function called the α-fractal function,

$$q_n(x) = \phi \circ L_n(x) - \alpha_n b_n(x), \quad \alpha_n \in (0, 1),\ n = 1, 2, \ldots, N-1, \tag{7.5}$$

where ϕ is the given continuous function and b is the base function equals to ϕ only at the initial point x_1 and end point x_N. The functional equation of α-fractal function is recapped as

$$\phi^\alpha(L_n(x)) = \phi(L_n(x)) + \alpha_n(\phi^\alpha - b)(x), \quad n = 1, 2, \ldots, N-1. \tag{7.6}$$

Now, the Weyl-Marchaud fractional derivative of ϕ^α can be explored by substituting the function q_n given in Eq. (7.5) in Theorem 7.1. Before entering the below theorem, it is essential to analyze the structure of α-fractal function since it is not only depending on the data points as in the case of linear and quadratic fractal function. Its functional equation involves both the given continuous function ϕ and base function b. Hence, in order to study its fractional derivative, it is necessary to impose certain additional constraints on both the fractional derivative of ϕ and the fractional derivative of b. That is, in the definition of α-fractal function, the base function obeys $b(x_1) = \phi(x_1), \quad b(x_N) = \phi(x_N), \quad b \neq \phi$. Similarly, while dealing the fractional derivative of α-fractal function, the new continuous function $D_{x_1}^{\lambda}\phi$ and new base function $D_{x_1}^{\lambda}b$ must satisfy the following condition,

$$(D_{x_1}^{\lambda}b)(x_1) = (D_{x_1}^{\lambda}\phi)(x_1), \quad (D_{x_1}^{\lambda}b)(x_N) = (D_{x_1}^{\lambda}\phi)(x_N), \quad (D_{x_1}^{\lambda}b) \neq (D_{x_1}^{\lambda}\phi).$$

The investigation of Weyl-Marchaud fractional derivative of ϕ^{α} is narrated in the below theorem with aforementioned conditions on fractional derivative of ϕ and fractional derivative of b.

Theorem 7.4 *Consider the α-fractal function, ϕ^{α} obeying the functional equation given in Eq. (7.6). Suppose the fractional derivative of ϕ and b defined by*

$$(\phi')^{\lambda}(x) = (D_{x_1}^{\lambda}\phi)(x) = \frac{\lambda}{\Gamma(1-\lambda)}\int_{x_1}^{x}\frac{\phi(x)-\phi(t)}{(x-t)^{1+\lambda}}dt,$$

$$(b')^{\lambda}(x) = (D_{x_1}^{\lambda}b)(x) = \frac{\lambda}{\Gamma(1-\lambda)}\int_{x_1}^{x}\frac{b(x)-b(t)}{(x-t)^{1+\lambda}}dt$$

obeys $\tilde{y}_1 = (\phi')^{\lambda}(x_1) = (b')^{\lambda}(x_1) = 0$ and $(\phi')^{\lambda}(x_N) = (b')^{\lambda}(x_N)$. Then, for some fixed $\lambda \in (0,1)$ such that $|\alpha_n| < |a_n^{\lambda}|$ for all $n = 1, 2, \ldots, N-1$, the function $(D_{x_1}^{\lambda}\phi^{\alpha})(x)$ is also an α-fractal function corresponding to new interpolation points $\{(x_n, \tilde{y}_n) : n = 1, 2, \ldots, N\}$. In addition, the new IFS $\{(L_n(x), \tilde{F}_n(x, y)) : n = 1, 2, \ldots, N-1\}$ determines the function $(D_{x_1}^{\lambda}\phi^{\alpha})(x)$, where $\tilde{F}_n(x, y) = a_n^{-\lambda}\alpha_n y + \tilde{q}_n(x)$, for $n = 1, 2, \ldots, N-1$,

$$\tilde{q}_n(x) = \tilde{y}_n + \phi_{n,\lambda}^{\alpha}(x) - a_n^{-\lambda}\alpha_n(b')^{\lambda}(x) + a_n^{-\lambda}(\phi')^{\lambda}(L_n(x)),$$

$$\tilde{y}_{n+1} = \sum_{j=1}^{n}\left(\phi_{j,\lambda}^{\alpha}(x_N) + a_j^{-\lambda}\alpha_j\tilde{y}_N - a_j^{-\lambda}\alpha_j(b')^{\lambda}(x_N) + a_j^{-\lambda}(\phi')^{\lambda}(L_j(x_N))\right),$$

$$\tilde{y}_N = \frac{\sum_{j=1}^{N-1}\left(\phi_{j,\lambda}^{\alpha}(x_N) - a_j^{-\lambda}\alpha_j(b')^{\lambda}(x_N) + a_j^{-\lambda}(\phi')^{\lambda}(L_j(x_N))\right)}{1-\sum_{j=1}^{N-1}a_j^{-\lambda}\alpha_j},$$

here, $\tilde{y}_{n+1}$, $n = 1, 2, \ldots, N-1$ is the new interpolation data associated with $D_{x_1}^{\lambda}\phi^{\alpha}$.

Proof By the definition of Weyl-Marchaud fractional derivative with $D_{x_1}^{\lambda}\phi^{\alpha}(x_1)$=0,

$$(D_{x_1}^{\lambda}\phi^{\alpha})(L_n(x)) = \frac{\lambda}{\Gamma(1-\lambda)}\int_{x_1}^{L_n(x)}\frac{\phi^{\alpha}(L_n(x)) - \phi^{\alpha}(t)}{(L_n(x)-t)^{1+\lambda}}dt$$

$$= \frac{\lambda}{\Gamma(1-\lambda)} \int_{x_1}^{x_n} \frac{\phi^\alpha(L_n(x)) - \phi^\alpha(t)}{(L_n(x)-t)^{1+\lambda}} dt + \frac{\lambda}{\Gamma(1-\lambda)} \int_{x_n}^{L_n(x)} \frac{\phi^\alpha(L_n(x)) - \phi^\alpha(t)}{(L_n(x)-t)^{1+\lambda}} dt.$$

Using change of variable in second term as $t = L_n(u)$, one can get

$$\begin{aligned}(D_{x_1}^\lambda \phi^\alpha)(L_n(x)) &= \frac{\lambda}{\Gamma(1-\lambda)} \int_{x_1}^{x_n} \frac{\phi^\alpha(x_n) - \phi^\alpha(t)}{(x_n - t)^{1+\lambda}} dt - \frac{\lambda}{\Gamma(1-\lambda)} \int_{x_1}^{x_n} \frac{\phi^\alpha(x_n) - \phi^\alpha(t)}{(x_{n-1} - t)^{1+\lambda}} dt \\ &+ \frac{\lambda}{\Gamma(1-\lambda)} \int_{x_1}^{x_n} \frac{\phi^\alpha(L_n(x)) - \phi^\alpha(t)}{(L_n(x)-t)^{1+\lambda}} dt + \frac{a_n\lambda}{\Gamma(1-\lambda)} \int_{x_1}^{x} \frac{\phi^\alpha(L_n(x)) - \phi^\alpha(L_n(u))}{(L_n(x)-L_n(u))^{1+\lambda}} dt.\end{aligned}$$

Given $\phi^\alpha(x)$ is a α-fractal function and the functional equation Eq. (7.6) provides

$$\begin{aligned}(D_{x_1}^\lambda \phi)(L_n(x)) &= \tilde{y}_n + \phi_{n,\lambda}^\alpha(x) \\ &+ \frac{a_n\lambda}{\Gamma(1-\lambda)} \int_{x_1}^{x} \frac{\phi \circ L_n(x) + \alpha_n \phi^\alpha(x) - \alpha_n b(x) - \phi \circ L_n(u) - \alpha_n \phi^\alpha(u) + \alpha_n b(u)}{(a_n x + b_n - a_n u - b_n)^{1+\lambda}} du \\ &= \tilde{y}_n + \phi_{n,\lambda}^\alpha(x) \\ &+ \frac{a_n\lambda}{\Gamma(1-\lambda)} \int_{x_1}^{x} \frac{\alpha_n(\phi^\alpha(x) - \phi^\alpha(u)) - \alpha_n(b(x) - b(u)) + \phi \circ L_n(x) - \phi \circ L_n(u)}{a_n^{1+\lambda}(x-u)^{1+\lambda}} du \\ &= \tilde{y}_n + \phi_{n,\lambda}^\alpha(x) + \frac{a_n^{-\lambda}\alpha_n\lambda}{\Gamma(1-\lambda)} \int_{x_1}^{x} \frac{\phi^\alpha(x) - \phi^\alpha(u)}{(x-u)^{1+\lambda}} du - \frac{a_n^{-\lambda}\alpha_n\lambda}{\Gamma(1-\lambda)} \int_{x_1}^{x} \frac{b(x) - b(u)}{(x-u)^{1+\lambda}} du \\ &+ \frac{a_n^{-\lambda}\lambda}{\Gamma(1-\lambda)} \int_{x_1}^{x} \frac{\phi \circ L_n(x) - \phi \circ L_n(u)}{(x-u)^{1+\lambda}} du \\ &= \tilde{y}_n + \phi_{n,\lambda}^\alpha(x) + a_n^{-\lambda}\alpha_n(D_{x_1}^\lambda \phi^\alpha)(x) - a_n^{-\lambda}\alpha_n(D_{x_1}^\lambda b)(x) + a_n^{-\lambda} D_{x_1}^\lambda(\phi \circ L_n)(x) \\ &= \tilde{y}_n + \phi_{n,\lambda}^\alpha(x) + a_n^{-\lambda}\alpha_n(D_{x_1}^\lambda \phi^\alpha)(x) - a_n^{-\lambda}\alpha_n(b')^\lambda(x) + a_n^{-\lambda}(\phi')^\lambda(L_n(x)) \\ &= a_n^{-\lambda}\alpha_n(D_{x_1}^\lambda \phi^\alpha)(x) + \tilde{q}_n(x) \\ &= \tilde{F}_n(x, (D_{x_1}^\lambda \phi^\alpha)(x)).\end{aligned}$$

Here

$$\tilde{q}_n(x) = \tilde{y}_n + \phi_{n,\lambda}^\alpha(x) - a_n^{-\lambda}\alpha_n(b')^\lambda(x) + a_n^{-\lambda}(\phi')^\lambda(L_n(x))$$

$$\phi_{n,\lambda}^\alpha(x) = \frac{\lambda}{\Gamma(1-\lambda)} \int_{x_1}^{x_n} \left(\frac{\phi^\alpha(L_n(x)) - \phi^\alpha(t)}{(L_n(x)-t)^{1+\lambda}} - \frac{\phi^\alpha(x_n) - \phi^\alpha(t)}{(x_n - t)^{1+\lambda}} \right) dt.$$

Hence, the Weyl-Marchaud fractional derivative of an α-fractal interpolation function determined by the IFS $\{L_n(x), \tilde{F}_n(x, y) : n = 1, 2, \ldots, N-1\}$ is also an α-fractal interpolation function. Let us consider

$$(D_{x_1}^\lambda \phi^\alpha)(L_n(x)) = \tilde{y}_n + \phi_{n,\lambda}^\alpha(x) + a_n^{-\lambda}\alpha_n(D_{x_1}^\lambda \phi^\alpha)(x) - a_n^{-\lambda}\alpha_n(b')^\lambda(x) + a_n^{-\lambda}(\phi')^\lambda(L_n(x)). \tag{7.7}$$

Let $x = x_N$ and $L_n(x_N) = x_{n+1}$. Then, Eq. (7.7) provides

$$\tilde{y}_{n+1} - \tilde{y}_n = \phi^{\alpha}_{j,\lambda}(x_N) + a_j^{-\lambda}\alpha_n \tilde{y}_N - a_j^{-\lambda}\alpha_n (b')^{\lambda}(x_N) + a_j^{-\lambda}(\phi')^{\lambda}(L_n(x_N)). \tag{7.8}$$

The equation $\tilde{y}_{n+1} = \tilde{y}_1 + \sum_{j=1}^{n}(\tilde{y}_{j+1} - \tilde{y}_j)$ gives

$$\tilde{y}_{n+1} = \sum_{j=1}^{n}\Big(\phi^{\alpha}_{j,\lambda}(x_N) + a_j^{-\lambda}\alpha_j \tilde{y}_N - a_j^{-\lambda}\alpha_j (b')^{\lambda}(x_N) + a_j^{-\lambda}(\phi')^{\lambda}(L_j(x_N))\Big).$$

Similarly, substituting $n = N - 1$ in the above equation, the end point $\tilde{y}_N$ is obtained as follows

$$\tilde{y}_N = \frac{\sum_{j=1}^{N-1}\Big(\phi^{\alpha}_{j,\lambda}(x_N) - a_j^{-\lambda}\alpha_j (b')^{\lambda}(x_N) + a_j^{-\lambda}(\phi')^{\lambda}(L_j(x_N))\Big)}{1 - \sum_{j=1}^{N-1} a_j^{-\lambda}\alpha_j}.$$

Observe

$$\begin{aligned}
\tilde{q}_n(x_1) &= \tilde{y}_n + \phi^{\alpha}_{n,\lambda}(x_1) - a_n^{-\lambda}\alpha_n (b')^{\lambda}(x_1) + a_n^{-\lambda}(\phi')^{\lambda}(L_n(x_1)) \\
&= \tilde{y}_n - a_n^{-\lambda}\alpha_n \tilde{y}_1, \\
\tilde{q}_n(x_N) &= \tilde{y}_n + \phi^{\alpha}_{n,\lambda}(x_N) - a_n^{-\lambda}\alpha_n (b')^{\lambda}(x_N) + a_n^{-\lambda}(\phi')^{\lambda}(L_n(x_N)).
\end{aligned}$$

By comparing the above equation with Eq. (7.8), one can get $\tilde{q}_n(x_N) = \tilde{y}_{n+1} - a_n^{-\lambda}\alpha_n \tilde{y}_N$. Note that

$$\begin{aligned}
\tilde{F}_n(x_1, \tilde{y}_1) &= \tilde{F}_n(x_1, 0) \\
&= a_n^{-\lambda}\alpha_n (D^{\lambda}_{x_1} f^{\alpha})(x_1) + \tilde{q}_n(x_1) \\
&= \tilde{q}_n(x_1) \\
&= \tilde{y}_n + \phi^{\alpha}_{n,\lambda}(x_1) - a_n^{-\lambda}\alpha_n (b')^{\lambda}(x_1) + a_n^{-\lambda}(\phi')^{\lambda}(L_n(x_1)) \\
&= \tilde{y}_n, \\
\tilde{F}_n(x_N, \tilde{y}_N) &= a_n^{-\lambda}\alpha_n (D^{\lambda}_{x_1} f^{\alpha})(x_N) + \tilde{q}_n(x_N) \\
&= \tilde{y}_n + \phi^{\alpha}_{n,\lambda}(x_N) + a_n^{-\lambda}\alpha_n (D^{\lambda}_{x_1}\phi^{\alpha})(x_N) - a_n^{-\lambda}\alpha_n (b')^{\lambda}(x_N) + a_n^{-\lambda}(\phi')^{\lambda}(L_n(x_N)) \\
&= (D^{\lambda}_{x_1}\phi^{\alpha})(L_n(x_N)) \\
&= \tilde{y}_{n+1}.
\end{aligned}$$

It is clear that the function $(D^{\lambda}_{x_1}\phi^{\alpha})(x)$ satisfies the endpoint conditions $\tilde{F}_n(x_1, \tilde{y}_1) = \tilde{y}_n$ and $\tilde{F}_n(x_N, \tilde{y}_N) = \tilde{y}_{n+1}$. Thus, the Weyl-Marchaud fractional derivative of a α-fractal function determined by the IFS $\{(L_n(x), \tilde{F}_n(x, y)) : n = 1, 2, \ldots, N - 1\}$ is also a α-fractal function which interpolate the new data set $\{(x_n, \tilde{y}_n) : n = 1, 2, \ldots, N\}$.

While investigating the fractional derivative of α-fractal function also, the fractional order constraint $|\alpha_n| < |a_n^\lambda|$ remains same, since the parameter a_n is associated with the contraction maps L_n and change in q_n has no impact with the constraint.

Remark 7.4 The Weyl-Marchaud fractional derivative of ϕ^α defined at the end point x_N, say $D_{x_N}^\lambda \phi$ can be demonstrated as an α-fractal function following the same argument of Theorem 7.4. The essential fact to be remembered is to predefine the necessary condition on the fractional derivative of ϕ and the fractional derivative of b.

Case (ii)
The hidden variable fractal interpolation function, **h** is a vector valued function having two components (h_1, h_2) obeying the following functional equations,

$$\begin{aligned} h_1\big(L_n(x)\big) &= \delta_n h_1(x) + \beta_n h_2(x) + p_n(x), \\ h_2\big(L_n(x)\big) &= \gamma_n h_2(x) + q_n(x). \end{aligned} \tag{7.9}$$

Here, the two continuous functions can be chosen as polynomial functions obeying the end point conditions discussed in Chap. 2. The interesting part in this fractal function is its scaling matrix, $A_n = \begin{bmatrix} \delta_n & \beta_n \\ 0 & \gamma_n \end{bmatrix}$, whose norm should be less than 1. This constraint on the matrix norm should be satisfied while investigating the Weyl-Marchaud fractional derivative of hidden variable fractal interpolation function. Also, the fractional derivative of h_1 and h_2 have to be evaluated separately for self-affine and non self-affine nature.

Theorem 7.5 *Consider the hidden variable fractal interpolation function* **h** *with contraction matrix* $A_n = \begin{bmatrix} \delta_n & \beta_n \\ 0 & \gamma_n \end{bmatrix}$. *Suppose* $\|A_n'\| < 1$, *where* $A_n' = a_n^{-\lambda} \begin{bmatrix} \delta_n & \beta_n \\ 0 & \gamma_n \end{bmatrix}$, *for all* $n = 1, 2, \ldots, N-1$, *then the function* $(D_{x_1}^\lambda \mathbf{h})(x) = ((D_{x_1}^\lambda h_1)(x), (D_{x_1}^\lambda h_2)(x))$ *is also a hidden variable fractal function. The new iterated function systems* $\{(L_n(x), \tilde{F}_{1n}(x, y, z)) : n = 1, 2, \ldots, N-1\}$ *and* $\{(L_n(x), \tilde{F}_{2n}(x, z)) : n = 1, 2, \ldots, N-1\}$ *generate the functions* $(D_{x_1}^\lambda h_1)(x)$ *and* $(D_{x_1}^\lambda h_2)(x)$ *respectively, where* $\tilde{F}_{1n}(x, y, z) = a_n^{-\lambda}\delta_n y + a_n^{-\lambda}\beta_n z + \tilde{p}_n(x)$ *and* $\tilde{F}_{2n}(x, z) = a_n^{-\lambda}\gamma_n z + \tilde{q}_n(x)$, *for* $n = 1, 2, \ldots, N-1$,

$$\tilde{p}_n(x) = \tilde{y}_{n-1} + h_{1(n,\lambda)}(x) + a_n^{-\lambda}(D_{x_1}^\lambda p_n)(x),$$

$$\tilde{q}_n(x) = \tilde{z}_{n-1} + h_{2(n,\lambda)}(x) + a_n^{-\lambda}(D_{x_1}^\lambda q_n)(x),$$

$$\tilde{y}_{n+1} = \sum_{j=1}^{n} \left(h_{1(j,\lambda)}(x_N) + a_j^{-\lambda}\delta_j \tilde{y}_N + a_j^{-\lambda}\beta_j (D_{x_1}^\lambda h_2)(x_N) + a_j^{-\lambda}(D_{x_1}^\lambda p_j)(x_N) \right),$$

$$\tilde{y}_N = \frac{\sum_{j=1}^{N-1} \left(h_{1(j,\lambda)}(x_N) + a_j^{-\lambda}\beta_j (D_{x_1}^\lambda h_2)(x_N) + a_j^{-\lambda}(D_{x_1}^\lambda p_j)(x_N) \right)}{1 - \sum_{j=1}^{N-1} a_j^{-\lambda}\delta_j}$$

$$\tilde{z}_{n+1} = \sum_{j=1}^{n} \left(h_{2(j,\lambda)}(x_N) + a_j^{-\lambda}\gamma_n \tilde{z}_N + a_j^{-\lambda}(D_{x_1}^{\lambda} q_j)(x_N) \right),$$

$$\tilde{z}_N = \frac{\sum_{j=1}^{N-1} \left(h_{2(j,\lambda)}(x_N) + a_j^{-\lambda}(D_{x_1}^{\lambda} q_j)(x_N) \right)}{1 - \sum_{j=1}^{N-1} a_j^{-\lambda}\gamma_j}$$

where, $\tilde{y}_{n+1}$ and $\tilde{z}_{n+1}$, for $n = 1, 2, \ldots, N-1$, are the new interpolation points associated with the functions $D_{x_1}^{\lambda} h_1$ and $D_{x_1}^{\lambda} h_2$.

Proof By the definition of Weyl-Marchaud fractional derivative given with $D_{x_1}^{\lambda} h_1(x_1) = 0$, we have

$$\begin{aligned}
(D_{x_1}^{\lambda} h_1)(L_n(x)) &= \frac{\lambda}{\Gamma(1-\lambda)} \int_{x_1}^{L_n(x)} \frac{h_1(L_n(x)) - h_1(t)}{(L_n(x) - t)^{1+\lambda}} dt \\
&= \frac{\lambda}{\Gamma(1-\lambda)} \int_{x_1}^{x_n} \frac{h_1(L_n(x)) - h_1(t)}{(L_n(x) - t)^{1+\lambda}} dt + \frac{\lambda}{\Gamma(1-\lambda)} \int_{x_n}^{L_n(x)} \frac{h_1(L_n(x)) - h_1(t)}{(L_n(x) - t)^{1+\lambda}} dt.
\end{aligned}$$

Using change of variable as $t = L_n(u)$ in second term provides

$$\begin{aligned}
(D_{x_1}^{\lambda} h_1)(L_n(x)) = {} & \frac{\lambda}{\Gamma(1-\lambda)} \int_{x_1}^{x_n} \frac{h_1(x_n) - h_1(t)}{(x_n - t)^{1+\lambda}} dt - \frac{\lambda}{\Gamma(1-\lambda)} \int_{x_1}^{x_n} \frac{h_1(x_n) - h_1(t)}{(x_n - t)^{1+\lambda}} dt \\
& + \frac{\lambda}{\Gamma(1-\lambda)} \int_{x_1}^{x_n} \frac{h_1(L_n(x)) - h_1(t)}{(L_n(x) - t)^{1+\lambda}} dt + \frac{a_n\lambda}{\Gamma(1-\lambda)} \int_{x_1}^{x} \frac{h_1(L_n(x)) - h_1(L_n(u))}{(L_n(x) - L_n(u))^{1+\lambda}} du.
\end{aligned}$$

Given $\mathbf{h}(x)$ is a hidden variable FIF and the functional equation in Eq. (7.9) yields,

$$\begin{aligned}
(D_{x_1}^{\lambda} h_1)(L_n(x)) = {} & \tilde{y}_n + h_{1(n,\lambda)}(x) + \\
& \frac{a_n\lambda}{\Gamma(1-\lambda)} \int_{x_1}^{x} \frac{\delta_n h_1(x) + \beta_n h_2(x) + p_n(x) - \delta_n h_1(u) - \beta_n h_2(u) - p_n(u)}{(a_n x + b_n - a_n u - b_n)^{1+\lambda}} du \\
= {} & \tilde{y}_n + h_{1(n,\lambda)}(x) + \frac{a_n\lambda}{\Gamma(1-\lambda)} \int_{x_1}^{x} \frac{\delta_n (h_1(x) - h_1(u))}{a_n^{1+\lambda}(x-u)^{1+\lambda}} du \\
& + \frac{a_n\lambda}{\Gamma(1-\lambda)} \int_{x_1}^{x} \frac{\beta_n (h_2(x) - h_2(u))}{a_n^{1+\lambda}(x-u)^{1+\lambda}} du + \frac{a_n\lambda}{\Gamma(1-\lambda)} \int_{x_1}^{x} \frac{p_n(x) - p_n(u)}{a_n^{1+\lambda}(x-u)^{1+\lambda}} du \\
= {} & \tilde{y}_n + h_{1(n,\lambda)}(x) + \frac{a_n^{-\lambda}\delta_n\lambda}{\Gamma(1-\lambda)} \int_{x_1}^{x} \frac{h_1(x) - h_1(u)}{(x-u)^{1+\lambda}} du \\
& + \frac{a_n^{-\lambda}\beta_n\lambda}{\Gamma(1-\lambda)} \int_{x_1}^{x} \frac{h_2(x) - h_2(u)}{(x-u)^{1+\lambda}} du + \frac{a_n^{-\lambda}\lambda}{\Gamma(1-\lambda)} \int_{x_1}^{x} \frac{p_n(x) - p_n(u)}{(x-u)^{1+\lambda}} du \\
= {} & \tilde{y}_n + h_{1(n,\lambda)}(x) + a_n^{-\lambda}\delta_n (D_{x_1}^{\lambda} h_1)(x) + a_n^{-\lambda}\beta_n (D_{x_1}^{\lambda} h_2)(x) + a_n^{-\lambda}(D_{x_1}^{\lambda} p_n)(x) \\
= {} & a_n^{-\lambda}\delta_n (D_{x_1}^{\lambda} h_1)(x) + a_n^{-\lambda}\beta_n (D_{x_1}^{\lambda} h_2)(x) + \tilde{p}_n(x) \\
= {} & \tilde{F}_{1n}(x, (D_{x_1}^{\lambda} h_1)(x), (D_{x_1}^{\lambda} h_2)(x)).
\end{aligned}$$

Here,

$$\tilde{p}_n(x) = \tilde{y}_n + h_{1(n,\lambda)}(x) + a_n^{-\lambda}(D_{x_1}^{\lambda} p_n)(x),$$
$$h_{1(n,\lambda)}(x) = \frac{\lambda}{\Gamma(1-\lambda)} \int_{x_1}^{x_n} \left(\frac{h_1(L_n(x)) - h_1(t)}{(L_n(x) - t)^{1+\lambda}} - \frac{h_1(x_n) - h_1(t)}{(x_n - t)^{1+\lambda}} \right) dt.$$

Similarly, from the definition of Weyl-Marchaud fractional derivative with $D_{x_1}^{\lambda} h_2(x_1) = 0$, it follows that

$$\begin{aligned}(D_{x_1}^{\lambda} h_2)(L_n(x)) &= \frac{\lambda}{\Gamma(1-\lambda)} \int_{x_1}^{L_n(x)} \frac{h_2(L_n(x)) - h_2(t)}{(L_n(x) - t)^{1+\lambda}} dt \\ &= \frac{\lambda}{\Gamma(1-\lambda)} \int_{x_1}^{x_n} \frac{h_2(L_n(x)) - h_2(t)}{(L_n(x) - t)^{1+\lambda}} dt + \frac{\lambda}{\Gamma(1-\lambda)} \int_{x_n}^{L_n(x)} \frac{h_2(L_n(x)) - h_2(t)}{(L_n(x) - t)^{1+\lambda}} dt.\end{aligned}$$

Using change of variable as $t = L_n(u)$ in second term provides

$$\begin{aligned}(D_{x_1}^{\lambda} h_2)(L_n(x)) = &\frac{\lambda}{\Gamma(1-\lambda)} \int_{x_1}^{x_n} \frac{h_2(x_n) - h_2(t)}{(x_n - t)^{1+\lambda}} dt - \frac{\lambda}{\Gamma(1-\lambda)} \int_{x_1}^{x_n} \frac{h_2(x_n) - h_2(t)}{(x_n - t)^{1+\lambda}} dt \\ &+ \frac{\lambda}{\Gamma(1-\lambda)} \int_{x_1}^{x_n} \frac{h_2(L_n(x)) - h_2(t)}{(L_n(x) - t)^{1+\lambda}} dt + \frac{a_n \lambda}{\Gamma(1-\lambda)} \int_{x_1}^{x} \frac{h_2(L_n(x)) - h_2(L_n(u))}{(L_n(x) - L_n(u))^{1+\lambda}} du.\end{aligned}$$

Given $\mathbf{h}(x)$ is a hidden variable FIF and the functional equation Eq. (7.9) yields,

$$\begin{aligned}(D_{x_1}^{\lambda} h_2)(L_n(x)) &= \tilde{z}_n + h_{2(n,\lambda)}(x) + \frac{a_n \lambda}{\Gamma(1-\lambda)} \int_{x_1}^{x} \frac{\gamma_n h_2(x) + q_n(x) - \gamma_n h_2(u) - q_n(u)}{(a_n x + b_n - a_n u - b_n)^{1+\lambda}} du \\ &= \tilde{z}_n + h_{2(n,\lambda)}(x) + \frac{a_n \lambda}{\Gamma(1-\lambda)} \int_{x_1}^{x} \frac{\gamma_n (h_2(x) - h_2(u))}{a_n^{1+\lambda}(x-u)^{1+\lambda}} du + \frac{a_n \lambda}{\Gamma(1-\lambda)} \int_{x_1}^{x} \frac{q_n(x) - q_n(u)}{a_n^{1+\lambda}(x-u)^{1+\lambda}} du \\ &= \tilde{z}_n + h_{2(n,\lambda)}(x) + \frac{a_n^{-\lambda} \lambda \gamma_n}{\Gamma(1-\lambda)} \int_{x_1}^{x} \frac{(h_2(x) - h_2(u))}{(x-u)^{1+\lambda}} du + \frac{a_n^{-\lambda} \lambda}{\Gamma(1-\lambda)} \int_{x_1}^{x} \frac{q_n(x) - q_n(u)}{(x-u)^{1+\lambda}} du \\ &= \tilde{z}_n + h_{2(n,\lambda)}(x) + a_n^{-\lambda} \gamma_n (D_{x_1}^{\lambda} h_2)(x) + a_n^{-\lambda} (D_{x_1}^{\lambda} q_n)(x) \\ &= a_n^{-\lambda} \gamma_n (D_{x_1}^{\lambda} h_2)(x) + \tilde{q}_n(x) \\ &= \tilde{F}_{2n}(x, (D_{x_1}^{\lambda} h_2)(x)).\end{aligned}$$

Here

$$\tilde{q}_n(x) = \tilde{z}_n + h_{2(n,\lambda)}(x) + a_n^{-\lambda}(D_{x_1}^{\lambda} q_n)(x)$$
$$h_{2(n,\lambda)}(x) = \frac{\lambda}{\Gamma(1-\lambda)} \int_{x_1}^{x_n} \left(\frac{h_2(L_n(x)) - h_2(t)}{(L_n(x) - t)^{1+\lambda}} - \frac{h_2(x_n) - h_2(t)}{(x_n - t)^{1+\lambda}} \right) dt.$$

Hence, the Weyl-Marchaud fractional derivative of a hidden variable FIF is also a hidden variable FIF. Let us consider

$$(D_{x_1}^{\lambda} h_1)(L_n(x)) = \tilde{y}_n + h_{1(n,\lambda)}(x) + a_n^{-\lambda}\delta_n (D_{x_1}^{\lambda} h_1)(x) + a_n^{-\lambda}\beta_n (D_{x_1}^{\lambda} h_2)(x) + a_n^{-\lambda}(D_{x_1}^{\lambda} p_n)(x). \tag{7.10}$$

Let $x = x_N$ and $L_n(x_N) = x_{n+1}$. Then, Eq. (7.10) gives

$$\tilde{y}_{n+1} - \tilde{y}_n = h_{1(n,\lambda)}(x_N) + a_n^{-\lambda}\delta_n \tilde{y}_N + a_n^{-\lambda}\beta_n (D_{x_1}^{\lambda} h_2)(x_N) + a_n^{-\lambda}(D_{x_1}^{\lambda} p_n)(x_N).$$

The equation $\tilde{y}_{n+1} = \tilde{y}_1 + \sum_{j=1}^{n}(\tilde{y}_{j+1} - \tilde{y}_j)$ provides

$$\tilde{y}_{n+1} = \sum_{j=1}^{n}\Big(h_{1(j,\lambda)}(x_N) + a_j^{-\lambda}\delta_j \tilde{y}_N + a_j^{-\lambda}\beta_j (D_{x_1}^{\lambda} h_2)(x_N) + a_j^{-\lambda}(D_{x_1}^{\lambda} p_j)(x_N)\Big).$$

Similarly, substituting $n = N-1$ in the above equation gives the end point $\tilde{y}_N$ as follows

$$\tilde{y}_N = \frac{\sum_{j=1}^{N-1}\Big(h_{1(j,\lambda)}(x_N) + a_j^{-\lambda}\beta_j (D_{x_1}^{\lambda} h_2)(x_N) + a_j^{-\lambda}(D_{x_1}^{\lambda} p_j)(x_N)\Big)}{1 - \sum_{j=1}^{N-1} a_j^{-\lambda}\delta_j}.$$

In order to find the new interpolation data $\{(x_n, \tilde{z}_n) : i = 1, 2, \ldots, N\}$, consider the following equation

$$(D_{x_1}^{\lambda} h_2)(L_n(x)) = \tilde{z}_n + h_{2(n,\lambda)}(x) + a_n^{-\lambda}\gamma_n (D_{x_1}^{\lambda} h_2)(x) + a_n^{-\lambda}(D_{x_1}^{\lambda} q_n)(x). \tag{7.11}$$

Let $x = x_N$ and $L_n(x_N) = x_{n+1}$. Then, Eq. (7.11) gives

$$\tilde{z}_{n+1} - \tilde{z}_n = h_{2(n,\lambda)}(x_N) + a_n^{-\lambda}\gamma_n \tilde{z}_N + a_n^{-\lambda}(D_{x_1}^{\lambda} q_n)(x_N).$$

The equation $\tilde{z}_{n+1} = \tilde{z}_1 + \sum_{j=1}^{n}(\tilde{z}_{j+1} - \tilde{z}_j)$ provides

$$\tilde{z}_{n+1} = \sum_{j=1}^{n}\Big(h_{2(j,\lambda)}(x_N) + a_j^{-\lambda}\gamma_j \tilde{z}_N + a_j^{-\lambda}(D_{x_1}^{\lambda} q_j)(x_N)\Big).$$

Similarly, substituting $n = N-1$ in the above equation gives the end point $\tilde{z}_N$ as follows

$$\tilde{z}_N = \frac{\sum_{j=1}^{N-1}\Big(h_{2(j,\lambda)}(x_N) + a_j^{-\lambda}(D_{x_1}^{\lambda} q_j)(x_N)\Big)}{1 - \sum_{j=1}^{N-1} a_j^{-\lambda}\gamma_j}.$$

The Weyl-Marchaud fractional derivative of a hidden variable FIF determined by the IFS $\{L_n(x), \tilde{F}_n(x, y, z) : n = 1, 2, \ldots, N-1\}$ is also a hidden variable fractal interpolation function which interpolates the new data set $\{(x_n, \tilde{y}_n, \tilde{z}_n) : n = 1, 2, \ldots, N\}$. □

Remark 7.5 The Weyl-Marchaud fractional derivative of $\mathbf{h}$ defined at the end point x_N, say $(D^{\lambda}_{x_N}\mathbf{h})(x) = ((D^{\lambda}_{x_N}h_1)(x), (D^{\lambda}_{x_N}h_2)(x))$ can be demonstrated as a hidden variable fractal function following the same argument of Theorem 7.5.

7.3 Numerical Illustration of Relation Between Fractional Order and Scaling Parameter

In this section, numerical examples are presented for different types of fractal interpolation functions and their Weyl-Marchaud fractional derivative with appropriate choice of fractional order λ. The graphical illustrations help to perceive the difference between the original fractal functions and their fractional derivative.

Example 7.1 For the data set $x=\{0, 0.3, 0.6, 0.9, 1.2, 1.5\}$ and $y=\{0, 0.25, -0.2, 0.3, -0.1, 0.3\}$, a linear fractal interpolation function is generated as an attractor of the IFS $\{(L_n(x), F_n(x, y))\}_{n=1}^{5}$, with the predefined free parameter $\alpha_n = 0.5$, where the maps L_n and F_n are defined as $L_n(x) = a_n x + b_n$ and $F_n(x, y) = 0.5y + c_n x + d_n$. The parameters a_n, b_n, c_n and d_n are estimated using the endpoint conditions of contractions L_n and F_n. Figure 7.1a shows the graph of linear fractal interpolation function f for the provided data set. To obtain the Weyl-Marchuad fractional derivative of linear fractal function f, the right choice of fractional order λ has to be made. Using the necessary condition $|\alpha_n| < |a_n^{\lambda}|$ stated in Theorem 7.2 and the given data set, two different fractional orders 0.18 and 0.3 are taken. The graph of the Weyl-Marchaud derivative of f with 0.18, the function $D^{0.18}f$ is plotted in Fig. 7.1b. The functions $D^{0.25}f$ and $D^{0.3}f$, with order 0.25 and 0.3, are plotted in Fig. 7.1c, d. Comparing f with $D^{0.3}f$, the latter is highly irregular at each sub-interval than the former. Also, the irregularity increases while increasing the fractional order from 0.18 to 0.3. The self-similarity of the original fractal function f is preserved by the different fractional-order fractal functions $D^{0.18}f$, $D^{0.25}f$ and $D^{0.3}f$.

Example 7.2 In this example, a quadratic fractal interpolation function is generated to approximate the data set $\{(0, 0), (0.3, 0.25), (0.6, -0.2), (0.9, 0.3), (1.2, -0.1), (1.5, 0.3)\}$. The constant scaling factor is chosen as 0.6 at each sub-interval. Figure 7.2a represents the graph of a quadratic fractal function g interpolating the given data. In addition, it is obtained as an attractor of the IFS $\{(L_n(x), F_n(x, y))\}_{n=1}^{5}$, where the maps L_n and F_n are defined as $L_n(x) = a_n x + b_n$ and $F_n(x, y) = 0.6y + x^2 + c_n x + d_n$. The parameter values a_n, b_n, c_n, d_n are estimated using the end point conditions of F_n. To obtain the Weyl-Marchuad fractional derivative of a quadratic fractal interpolation function g, the two fractional orders 0.1 and 0.3 are chosen. The two choices of fractional order satisfies the constraint $|\alpha_n| < |a_n^{\lambda}|$ stated in Theorem 7.3 with respect to the given interpolation data and scaling factor 0.6. The graphical representations of $D^{0.1}g$, $D^{0.23}g$ and $D^{0.3}g$ are shown in Figs. 7.2b, 7.2(c) and 7.2(d), respectively. The plots g and $D^{0.1}g$ may look similar but when observed at sub-intervals, the variation can be noticed. That is, the irregular pattern

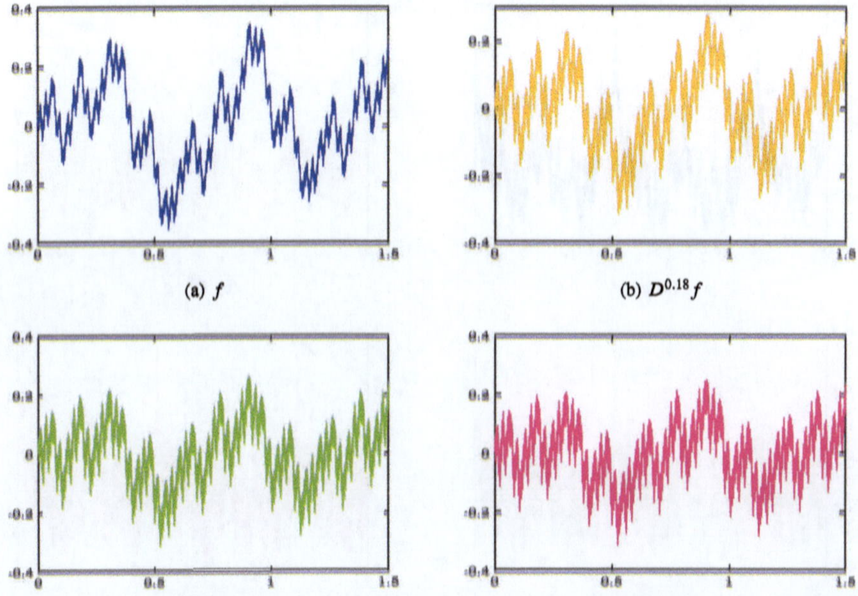

(a) f (b) $D^{0.18}f$ (c) $D^{0.25}f$ (d) $D^{0.3}f$

Fig. 7.1 Linear fractal function and its Weyl-Marchaud fractional derivative with scaling factor $\alpha_n = 0.5, \forall n = 1, 2, 3, 4, 5$: **a** f, **b** $D^{0.18}f$, **c** $D^{0.25}f$ and **d** $D^{0.3}f$

is dense in the fractional derivative $D^{0.1}g$ than the original quadratic fractal function g. Besides the irregularity increases further when the fractional order is increased to be 0.3, as depicted in Fig. 7.2c.

Example 7.3 In this example, graphs of α-fractal function and its Weyl-Marchaud fractional derivative are generated to approximate the given continuous function $\phi(x) = x^2 + \frac{x}{2}$ on the interval [0, 1]. The base function $b(x) = \frac{3x}{2}$ is chosen such that it only equals to ϕ at the end points $x = 0$ and $x = 1$. The scaling factor α_n is chosen as $\alpha_n = 0.45$ for each i. For $x = \{0, 0.25, 0.5, 0.75, 1\}$, $\phi(x)$ is generated and it is approximated using ϕ^α as shown in Fig. 7.3a. The two distinct fractional orders 0.2 and 0.4 are appropriately selected obeying the necessary condition $|\alpha_n| < |a_n^\lambda|$. The graph of $D^{0.2}\phi^\alpha$ is demonstrated in Fig. 7.3b and the graphs of $D^{0.32}\phi^\alpha$ and $D^{0.4}\phi^\alpha$ are demonstrated in Fig. 7.3c, d. From the two plots ϕ^α and $D^{0.2}\phi^\alpha$, it is observed that the fractional derivative has more complexity preserving the self-similarity of ϕ^α. Also, the irregular pattern varies as fractional order gradually increases from 0.2 to 0.4. The new fractal functions $D^{0.2}\phi^\alpha$, $D^{0.32}\phi^\alpha$ and $D^{0.4}\phi^\alpha$ are also nowhere differentiable as the original fractal function ϕ^α.

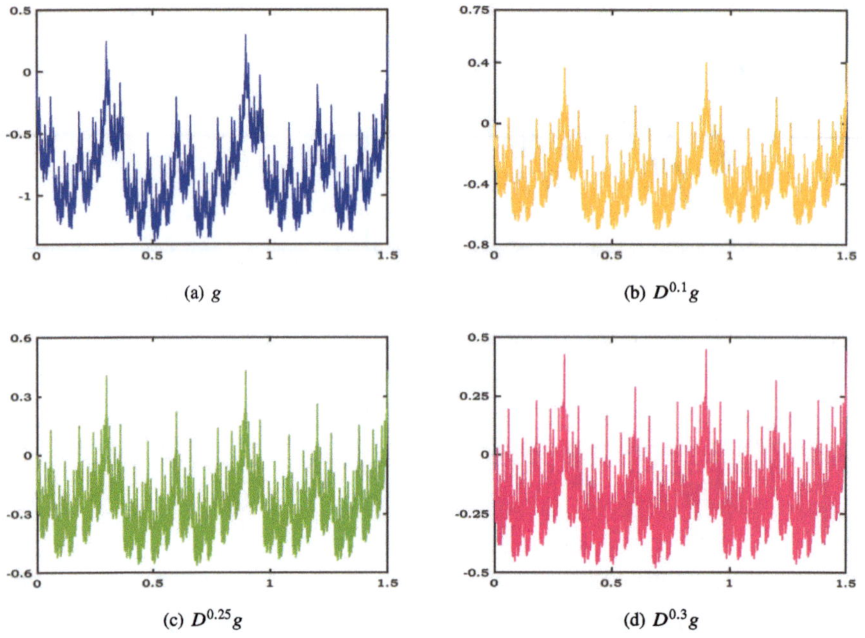

Fig. 7.2 Quadratic fractal function and its Weyl-Marchaud fractional derivative with scaling factor $\alpha_n = 0.5, \forall n = 1, 2, 3, 4, 5$: **a** g, **b** $D^{0.1}g$, **c** $D^{0.23}g$ and **d** $D^{0.3}g$

Example 7.4 Consider the generalized interpolation data $\{(0, 0, 0), (0.3, -0.5, -0.7)(0.7, 0.5, 0.3), (1, -1, -1)\}$ having hidden variables $\{0, -0.7, 0.3, -1\}$. The parameters involved in the contraction matrix are chosen as $\delta = (0.5, 0.5, 0.5)$, $\beta = (0.2, 0.2, 0.2)$, $\gamma = (0.3, 0.3, 0.3)$ such that $\|\delta_n\| < 1$ and $\|\beta_n\| + \|\gamma_n\| < 1$. Choosing the continuous functions p_n and q_n to be linear, we obtain the following functional equations

$$\begin{aligned} h_1\big(L_n(x)\big) &= \delta_n h1(x) + \beta_n h_2(x) + c_n x + d_n, \\ h_2\big(L_n(x)\big) &= \gamma_n h_2(x) + e_n x + d_n, \end{aligned} \tag{7.12}$$

where, $L_n(x) = a_n x + b_n$. Utilizing the endpoint conditions and provided scaling parameters, the co-efficients $a_n, b_n, c_n, d_n, e_n, f_n$ involved in the above equations can be estimated. Figure 7.4a, b represent the graphs of self-affine h_1 and non self-affine fractal functions h_2 respectively. The Weyl-Marchaud fractional derivative of h_1 and h_2 with fractional orders 0.25 are respectively provided in Fig. 7.4c, d. To analyse the role of fractional order, the graphs of fractional derivative are also illustrated with order 0.4 in Fig. 7.4e, f. Both the fractional orders 0.25 and 0.4 are chosen such that the matrix norm $\|A'_n\| < 1$, where $A'_n = a_n^{-\lambda} \begin{bmatrix} \delta_n & \beta_n \\ 0 & \gamma_n \end{bmatrix}$, for all $n = 1, 2, 3, 4$.

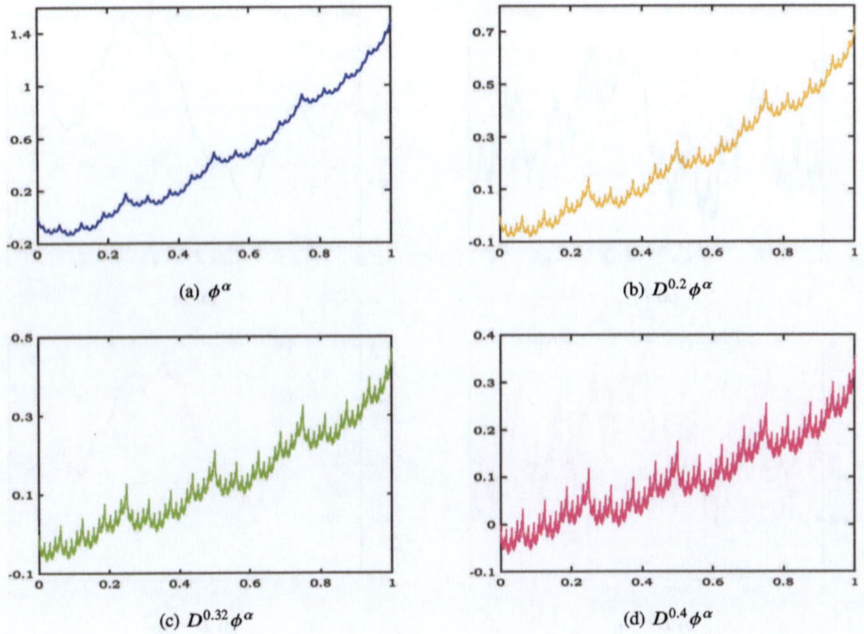

(a) ϕ^{α} (b) $D^{0.2}\phi^{\alpha}$ (c) $D^{0.32}\phi^{\alpha}$ (d) $D^{0.4}\phi^{\alpha}$

Fig. 7.3 α-fractal function and its Weyl-Marchuad fractional derivative with scaling factor $\alpha_n = 0.5, \forall n = 1, 2, 3, 4, 5$: **a** ϕ^{α}, **b** $D^{0.2}\phi^{\alpha}$, **c** $D^{0.32}\phi^{\alpha}$ and **d** $D^{0.4}\phi^{\alpha}$

7.4 Relation Between Fractional Order and Fractal Scalings

In [7], Ruan has quite modified the dimension formula of Barnsley [11] and has estimated the box-dimension of a linear fractal interpolation function based on the scaling factor α_n and continuous function $q_n(x)$. The author has demonstrated that if $\sum_{n=1}^{N-1} |\alpha_n| > 1$ and q_n is a linear polynomial with box-dimension $\dim_B(\mathcal{G}(q_n)) = 1$ for any $1 \leq n \leq N-1$, then the box-dimension of a linear FIF f is given by

$$\dim_B(\mathcal{G}(f)) = D(a_n, \alpha_n) \text{ or } 1, \tag{7.13}$$

where $D(a_n, \alpha_n)$ is the unique solution s of $\sum_{n=1}^{N-1} a_n^{s-1}|\alpha_n| = 1$. Based on this result, the box dimension of the fractional integral of linear FIF $I_{x_1}^{\nu} f$ is estimated in terms of fractional order ν as follows.

Theorem 7.6 *Suppose* $\sum_{n=1}^{N-1} |\alpha_n| > 1$ *and* $\dim_B(\mathcal{G}(f)) = D(a_n, \alpha_n)$, *then for any* $0 < \nu < D(a_n, \alpha_n)$,

$$\dim_B(\mathcal{G}(I_{x_1}^{\nu} f)) > 1,$$

meanwhile if $\dim_B(\mathcal{G}(I_{x_1}^{\nu} q_n)) = 1$ *for* $1 \leq n \leq N-1$, *then*

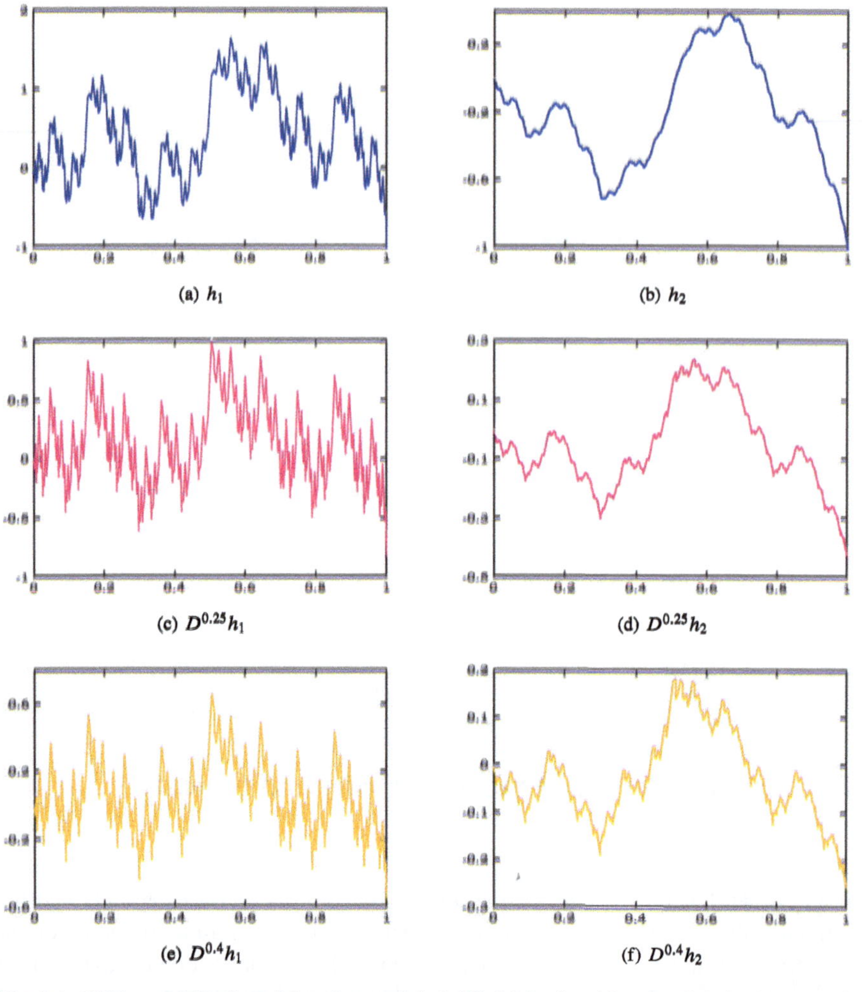

Fig. 7.4 Hidden variable fractal functions and their Weyl-Marchaud fractional derivative: **a** h_1, **b** h_2, **c** $D^{0.25}h_1$, **d** $D^{0.25}h_2$, **e** $D^{0.4}h_1$ and **f** $D^{0.4}h_2$

$$\dim_B(\mathcal{G}(I_{x_1}^{\nu} f)) = \dim_B(\mathcal{G}(f)) - \nu.$$

Theorem 7.6 clarifies that the fractal dimension of the fractional integral of linear FIF is the subtraction of fractional order ν from the fractal dimension of original linear FIF f [7]. In a deeper sense, the fractal dimension decreases for the fractional integral of FIF since the irregularity decreases than the original FIF. In Ref. [12], Peng has established a linear relation between the fractional order of Weyl-Marchaud derivative and the fractal dimension of linear fractal interpolation function using the following theorem.

Theorem 7.7 *[12] Let f be the linear fractal interpolation function. If $\sum_{n=1}^{N-1} |\alpha_n| > 1$ and $\dim_B(\mathcal{G}(f)) = D(a_n, \alpha_n)$, then*

$$\dim_B(\mathcal{G}(D_{x_1}^{\lambda} f)) = \dim_B(\mathcal{G}(f)) + \lambda \text{ holds,}$$

for any $\lambda \in (0, 2 - D(a_n, \alpha_n))$, whenever $\dim_B(\mathcal{G}(D_{x_1}^{\lambda} f)) > 1$.

Comparing Theorems 7.7 and 7.6, the fractal dimension of fractional derivative of linear FIF increases with respect to fractional order λ whereas the fractal dimension of fractional integral of linear FIF decreases. This relationship helps to understand that the fractal dimension depends on the fractional order and it quantitatively measures the irregularity of graph of fractional derivative and fractional integral of fractal interpolation functions.

Example 7.5 The fractal dimensions of the fractional integral and fractional derivative of linear FIF f interpolating the data $\{(0, 0), (0.2, 0.2), (0.4, -0.2), (0.6, 0.3), (0.8, -0.1), (1, 0.5)\}$ with scale vector $\alpha = (0.4, 0.4, 0.4, 0.4, 0.4)$ are estimated in this example. Their corresponding graphical illustrations are provided in Fig. 7.5a, b, respectively with orders $\lambda = 0.2, 0.38$ and $\nu = 0.3, 0.65$. Using Theorems 7.6 and 7.7, the box-dimension of $\mathcal{G}(f)$ is determined to be

$$\dim_B(\mathcal{G}(f)) = D(a_n, \alpha_n) = 1.5.$$

Then, using the relation $\dim_B(\mathcal{G}(I^{\nu} f)) = \dim_B(\mathcal{G}(f)) - \nu$, the fractal dimension of graphs of the functions $I^{0.3} f$ and $I^{0.65} f$ are estimated as

$$\dim_B(\mathcal{G}(I^{0.3} f)) = \dim_B(\mathcal{G}(f)) - 0.3 = 1.2,$$
$$\dim_B(\mathcal{G}(I^{0.65} f)) = \dim_B(\mathcal{G}(f)) - 0.65 = 0.85.$$

Similarly, using the relation $\dim_B(\mathcal{G}(D^{\lambda} f)) = \dim_B(\mathcal{G}(f)) + \lambda$, the fractal dimension of graphs of the functions $D^{0.2} f$ and $D^{0.38} f$ are estimated as

$$\dim_B(\mathcal{G}(D^{0.2} f)) = \dim_B(\mathcal{G}(f)) + 0.2 = 1.7,$$
$$\dim_B(\mathcal{G}(D^{0.38} f)) = \dim_B(\mathcal{G}(f)) + 0.38 = 1.88.$$

The dimension results support the graphical illustrations for observing the roughness of the functions. That is, as the fractional integral order increases, the irregularity of the graphs decrease, in contrary, when the fractional derivative order is increased, irregularity of the graphs increase successively.

In the direction of Peng [12], the fractal dimension of Weyl-Marchaud fractional derivative of quadratic FIF is estimated [10]. The upper box-dimension of Riemann-Liouville fractional integral of linear FIF is discussed by Liang, et al., in the literature [8]. In 2023, authors have investigated the box-dimension for the generalized

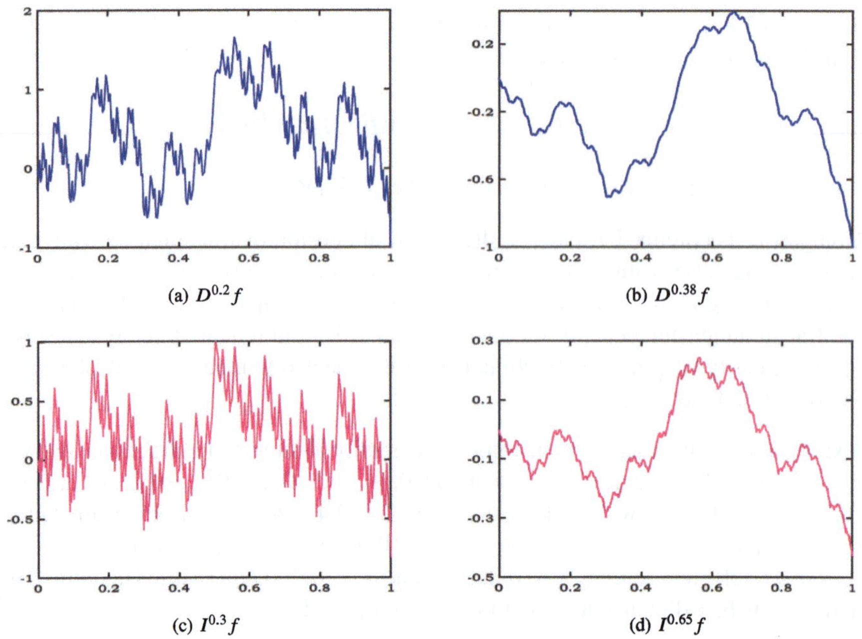

Fig. 7.5 Weyl-Marchaud fractional derivative and Riemann-Liouville fractional integral: (a) $D^{0.2}f$, (b) $D^{0.38}f$, (c) $I^{0.3}f$ and (d) $I^{0.65}f$

affine fractal interpolation function (FIF with variable scalings) using the sum function of variable scaling factors $\gamma(x) = \sum_{n=1}^{N-1} \alpha_n(x)$ in the article [13]. It is demonstrated that the dimension results are applicable to the classical fractal functions like Weierstrass function. Ruan [14] has estimated the box-dimension of recurrent FIF generated using recurrent IFS. For more related works on fractal dimension of different fractal interpolation functions, consult [15–20].

7.5 Summary

This chapter has introduced new fractal functions by differentiating types of fractal interpolation functions with Weyl-Marchaud fractional derivative. It highlights the choice of fractional order depending on the provided fractal scaling factors. Numerical examples are presented to observe how the fractional derivative of fractal functions evolve on increasing the fractional order. For all the different types of fractal functions, their fractional derivative shows more complexity maintaining self-similar property. In general, fractal interpolation functions are developed to approximate naturally existing complex functions. The higher irregularity of fractional derivative of fractal functions ensure that they can be effectively used to approximate highly irregular physical phenomena such as stock price, fluctuations in gold price and ECG signals.

References

1. F.B. Tatom, The relationship between fractional calculus and fractals. Fractals **3**(01), 217–229 (1995)
2. Y.-S. Liang, Q. Zhang, A type of fractal interpolation functions and their fractional calculus. Fractals **24**(02), 1650026 (2016)
3. X.E. Wu, J.H. Du, Box dimension of hadamard fractional integral of continuous functions of bounded and unbounded variation. Fractals **25**(03), 1750035 (2017)
4. Y.S. Liang, W.Y. Su, A geometric based connection between fractional calculus and fractal functions. Acta Math. Sin., Engl. Ser. **40**(2), 537–567 (2024)
5. K. Yao, Y.S. Liang, F. Zhang, On the connection between the order of the fractional derivative and the hausdorff dimension of a fractal function. Chaos Solitons Fractals **41**(5), 2538–2545 (2009)
6. K. Yao, W.Y. Su, S.P. Zhou, The fractional derivatives of a fractal function. Acta Math. Sin. **22**(3), 719–722 (2006)
7. H.-J. Ruan, W.-Y. Su, K. Yao, Box dimension and fractional integral of linear fractal interpolation functions. J. Approx. Theory **161**(1), 187–197 (2009)
8. Y.S. Liang, H.X. Wang, Upper box dimension of Riemann-Liouville fractional integral of fractal functions. Fractals **29**(01), 2150015 (2021)
9. A. Gowrisankar, M.G.P. Prasad, Riemann-liouville calculus on quadratic fractal interpolation function with variable scaling factors. J. Anal. **27**(2), 347–363 (2019)
10. T.M.C. Priyanka, A. Gowrisankar, Analysis on wely-marchaud fractional derivative for types of fractal interpolation function with fractal dimension. Fractals **29**(07), 2150215 (2021)
11. M.F. Barnsley, *Fractals Everywhere* (Academic Press, 2014)
12. W.L. Peng, K. Yao, X. Zhang, J. Yao, Box dimension of weyl-marchaud fractional derivative of linear fractal interpolation functions. Fractals **27**(04), 1950058 (2019)
13. L. Jiang, H.-J. Ruan, Box dimension of generalized affine fractal interpolation functions. J. Fractal Geom. **10**(3), 279–302 (2023)
14. H.-J. Ruan, J.-C. Xiao, B. Yang, Existence and box dimension of general recurrent fractal interpolation functions. Bull. Aust. Math. Soc. **103**(2), 278–290 (2021)
15. V. Agrawal, T. Som, Fractal dimension of α-fractal function on the Sierpiński gasket. Eur. Phys. J. Spec. Top. **230**(21), 3781–3787 (2021)
16. S. Jha, S. Verma, Dimensional analysis of α-fractal functions. Results Math. **76**(4), 186 (2021)
17. M. Verma, A. Priyadarshi, Dimensions of new fractal functions and associated measures. Numer. Algorithms **94**(2), 817–846 (2023)
18. A. Hossain, Md.N. Akhtar, M.A. Navascués, Fractal dimension of fractal functions on the real projective plane. Fractal Fract. **7**(7), 510 (2023)
19. S. Verma, P. Viswanathan, A revisit to α-fractal function and box dimension of its graph. Fractals **27**(06), 1950090 (2019)
20. Y.S. Liang, Approximation by the linear fractal interpolation functions with the same fractal dimension. Eur. Phys. J. Spec. Top. **232**(7), 1071–1076 (2023)

Chapter 8
Fractal Splines and Fractional Derivative

8.1 Introduction

This chapter interestingly discusses the existence of "differentiable" fractal interpolation functions, in contrary to the above discussed non-differentiable fractal functions, and their fractional calculus.[1] Barnsley has discussed the calculus of fractal interpolation functions in Ref. [1] and he has demonstrated that the FIFs can be indefinitely integrated such that the resultant functions are again attractors for a new IFS. If f is the fractal interpolation function associated with the IFS $\{(L_n(x), F_n(x, y)) : n = 1, 2, \ldots, N-1\}$ and

$$\hat{f}(x) = \hat{y}_1 + \int_{x_1}^{x} f(t)dt,$$

then $\hat{f}$ is also a FIF generated by $\{(L_n(x), \hat{F}_n(x, \hat{y})) : n = 1, 2, \ldots, N-1\}$, where $\hat{F}_n(x, \hat{y}) = \alpha_n a_n \hat{y} + \hat{q}_n(x)$ (refer Theorem 1 in [1]). Noticing the above result, there is no additional condition imposed on either the scaling factor or given data set yet the integral of FIF is shown to be again a FIF. The integration of FIF is facilitated as the fractal functions are continuous everywhere though non-differentiable. In contrast to the general non-differentiable fractal interpolation functions, Barnsley has introduced the differentiable fractal interpolation function by incorporating certain additional conditions on α_n and q_n, see [1]. The most valuable theorem assuring the existence of such a differentiable fractal interpolation function is as provided as follows.

[1] Here, we wish to quote the lines by Barnsley in Ref. [1], '*The differentiable examples cannot actually be fractals. We retain the name fractal interpolation function because of the flavor of the scaling in the definition* $f(L_n(x)) = F_n(x, f(x))$ *and because some derivative of these functions are typically fractal*'.

A. Gowrisankar et al., *Fractal Functions Communication with Fractional Calculus*,
https://doi.org/10.1007/978-3-031-96993-5_8

Theorem 8.1 *Consider the mappings* $L_n(x) = a_n x + b_n$ *and* $F_n(x, y) = \alpha_n y + q_n(x)$ *for* $n = 1, 2, \ldots, N-1$ *with* $x_n < x_{n+1}$ *satisfying the Eqs. (2.1) and (2.2), respectively. If the scaling factors obey the constraint* $|\alpha_n| < a_n^p$, *for* $p \geq 0$ *and* $q_n \in C^p([x_1, x_N])$ *with*

$$F_{n,r}(x, y) = \frac{\alpha_n y + q_n^{(r)}(x)}{a_n^r}, \; (q_n^{(0)} = q_n), \tag{8.1}$$

$$y_{1,r} = \frac{q_1^r(x_1)}{a_1^r - \alpha_1}, \; y_{N,r} = \frac{q_{n-1}^r(x_N)}{a_{n-1}^r - \alpha_{n-1}}, \; r = 1, 2 \ldots, p. \tag{8.2}$$

Suppose

$$F_{n-1,r}(x_N, y_{N,r}) = F_{n,r}(x_1, y_{1,r}), \; n = 2, 3, \ldots, N-1, \; r = 1, 2 \ldots, p, \tag{8.3}$$

then $g \in C^p[x_1, x_N]$ *is a FIF generated by the IFS* $\{(L_n(x), F_n(x, y)) : n = 1, 2, \ldots, N-1\}$ *and* $g^{(r)}$ *is the associated* r*th derivative of* g *determined by the IFS* $\{(L_n(x), F_{n,r}(x, y)) : n = 1, 2, \ldots, N-1, \; r = 1, 2 \ldots, p\}$.

The above stated Barnsley-Harrington Theorem has defined a special differentiable fractal interpolation function namely, spline fractal interpolation function generalizing the spline interpolation function. For instance, consider q_n to be a cubic polynomial and $g \in C^2$, for $p = 2$. If the vertical scaling factor α_n satisfies $|\alpha_n| < a_n^2$, then the resulting fractal function is referred as a cubic spline fractal interpolation function.

In [2], Navascués who pioneered the α-fractal function ϕ^α approximating given continuous function ϕ, has explored its rth derivative using the Barnsley-Harrington theory (see Theorem 8.1). As the base function plays a significant role in defining the α-fractal function, sufficient conditions have to be indulged in order to satisfy the hypotheses of Theorem 8.1. Therefore, take $q_n \in C^p(I)$ and $|\alpha_n| < a_n^p$, for $n = 1, 2, \ldots, N-1$. For the special choice,

$$q_n(x) = \phi \circ L_n(x) - \alpha_n b(x)$$

with $L_n(x) = a_n x + b_n$ and $L_n'(x) = a_n$, the prescribed assumption in Theorem 8.1,

$$F_{n,r}(x, y) = \frac{\alpha_n y + q_n^{(r)}(x)}{a_n^r}, \; r = 1, 2, \ldots, p, \tag{8.4}$$

implies that

$$q_n^{(r)}(x) = a_n^r \phi^{(r)} \circ L_n(x) - \alpha_n b^{(r)}(x). \tag{8.5}$$

Also, in order to satisfy the below condition

$$F_{n-1,r}(x_N, y_{N,r}) = F_{n,r}(x_1, y_{1,r}), \; n = 2, 3, \ldots, N-1, \; r = 1, 2, \ldots, p,$$

the base function $b \in C^p(I)$ is chosen in such a way that

$$b^r(x_1) = \phi^r(x_1),$$
$$b^r(x_N) = \phi^r(x_N).$$

Hence, using the theorem of Barnsley, the IFS generating the rth derivative of an α-fractal function $(\phi^\alpha)^{(r)}$ of g is determined by substituting Eq. (8.5) in Eq. (8.4) as below

$$L_n(x) = a_n x + b_n,$$
$$F_{n,r}(x, y) = \frac{\alpha_n y}{a_n^r} + \phi^{(r)} \circ L_n(x) - \frac{\alpha_n b^{(r)}(x)}{a_n^r}, \; n = 1, 2, \ldots, N-1, \; r = 1, 2, \ldots, p.$$

Here, $q_{n,r}(x) = \phi^{(r)} \circ L_n(x) - \frac{\alpha_n b^{(r)}(x)}{a_n^r}$ is the map associated with $F_{n,r}(x, y)$ so that the rth derivative of ϕ^α, $(\phi^\alpha)^{(r)}$ agrees with the rth derivative of g in relation to the scaling factor $a_n^{-r}\alpha$ and base function $b^{(r)}$. That is,

$$(\phi_b^\alpha)^{(r)} = (\phi^{(r)})_{b^{(r)}}^{a_n^{-r}\alpha}, \; r = 1, 2, \ldots, p.$$

The fractal function $(\phi_b^\alpha)^{(r)}$ interpolates $\phi^{(r)}$ at all the points, for each $r = 1, 2, \ldots, p$. For the detailed study on α-fractal splines and fractal cubic splines, the readers are suggested to visit [2, 3].

Remark 8.1 The necessary conditions for the existence of α-fractal spline has been concretely discussed in [2], yet its functional equation is not provided explicitly. Later in [4], taking the uniform partition $a_n = \frac{1}{N}$ and constant scaling factor $\alpha_n = \alpha$, an IFS defining the rth derivative of α-fractal function is provided as

$$L_n(x) = \frac{1}{N}x + b_k,$$
$$F_{n,r}(x, y) = N^r \alpha y + N^r q_n^{(r)}(x), \; r = 1, 2, \ldots, p.$$

From the above described IFS theory, one can get the functional equation of first derivative of α-fractal function as

$$(\phi^\alpha)'(L_n(x)) = \alpha N((\phi^\alpha)'(x) - b(x)) + \phi'(L_n(x)), \; \forall\, x \in I.$$

In a similar fashion, for any $r \in \mathbb{N}_p$, the respective rth derivative of α-fractal function can be obtained.

Example 8.1 Let $\{(0, 0), (0.25, -0.125), (0.5, 0.5), (0.75, 1.875), (1, 4)\}$ be the data set obtained using the first derivative of the continuous function $\phi(x) = 2x^3 - x^2$ defined on the closed interval $[0, 1]$. The graph of the prescribed function g is illustrated in Fig. 8.1a. The α-fractal function ϕ^α is generated in Fig. 8.1b with respect

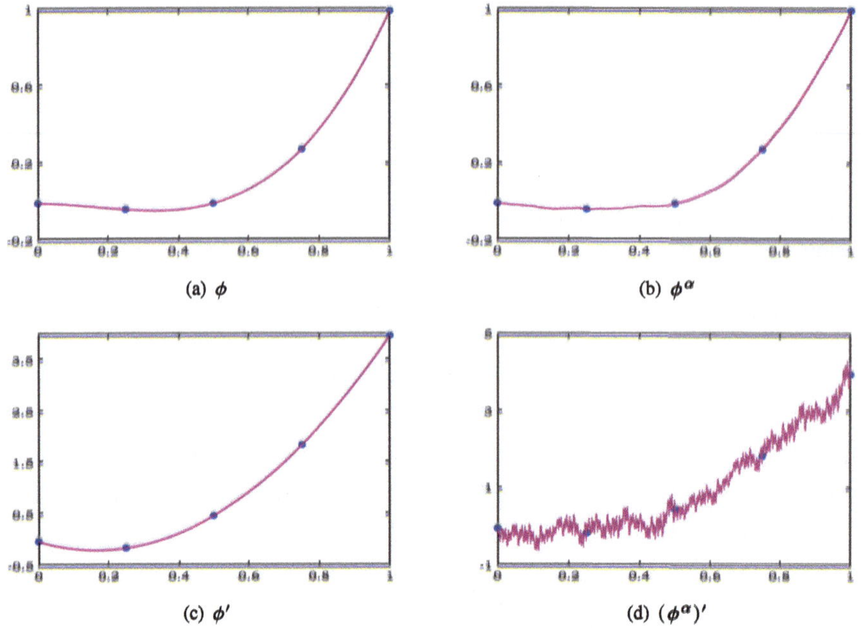

(a) ϕ (b) ϕ^{α} (c) ϕ' (d) $(\phi^{\alpha})'$

Fig. 8.1 Graphical illustrations of **a** Classical interpolant g, **b** α-fractal interpolant, **c** Classical derivative ϕ' and **d** C^1-α-fractal approximant of $\phi'(x) = 6x^2 - 2x$ with base function $b'(x) = 4x^3$

to the base function $b(x) = x^4$ and scale vector $\alpha = \{0.2, -0.23, 0.18, -0.24\}$. The first derivatives of g and ϕ^{α} are also generated in Fig. 8.1c, d, respectively. The corresponding base function $b'(x) = 4x^3$ is verified to satisfy the necessary constraints $b'(0) = \phi'(0)$ and $b'(1) = \phi'(1)$.

8.2 Caputo-Fractional Derivative for $C^1 - \alpha$-Fractal Function

In 2009, Ruan explored the fractional integral of a linear fractal interpolation function and also estimated the box dimension of its graph [5]. In [6], Pan has investigated the Riemann-Liouville fractional integral of an affine FIF defined on the interval $[0, b] (b > 0)$ corresponding to the IFS consisting $L_n(x) = \frac{1}{N}t + \frac{N-1}{N}b,\ F_n(x, y) = c_n x + d_n y + f_n$. In [7], the authors have discussed the Weyl-Marchaud fractional derivative of a FIF assuming its $q_n(x)$ to be a linear function. However, the generation of new data points from the newly obtained IFS is not discussed in any of the aforementioned works. Keeping in mind, the adequate formulae for finding the new data points have been explicitly provided by Gowrisankar in [8] while discussing the Riemann-Liouville fractional integral of a quadratic FIF by choosing

$q_n(x) = x^2 + c_n x + d_n$. In this direction, the fractional calculus methods of various FIFs have been investigated providing the explicit expressions for new data points. Among the different fractional calculus methods like Riemann-Liouville, Weyl-Marchaud and so forth, the Caputo-fractional derivative of a continuous function remains unique as it can be estimated only if the function is atleast one-time differentiable. While solving many real world problems involving fractional differential equations having defined the arbitrary function at the origin to be constant, the Riemann-Liouville derivative loses its advantage since its derivative is not zero for a constant function. This issue is overcome by the Caputo-fractional derivative demanding higher conditions for differentiability.

In light of the aforementioned property for the necessity of first order derivative, here we discuss the Caputo-fractional derivative of a C^1-α-fractal function and interprets its resultant function. In addition, a fractal operator is proposed to yield the Caputo-fractional derivative of a C^1-α-fractal function for any given differentiable continuous function. With an aim to discuss the operator's bound, the weighted space of continuous functions is employed rather than the space of continuous functions.

Definition 8.1 The Caputo-fractional derivative of a C^1-α-FIF defined at the initial point $x = x_1$ is expressed by

$$({}^C D_{x_1}^{\upsilon(x)} \phi^\alpha)(x) = \frac{1}{\Gamma(1-\upsilon(x))} \int_{x_1}^{x} (x-u)^{-\upsilon(x)} (\phi^\alpha)'(u) du,$$

with $({}^C D_{x_1}^{\upsilon(x)} \phi^\alpha)(x_1) = \tilde{y}_1 = 0$ and $({}^C D_{x_1}^{\upsilon(x)} \phi^\alpha)(x_N) = \tilde{y}_N$. Here $\upsilon : [x_1, x_N] \rightarrow (0, 1)$ is the variable fractional order.

The following theorem interprets on the resultant function of the Caputo-fractional derivative of a α-fractal function pre-defined at the initial point $x = x_1$.

Theorem 8.2 *Let ϕ^α be the α-fractal function corresponding to $\phi \in C^1(I)$ and $(\phi^\alpha)'$ be its first order derivative associated with the IFSs $\{(L_n(x), F_n(x, y)) : n = 1, 2, \dots, N-1\}$ and $\{(L_n(x), F_{n,1}(x, y)) : n = 1, 2, \dots, N-1\}$, respectively, where*

$$F_n(x, y) = \alpha_n y + \phi \circ L_n(x) - \alpha_n b(x)$$

and

$$F_{n,1}(x, y) = a_n^{-1} \alpha_n y + \phi' \circ L_n(x) - a_n^{-1} \alpha_n b'(x)$$

for all $n = 1, 2, \dots, N-1$. Then, under the constraints

$$({}^C D_{x_1}^{\upsilon(x)} \phi)(x_1) = ({}^C D_{x_1}^{\upsilon(x)} b)(x_1), \ ({}^C D_{x_1}^{\upsilon(x)} \phi)(x_N) = ({}^C D_{x_1}^{\upsilon(x)} b)(x_N),$$

the Caputo-fractional derivative $({}^C D_{x_1}^{\upsilon(x)} \phi^\alpha)$ of ϕ^α is also an α-fractal function corresponds to a new IFS $\{(L_n(x), \tilde{F}_n(x, \tilde{y})) : n = 1, 2, \dots, N-1\}$, where $\tilde{F}_n(x, \tilde{y}) = a_n^{-\upsilon(x)} \alpha_n \tilde{y} + \tilde{q}_n(x)$,

$$\tilde{q}_n(x) = \tilde{y}_n + \phi^{\alpha}_{n,v}(x) + a_n^{-v(x)(x)}({}^C D^{v(x)}_{x_1}\phi \circ L_n)(x) - a_n^{-v(x)}\alpha_n({}^C D^{v(x)}_{x_1} b)(x),$$

$$\tilde{y}_n = \frac{1}{\Gamma(1-v(x))}\int_{x_1}^{x_n}(x_n-u)^{-v(x)}(\phi^{\alpha})'(u)du,$$

$$\phi^{\alpha}_{n,v}(x) = \frac{1}{\Gamma(1-v(x))}\int_{x_1}^{x_n}\{(L_n(x)-u)^{-v(x)} - (x_n-u)^{-v(x)}\}(\phi^{\alpha})'(u)du,$$

$$\tilde{y}_{n+1} = \sum_{j=1}^{n}\{\phi^{\alpha}_{j,v}(x_N) + a_j^{-v(x)}\alpha_j({}^C D^{v(x)}_{x_1}\phi^{\alpha})(x_N) + a_j^{-v(x)}({}^C D^{v(x)}_{x_1}\phi \circ L_j)(x_N) - a_j^{-v(x)}\alpha_j({}^C D^{v(x)}_{x_1} b)(x_N)\},$$

$$\tilde{y}_N = \frac{\sum_{j=1}^{N-1}\{\phi^{\alpha}_{j,v}(x_N) + a_j^{-v(x)}({}^C D^{v(x)}_{x_1}\phi \circ L_j)(x_N) - a_j^{-v(x)}\alpha_j({}^C D^{v(x)}_{x_1} b)(x_N)\}}{1-\sum_{j=1}^{N-1} a_j^{-v(x)}\alpha_j}.$$

Proof Consider

$$\begin{aligned}({}^C D^{v(x)}_{x_1}\phi^{\alpha})(L_n(x)) &= \frac{1}{\Gamma(1-v(x))}\int_{x_1}^{L_n(x)}(L_n(x)-u)^{-v(x)}(\phi^{\alpha})'(u)du\\ &= \frac{1}{\Gamma(1-v(x))}\int_{x_1}^{x_n}(x_n-u)^{-v(x)}(\phi^{\alpha})'(u)du + \frac{1}{\Gamma(1-v(x))}\int_{x_n}^{L_n(x)}(L_n(x)-u)^{-v(x)}(\phi^{\alpha})'(u)du\\ &+ \frac{1}{\Gamma(1-v(x))}\int_{x_1}^{x_n}\{(L_n(x)-u)^{-v(x)} - (x_n-u)^{-v(x)}\}(\phi^{\alpha})'(u)du.\end{aligned}$$

Taking,

$$\tilde{y}_n = \frac{1}{\Gamma(1-v(x))}\int_{x_1}^{x_n}(x_n-u)^{-v(x)}(\phi^{\alpha})'(u)du,$$

$$\phi^{\alpha}_{n,v}(x) = \frac{1}{\Gamma(1-v(x))}\int_{x_1}^{x_n}\{(L_n(x)-u)^{-v(x)} - (x_n-u)^{-v(x)}\}(\phi^{\alpha})'(u)du,$$

and performing change of variable $u = L_n(y)$ in the second integral provides

$$\begin{aligned}({}^C D^{v(x)}_{x_1}\phi^{\alpha})(L_n(x)) &= \tilde{y}_n + \phi^{\alpha}_{n,v}(x) + \frac{a_n}{\Gamma(1-v(x))}\int_{x_1}^{x}(L_n(x)-L_n(y))^{-v(x)}(\phi^{\alpha})'(L_n(y))dy\\ &= \tilde{y}_n + \phi^{\alpha}_{n,v}(x)\\ &+ \frac{a_n^{1-v}}{\Gamma(1-v(x))}\int_{x_1}^{x}((x-y)^{-v(x)}\{a_n^{-1}\alpha_n(\phi^{\alpha})'(y) + \phi'(L_n(y)) - a_n^{-1}\alpha_n b'(y)\}dy\\ &= \tilde{y}_n + \phi^{\alpha}_{n,v}(x) + \frac{a_n^{-v(x)}\alpha_n}{\Gamma(1-v(x))}\int_{x_1}^{x}(x-y)^{-v(x)}(\phi^{\alpha})'(y)dy + \frac{a_n^{1-v}}{\Gamma(1-v(x))}\\ &\times \int_{x_1}^{x}(x-y)^{-v(x)}\phi' \circ L_n(y)dy - \frac{a_n^{-v(x)}\alpha_n}{\Gamma(1-v(x))}\int_{x_1}^{x}(x-y)^{-v(x)}b'(y)dy\\ &= \tilde{y}_n + \phi^{\alpha}_{n,v}(x) + \frac{a_n^{-v(x)}\alpha_n}{\Gamma(1-v(x))}\int_{x_1}^{x}(x-y)^{-v(x)}(\phi^{\alpha})'(y)dy + \frac{a_n^{1-v}}{\Gamma(1-v(x))}\\ &\times \int_{x_1}^{x}(x-y)^{-v(x)}\frac{(\phi \circ L_n)'(y)}{L_n'(y)}dy - \frac{a_n^{-v(x)}\alpha_n}{\Gamma(1-v(x))}\int_{x_1}^{x}(x-y)^{-v(x)}b'(y)dy\\ &= \tilde{y}_n + \phi^{\alpha}_{n,v}(x) + a_n^{-v(x)}\alpha_n({}^C D^{v(x)}_{x_1}\phi^{\alpha})(x) + a_n^{1-v}({}^C D^{v(x)}_{x_1}\phi)(L_n(x))\\ &- a_n^{-v(x)}\alpha_n {}^C D^{v(x)}_{x_1} b(x)\\ &= \tilde{y}_n + \phi^{\alpha}_{n,v}(x) + a_n^{-v(x)}\alpha_n({}^C D^{v(x)}_{x_1}\phi^{\alpha})(x) + a_n^{-v(x)}({}^C D^{v(x)}_{x_1}\phi \circ L_n)(x)\\ &- a_n^{-v(x)}\alpha_n({}^C D^{v(x)}_{x_1} b)(x)\end{aligned}$$

$$= a_n^{-v(x)} \alpha_n ({}^C D_{x_1}^{v(x)} \phi^\alpha)(x) + \tilde{q}_n(x)$$
$$= \tilde{F}_n(t, ({}^C D_{x_1}^{v(x)} \phi^\alpha)(x)).$$

Here $\tilde{q}_n(x) = \tilde{y}_n + \phi_{n,v}^\alpha(x) + a_n^{-v(x)}({}^C D_{x_1}^{v(x)} \phi \circ L_n)(x) - a_n^{-v(x)} \alpha_n ({}^C D_{x_1}^{v(x)} b)(x)$. The new data points $\{(x_n, \tilde{y}_n) : n = 1, 2, \ldots, N\}$ are obtained by taking $x = x_N$ and $L_n(x_n) = x_{N+1}$ in the below equation

$$({}^C D_{x_1}^{v(x)} \phi)(L_n(x)) = \tilde{y}_n + \phi_{n,v}^\alpha(x) + a_n^{-v(x)} \alpha_n ({}^C D_{x_1}^{v(x)} \phi^\alpha)(x) + a_n^{-v(x)} ({}^C D_{x_1}^{v(x)} \phi \circ L_n)(x)$$
$$- a_n^{-v(x)} \alpha_n ({}^C D_{x_1}^{v(x)} b)(x).$$

Hence,

$$\tilde{y}_{n+1} - \tilde{y}_n = \phi_{n,v}^\alpha(x_N) + a_n^{-v(x)} \alpha_n ({}^C D_{x_1}^{v(x)} \phi^\alpha)(x_N) + a_n^{-v(x)} ({}^C D_{x_1}^{v(x)} \phi \circ L_n)(x_N) - a_n^{-v(x)} \alpha_n ({}^C D_{x_1}^{v(x)} b)(x_N).$$

The equation $\tilde{y}_{n+1} = \tilde{y}_1 + \sum_{j=1}^{n} (\tilde{y}_{j+1} - \tilde{y}_j)$ determines the new interpolants $\tilde{y}_n$ for all $n = 2, 3, \ldots, N-1$, thus

$$\tilde{y}_{n+1} = \sum_{j=1}^{n} \{\phi_{j,v}^\alpha(x_N) + a_j^{-v(x)} \alpha_j ({}^C D_{x_1}^{v(x)} \phi^\alpha)(x_N) + a_j^{-v(x)} ({}^C D_{x_1}^{v(x)} \phi \circ L_j)(x_N) - a_j^{-v(x)} \alpha_j ({}^C D_{x_1}^{v(x)} b)(x_N)\}.$$

The new interpolant at $x = x_N$ is obtained by substituting $n = N - 1$ in the previous equation,

$$\tilde{y}_N = \frac{\sum_{j=1}^{N-1} \{\phi_{j,v}^\alpha(x_n) + a_j^{-v(x)} ({}^C D_{x_1}^{v(x)} \phi \circ L_j)(x_n) - a_j^{-v(x)} \alpha_j ({}^C D_{x_1}^{v(x)} b)(x_n)\}}{1 - \sum_{j=1}^{N-1} a_j^{-v(x)} \alpha_j}.$$

The endpoint conditions of $\tilde{F}_n$ are verified below

$$\tilde{F}_n(x_1, ({}^C D_{x_1}^{v(x)} \phi^\alpha)(x_1)) = \tilde{y}_n,$$
$$\tilde{F}_n(x_N, ({}^C D_{x_1}^{v(x)} \phi^\alpha)(x_N)) = a_n^{-v(x)} \alpha_n ({}^C D_{x_1}^{v(x)} \phi^\alpha)(x_N) + \tilde{q}_n(x_N)$$
$$= ({}^C D_{x_1}^{v(x)} \phi^\alpha)(L_n(x_N))$$
$$= \tilde{y}_{n+1}.$$

Therefore, the newly obtained Caputo-fractional derivative of a C^1-α-fractal function, $({}^C D_{x_1}^{v(x)} \phi^\alpha)$ is again an α-fractal function corresponding to the new data set $\{(x_n, \tilde{y}_n) : n = 1, 2, \ldots, N\}$. □

Remark 8.2 Suppose the value of the Caputo-fractional derivative of $g \in C^1(I)$ is known at the end point x_n instead of $x = x_1$, then an analogue of the above theorem can be obtained by defining

$$({}^C D_{x_N}^{v(x)} \phi^\alpha)(x) = \frac{-1}{\Gamma(1 - v(x))} \int_x^{x_N} (u - x)^{-v(x)} (\phi^\alpha)'(u) du,$$

with $({}^C D^{v(x)}_{x_N} \phi^\alpha)(x_1) = \tilde{y}_1$ and $({}^C D^{v(x)}_{x_N} \phi^\alpha)(x_N) = \tilde{y}_N = 0$. The function $({}^C D^{v(x)}_{x_N} \phi^\alpha)$ corresponds to the IFS $\{(L_n(x), \tilde{F}_n(x, \tilde{y})) : n = 1, 2, \ldots, N-1\}$, where $\tilde{F}_n(x, \tilde{y}) = a_n^{-v(x)} \alpha_n \tilde{y} + \tilde{q}_n(x)$,

$$\tilde{q}_n(x) = \tilde{y}_{n+1} - \phi^\alpha_{n,v}(x) + a_n^{-v(x)} ({}^C D^{v(x)}_{x_n} \phi \circ L_n)(x) - a_n^{-v(x)} \alpha_n ({}^C D^{v(x)}_{x_n} b)(x),$$

$$\tilde{y}_{n+1} = \frac{-1}{\Gamma(1 - v(x))} \int_{x_{n+1}}^{x_N} (u - x_{n+1})^{-v(x)} (\phi^\alpha)'(u) du,$$

$$\phi^\alpha_{n,v}(x) = \frac{1}{\Gamma(1 - v(x))} \int_{x_{n+1}}^{x_N} \{(u - L_n(x))^{-v(x)} - (u - x_{n+1})^{-v(x)}\} (\phi^\alpha)'(u) du,$$

$$\tilde{y}_n = -\sum_{j=n}^{N-1} \Big\{ \phi^\alpha_{j,v}(x_1) - a_j^{-v(x)} \alpha_j ({}^C D^{v(x)}_{x_n} \phi^\alpha)(x_1) - a_j^{-v(x)} ({}^C D^{v(x)}_{x_n} \phi \circ L_j)(x_1)$$
$$+ a_j^{-v(x)} \alpha_j ({}^C D^{v(x)}_{x_n} b)(x_1) \Big\},$$

$$\tilde{y}_1 = -\frac{\sum_{j=1}^{N-1} \{\phi^\alpha_{j,v}(x_1) - a_j^{-v(x)} ({}^C D^{v(x)}_{x_n} \phi \circ L_j)(x_1) + a_j^{-v(x)} \alpha_j ({}^C D^{v(x)}_{x_n} b)(x_1)\}}{1 - \sum_{j=1}^{N-1} a_j^{-v(x)} \alpha_j}.$$

From this instant, the notation ${}^C D^{v(x)}_{x_1}$ is simply used as ${}^C D^{v(x)}$ for the convenience. Whenever an α-fractal function is investigated for any continuous function, its associated fractal operator is explored to obtain the fractal version for the invoked continuous function. In this direction, the following subsection presents a fractal operator associated with the Caputo-fractional derivative of an α-fractal function.

8.2.1 Fractal Operator on the Weighted Space

Fractal interpolation operator, $\mathcal{F}$ is proposed in [9] with an aim to fractalize any given continuous function using the family of non-smooth fractal interpolants. The fractal operators discussed so far in the literature are defined on the spaces $C(I)$, $C^n(I)$, $\mathcal{L}ip(I)$ (see, for instance [9–11]). The bounds for the types of such fractal operators are also estimated using the uniform norm associated with the embedded space. In this subsection, a new kind of fractal operator is proposed to associate the given C^1- function to the Caputo-fractional derivative of its fractal version. Hence, in order to estimate the bound of the proposed operator we consider the weighted space, since the Caputo-fractional derivative is bounded in the weighted space rather than the continuous space of functions, for additional information on the weighted space of continuous functions and bound of ${}^C D^v$, the readers are referred to see [12].

To discuss the bound of the fractal operator associated with Caputo-fractional derivative, the variable fractional order $v(x)$ is particularly taken as a constant fractional order v. The space of continuous functions defined on I reserves the notation $C(I)$ with the norm $\|\phi\|_C = \|\phi\|_\infty = \max_{x \in I}\{|\phi(x)|\}$. In order to define the bound for $({}^C D^v \phi)$, we consider the weighted space, $C_\gamma(I)$, where $\gamma \in \mathbb{R}$, which consists of functions given on $(x_1, x_N]$ such that $(x - x_1)^\gamma g(x) \in C(I)$ and

$$\|\phi\|_{C_\gamma} = \|(x - x_1)^\gamma g(x)\|_C.$$

Now the fractal operator which assigns the Caputo-fractional derivative of α-fractal function for each prescribed C^1-function is given by

$$\tilde{\mathcal{F}}^v : C^1(I) \to C_{x_1}(I)$$
$$g \longmapsto {}^C D^v \phi^\alpha$$

and $\tilde{\mathcal{F}}^v(\phi)$ is defined by

$$\tilde{\mathcal{F}}^v(\phi(x)) = \tilde{y}_n + \phi^\alpha_{n,v}(L_n^{-1}(x)) + a_n^{-v}({}^C D^v \phi)x + a_n^{-v}\alpha_n(({}^C D^v \phi^\alpha) - ({}^C D^v(\phi \circ c))) \circ L_n^{-1}(x). \quad (8.6)$$

The analytical properties such as linearity and boundedness, and one of the algebraic properties known as semi-group property of the operator $\tilde{\mathcal{F}}^v(\phi)$ are investigated below.

Theorem 8.3 *$\tilde{\mathcal{F}}^v$ is a linear operator and its bound is given by*

$$\|\tilde{\mathcal{F}}^v(\phi)\| \leq \left(\frac{2K_v}{1 - |\alpha|_\infty}\right)\|\phi\|_\infty,$$

where $K_v = \frac{(x_N - x_1)^{1-v}}{|\Gamma(1-v)|(2-v)}$.

Proof For the functions $\phi, \psi \in C^1(I)$, let $r, s \in \mathbb{R}$.

$$\tilde{\mathcal{F}}^v(r\phi) = r\tilde{y}_n + r\phi^\alpha_{n,v}(L_n^{-1}(x)) + a_n^{-v}r({}^C D^v \phi)t + a_n^{-v}\alpha_n(r({}^C D^v \phi^\alpha) - r({}^C D^v(\phi \circ c))) \circ L_n^{-1}(x),$$
$$\tilde{\mathcal{F}}^v(s\psi) = s\tilde{y}_n + s\psi^\alpha_{n,v}(L_n^{-1}(x)) + a_n^{-v}s({}^C D^v \psi)t + a_n^{-v}\alpha_n(s({}^C D^v \psi^\alpha) - s({}^C D^v(\psi \circ c))) \circ L_n^{-1}(x).$$

It can be trivially verified that

$$\tilde{\mathcal{F}}^v(r\phi + s\psi) = r\tilde{\mathcal{F}}^v(\phi) + s\tilde{\mathcal{F}}^v(\psi).$$

Hence, $\tilde{\mathcal{F}}^v$ is a linear operator.

$$\begin{aligned}\|\tilde{\mathcal{F}}^v(\phi)\|_{C_{x_1}} &= |\tilde{y}_n + \phi^\alpha_{n,v}(L_n^{-1}(x)) + a_n^{-v}({}^C D^v \phi)t + a_n^{-v}\alpha_n(({}^C D^v \phi^\alpha) - ({}^C D^v(\phi \circ c))) \circ L_n^{-1}(x)| \\ &= \|\tilde{y}_n + \phi^\alpha_{n,v}(L_n^{-1})\|_{C_{x_1}} + |a^{-v}|_\infty \|{}^C D^v \phi\|_{C_{x_1}} + |a^{-v}|_\infty |\alpha|_\infty \\ &\quad \times \|(({}^C D^v \phi^\alpha) - ({}^C D^v(\phi \circ c))) \circ L_n^{-1}\|_{C_{x_1}},\end{aligned}$$

where $|a^{-v}|_\infty = \max_{1 \leq n \leq N-1}\{a_n^{-v}\}$ and $|\alpha|_\infty = \max_{1 \leq n \leq N-1}\{\alpha_n\}$. Since the Caputo-fractional derivative is bounded from the space C^1 to the space C_{x_1} by $K_v\|\phi\|_\infty$ with $K_v = \frac{(x_N - x_1)^{1-v}}{|\Gamma(1-v)|(2-v)}$, [12]. Also note that

$$\|\tilde{y}_n + \phi^\alpha_{n,v}(L_n^{-1})\|_{C_{x_1}} \leq \frac{(x_{n+1} - x_n)^{1-v}}{|\Gamma(1-v)|(2-v)}\|\phi\|_\infty \leq K_v\|\phi\|_\infty.$$

It follows that

$$\|\tilde{\mathcal{F}}^{v}(\phi)\|_{C_{x_1}} \leq K_v\|\phi\|_\infty + |a^{-v}|_\infty K_v\|\phi\|_\infty + |a^{-v}|_\infty|\alpha|_\infty K_v\|(\phi^\alpha - \phi \circ c)\|_\infty$$

It is known that $|a^{-v}| < 1$, thus

$$\begin{aligned}\|\tilde{\mathcal{F}}^{v}(\phi)\|_{C_{x_1}} &\leq 2K_v\|\phi\|_\infty + |\alpha|_\infty K_v\|(\phi^\alpha - \phi \circ c)\|_\infty \\ &\leq 2K_v\|\phi\|_\infty + |\alpha|_\infty K_v\|\phi^\alpha\|_\infty + |\alpha|_\infty K_v\|\phi \circ c\|_\infty.\end{aligned}$$

The fractal interpolation operator, $\mathcal{F}$ discussed in [9] provides the bound of $\|\phi^\alpha\|_\infty$ to be $\frac{1+|\alpha|_\infty}{1-|\alpha|_\infty}\|\phi\|_\infty$. Then,

$$\|\tilde{\mathcal{F}}^{v}(\phi)\|_{C_{x_1}} \leq 2K_v\|\phi\|_\infty + |\alpha_n|_\infty K_v\frac{1+|\alpha|_\infty}{1-|\alpha|_\infty}\|\phi\|_\infty + |\alpha|_\infty K_v\|\phi \circ c\|_\infty$$

Since $\|\phi \circ c\|_\infty = \|\phi\|_\infty$,

$$\|\tilde{\mathcal{F}}^{v}(\phi)\|_{C_{x_1}} \leq \left(\frac{2K_v}{1-|\alpha|_\infty}\right)\|\phi\|_\infty.$$

Therefore, the operator $\tilde{\mathcal{F}}^{v}$ is bounded by $\left(\frac{2K_v}{1-|\alpha|_\infty}\right)\|\phi\|_\infty$, where $K_v = \frac{(x_N-x_1)^{1-v}}{|\Gamma(1-v)|(2-v)}$. □

The effect of composition operation on the Caputo-fractional derivative is investigated as follows. The composite of Caputo-fractional derivatives of $g \in C^1(I)$ with respect to the fractional orders $v_1, v_2 > 0$ is given by

$${}^{C}D^{v_1}({}^{C}D^{v_2}\phi) = {}^{C}D^{v_1+v_2}\phi.$$

Next theorem explores the semi-group property of the operator $\tilde{\mathcal{F}}^{v}$ through the composition operation by considering the set of all fractal operators $\{\tilde{\mathcal{F}}^{v} : C^1(I) \to C_{x_1}(I) |\ v > 0\}$.

Theorem 8.4 *The composite of the set of all fractal operators* $\{\tilde{\mathcal{F}}^{v} : C^1(I) \to C_{x_1}(I) : v > 0\}$ *forms a semi-group. For* $v_1, v_2 > 0$,

$$\mathcal{F}^{v_1}\mathcal{F}^{v_2}(\phi) = \mathcal{F}^{v_1+v_2}(\phi).$$

Proof For a fixed function $\phi \in C^1(I)$ and $v_1, v_2 > 0$.

$$\begin{aligned}\mathcal{F}^{v_1}\mathcal{F}^{v_2}(\phi) &= \tilde{y}_{n,v_1+v_2} + \phi^{\alpha}_{n,v_1+v_2}(L_n^{-1}(x)) + \frac{a_n^{-(v_1+v_2)}}{\Gamma(1-v_1)\Gamma(1-v_2)}\int_{x_1}^{x}(x-y)^{-v_1}\int_{x_1}^{y}(y-t)^{-v_2}\phi'(t)dtdy \\ &+ \frac{a_n^{-(v_1+v_2)}\alpha_n}{\Gamma(1-v_1)\Gamma(1-v_2)}\int_{x_1}^{x}(x-y)^{-v_1}\int_{x_1}^{y}(y-t)^{-v_2}((\phi^\alpha)' - b')\circ L_n^{-1}(t)dtdy \\ &= \tilde{y}_{n,v_1+v_2} + \phi^{\alpha}_{n,v_1+v_2}(L_n^{-1}(x)) + \frac{a_n^{-(v_1+v_2)}}{\Gamma(1-v_1)\Gamma(1-v_2)}\int_{x_1}^{x}\phi'(t)\int_{t}^{x}(x-y)^{-v_1}(x-y)^{-v_2}dydt\end{aligned}$$

$$+\frac{a_n^{-(v_1+v_2)}\alpha_n}{\Gamma(1-v_1)\Gamma(1-v_2)}\int_{x_1}^{x}((\phi^\alpha)'-b')\circ L_n^{-1}(t)\int_t^x (x-y)^{-v_1}(y-t)^{-v_2}\,dydt.$$

On substituting $y=t+v(x-t)$ and simplifying the integrals in the above equation, it is seen that

$$\begin{aligned}\mathcal{F}^{v_1}\mathcal{F}^{v_2}(\phi) &= \tilde{y}_{n,v_1+v_2}+\phi^\alpha_{n,v_1+v_2}(L_n^{-1}(x))+\frac{a_n^{-(v_1+v_2)}}{\Gamma(1-v_1)\Gamma(1-v_2)}\int_{x_1}^{x}\phi'(t)(x-t)^{1-(v_1+v_2)}\\ &\times\int_0^1(1-v)^{-v_1}v^{-v_2}\,dvdt+\frac{a_n^{-(v_1+v_2)}\alpha_n}{\Gamma(1-v_1)\Gamma(1-v_2)}\int_{x_1}^{x}((\phi^\alpha)'-b')\circ L_n^{-1}(t)(x-t)^{1-(v_1+v_2)}\\ &\times\int_0^1(1-v)^{-v_1}v^{-v_2}\,dvdt.\end{aligned}$$

Since $\frac{\Gamma(1-v_1)\Gamma(1-v_2)}{\Gamma(1-v_1+1-v_2)}=\int_0^1(1-v)^{-v_1}v^{-v_2}dv$, thus

$$\begin{aligned}\mathcal{F}^{v_1}\mathcal{F}^{v_2}(\phi) &= \tilde{y}_{n,v_1+v_2}+\phi^\alpha_{n,v_1+v_2}(L_n^{-1}(x))+\frac{a_n^{-(v_1+v_2)}}{\Gamma(2-(v_1+v_2))}\int_{x_1}^{x}\phi'(y)(x-t)^{1-(v_1+v_2)}dt\\ &+\frac{a_n^{-(v_1+v_2)}\alpha_n}{\Gamma(2-(v_1+v_2))}\int_{x_1}^{x}((\phi^\alpha)'-b')\circ L_n^{-1}(t)(x-t)^{1-(v_1+v_2)}dt\\ &= \tilde{y}_{n,v_1+v_2}+\phi^\alpha_{n,v_1+v_2}(L_n^{-1}(x))+a_n^{-(v_1+v_2)}{}^{C}D_{x_1}^{v_1+v_2}\phi(x)-a_n^{-(v_1+v_2)}\alpha_n\\ &\times{}^{C}D_{x_1}^{v_1+v_2}(\phi^\alpha-b)\circ(L_n^{-1}(x))\\ &= \mathcal{F}^{v_1+v_2}(\phi).\end{aligned}$$

Hence, the set of all fractal operators forms a semi-group under the composition operation. □

8.3 Caputo-Fractional Derivative of Hermite Fractal Spline

Let $\{(x_n,y_n,d_n): n=1,2,\ldots,N\}$ be the given Hermite data set with partition $\Delta: a=x_1<x_2<\cdots<x_N=b$ spacing on $[x_1,x_N]$, where y_n and d_n are the corresponding function values and derivative values at the knots x_n. Based on the Theorem 8.1, for each $n=1,2,\ldots,N-1$, the scaling parameter α_n is chosen such that $|\alpha_n|<la_n$ with $0<l<1$, where $a_n=\frac{x_{n+1}-x_n}{x_N-x_1}$. Let $\mathcal{G}=\{g\in C^1(I,\mathbb{R})|\ g(x_1)=y_1, g(x_N)=y_N, g^{(1)}(x_1)=d_1, g^{(1)}(x_N)=d_N\}$ be the complete metric space with respect to the metric ρ induced by the C^1 norm, defined by $\|g\|_{C^1}=\max\{\|g\|_\infty,\|g^{(1)}\|_\infty\}$. The Read-Bajraktarević operator $\mathcal{T}$ is defined as a self-map on $\mathcal{G}$ by

$$(\mathcal{T}g)(x)=\alpha_n g(L_n^{-1}(x))+q_n(L_n^{-1}(x)),$$

where L_n is the contraction map as in Eq. (2.1) and q_n is a suitable cubic polynomial obeying $q_n(x_1)=y_n-\alpha_n y_1$ and $q_n(x_N)=y_{n+1}-\alpha_n y_N$. By the Banach contraction principle, the operator T admits a unique fixed point $(\mathcal{T}f=)f$ and satisfies the following functional equation,

$$f(L_n(x)) = \alpha_n f(x) + q_n(x),\ x \in I,\ n = 1, 2, \ldots, N-1.$$

Moreover, the derivative f' is the unique fixed point of the operator,

$$(T^* g^*)(x) = \frac{\alpha_n g^*(L_n^{-1}(x)) + q_n'(L_n^{-1}(x))}{a_n}$$

defined on $(\mathcal{G}^*, \rho^*)$ where ρ^* is the metric associated with $\mathcal{G}^* = \{g^* \in C(I, \mathbb{R}) | g^*(x_1) = d_1, g^*(x_N) = d_N\}$. For more details on C^1-cubic Hermite fractal interpolation function, visit [13]. Consider the cubic polynomial q_n of the form,

$$q_n(x) = A_n \left(\frac{x - x_1}{x_N - x_1}\right)^3 + B_n \left(\frac{x - x_1}{x_N - x_1}\right)^2 + C_n \left(\frac{x - x_1}{x_N - x_1}\right) + D_n, \tag{8.7}$$

where A_n, B_n, C_n, D_n are the unknown parameters, determined employing the Eq. (2.2)), and $F_{n,1}(x_1, w_1) = w_n$, $F_{n,1}(x_N, w_N) = w_{n+1}$ as follows

$$\begin{aligned} A_n &= h_n(d_{n+1} + d_n) - \alpha_n(d_N + d_1)(y_N - y_1) - 2(w_{n+1} - w_n) + 2(w_N - w_1), \\ B_n &= -h_n(2d_n + d_{n+1}) + 3(w_{n+1} - w_n) - \alpha_n[-(y_N - y_1)(2d_1 + d_N) + 3(w_N - w_1)], \\ C_n &= h_n d_n - \alpha_n d_1(y_N - y_1), \\ D_n &= w_n - \alpha_n w_1, \end{aligned} \tag{8.8}$$

where $h_n = x_{n+1} - x_n$. The following is the functional equation associated with the C^1-cubic Hermite FIF obtained using the above parameter values and $\theta = \frac{x - x_1}{x_N - x_1}$,

$$\begin{aligned} f(L_n(x)) = {} & \alpha_n f(x) + \{h_n(d_{n+1} + d_n) - \alpha_n(d_N + d_1)(y_N - y_1) - 2(w_{n+1} - w_n) + 2(w_N - w_1)\}\theta^3 \\ & + \{-h_n(2d_n + d_{n+1}) + 3(w_{n+1} - w_n) - \alpha_n[-(y_N - y_1)(2d_1 + d_N) + 3(w_N - w_1)]\}\theta^2 \\ & + \{h_n d_n - \alpha_n d_1(y_N - y_1)\}\theta + w_n - \alpha_n w_1. \end{aligned} \tag{8.9}$$

Meanwhile, from (8.7),

$$q_n'(x) = \frac{3A_n\theta^2 + 2B_n\theta + C_n}{x_N - x_1},$$

the functional relation of f' is given by

$$\begin{aligned} f'(L_n(x)) = {} & \frac{\alpha_n}{a_n} f(x) + \Big\{3\{h_n(d_{n+1} + d_n) - \alpha_n(d_N + d_1)(y_N - y_1) - 2(w_{n+1} - w_n) \\ & + 2(w_N - w_1)\}\theta^2 + 2\{-h_n(2d_n + d_{n+1}) + 3(w_{n+1} - w_n) - \alpha_n[-(y_N - y_1)(2d_1 + d_N) \\ & + 3(w_N - w_1)]\}\theta + h_n d_n - \alpha_n d_1(y_N - y_1)\Big\} \Big/ a_n(x_N - x_1). \end{aligned}$$

Note that if the scaling factor α_n is taken as zero, for all $n = 1, 2, \ldots, N-1$, then the C^1-cubic Hermite FIF in (8.9) reduces to the C^1-Classical cubic Hermite interpolation function.

Now, let us discuss the variable order Caputo-fractional derivative of a C^1-cubic Hermite FIF. The variable order Caputo-fractional derivative of the C^1-cubic Hermite FIF f defined at the initial point $x = x_1$ is given by

$$D_{x_1}^{v(x)} f(x) = \frac{1}{\Gamma(1-v(x))} \int_{x_1}^{x} (x-s)^{-v(x)} f'(s)ds,$$

with $D_{x_1}^{v(x)} f(x_1) = 0$, where $v : [x_1, x_N] \to (0, 1)$.

Theorem 8.5 *Let f be a C^1-cubic Hermite FIF with cubic polynomial $q_n(x) = ax^3 + bx^2 + cx + d$, $n = 1, 2, \ldots, N-1$ and f' be its derivative associated with the IFS $\{(L_n(x), F_{n,1}(x, y) : n = 1, 2, \ldots, N-1\}$. If $\sum_{k=1}^{N} \alpha_k a_k^{-v(x)} < 1$, then the function $D_{x_1}^{v(x)} g$ is a FIF corresponding to the new data set $\{(x_n, \tilde{x}_n) : n = 1, 2, \ldots, N\}$, where $\tilde{y}_n$ represents the Caputo-fractional derivative of f at the knot x_n and its corresponding IFS is given by $\{(L_n(x), \tilde{F}_n(x, \tilde{y})) : n = 1, 2, \ldots, N-1\}$.*

Proof The variable order Caputo-fractional derivative of the prescribed fractal function f is given by

$$\begin{aligned}
D_{x_1}^{v(x)} f(L_n(x)) &= \frac{1}{\Gamma(1-v(x))} \int_{x_1}^{L_n(x)} (L_n(x)-s)^{-v(x)} f'(s)ds \\
&= \frac{1}{\Gamma(1-v(x))} \int_{x_1}^{x_n} (x_n-s)^{-v(x)} f'(s)ds + \frac{1}{\Gamma(1-v(x))} \int_{x_n}^{L_n(x)} (L_n(x)-s)^{-v(x)} f'(s)ds \\
&+ \frac{1}{\Gamma(1-v(x))} \int_{x_1}^{x_n} \{(L_n(x)-s)^{-v(x)}) - (x_n-s)^{-v(x)}\} f'(s)ds.
\end{aligned}$$

Performing change of variable $s = L_n(u)$ in the second term provides

$$\begin{aligned}
&= \tilde{y}_n + f_{n,v}(x) + \frac{a_n}{\Gamma(1-v(x))} \int_{x_1}^{x} (L_n(x) - L_n(u))^{-v(x)} f'(L_n(u))du \\
&= \tilde{y}_n + f_{n,v}(x) + \frac{a_n^{1-v(x)}}{\Gamma(1-v(x))} \int_{x_1}^{x} (x-u)^{-v(x)} \frac{\alpha_n f'(u) + q_n'(u)}{a_n} du \\
&= \tilde{y}_n + f_{n,v}(x) + \frac{\alpha_n a_n^{-v(x)}}{a_n^{v(x)} \Gamma(1-v(x))} \int_{x_1}^{x} (x-u)^{-v(x)} f'(u)du \\
&+ \frac{a_n^{-v(x)}}{\Gamma(1-v(x))} \int_{x_1}^{x} (x-u)^{-v(x)} (3au^2 + 2bu + c)du \\
&= \tilde{y}_n + f_{n,v}(x) + \alpha_n a_n^{-v(x)} D_{x_1}^{v(x)} f(x) + \frac{a_n^{-v(x)}}{\Gamma(1-v(x))} \left\{ \frac{(x-x_1)^{1-v(x)}}{1-v(x)} (3ax_1^2 + 2bx_1 + c) \right. \\
&\left. + \frac{(x-x_1)^{2-v(x)}}{(1-v(x))(2-v(x))} (6ax_1 + 2b) + \frac{6a(x-x_1)^{3-v(x)}}{(1-v(x))(2-v(x))(3-v(x))} \right\} \\
&= \alpha_n a_n^{-v(x)} D_{x_1}^{v(x)} f(x) + \tilde{q}_n(x) \\
&= \tilde{F}_n(x, D_{x_1}^{v(x)} f(x)).
\end{aligned}$$

The notations employed in the above derivations for simplification are described in the following,

$$\tilde{q}_n(x) = \tilde{x}_n + f_{n,v}(x) + \frac{a_n^{-v(x)}}{\Gamma(1-v(x))}\left\{\frac{(x-x_1)^{1-v(x)}}{1-v(x)}(3ax_1^2+2bx_1+c)\right.$$
$$\left.+\frac{(x-x_1)^{2-v(x)}}{(1-v(x))(2-v(x))}(6ax_1+2b)+\frac{6a(x-x_1)^{3-v(x)}}{(1-v(x))(2-v(x))(3-v(x))}\right\},$$
$$f_{n,v}(x) = \frac{1}{\Gamma(1-v(x))}\int_{x_1}^{x_n}\{(L_n(x)-s)^{-v(x)}) - (x_n-s)^{-v(x)}\}f'(s)ds.$$

In order to get the new data points $\{(x_n, \tilde{y}_n) : n = 1, 2, \ldots, N\}$, take $x = x_N$ and $L_n(x_N) = x_{n+1}$ in the below equation,

$$D_{x_1}^{v(x)} f(L_n(x)) = \tilde{y}_n + f_{n,v}(x) + \alpha_n a_n^{-v(x)} D_{x_1}^{v(x)} f(x) + \frac{a_n^{-v(x)}}{\Gamma(1-v(x))}\left\{\frac{(x-x_1)^{1-v(x)}}{1-v(x)}(3ax_1^2+2bx_1+c)\right.$$
$$\left.+\frac{(x-x_1)^{2-v(x)}}{(1-v(x))(2-v(x))}(6ax_1+2b)+\frac{6a(x-x_1)^{3-v(x)}}{(1-v(x))(2-v(x))(3-v(x))}\right\}$$

and using the below equation,

$$\tilde{y}_{n+1} = \tilde{y}_1 + \sum_{k=1}^{n}(\tilde{y}_{k+1} - \tilde{y}_k),$$

for all $n = 1, 2, 3, \ldots, N-1$, one can obtain

$$\tilde{y}_{n+1} = \sum_{k=1}^{n}\left\{f_{k,v}(x_N) + \alpha_k a_k^{-v(x_N)} D^{v(x_N)} f(x_N) + \frac{a_k^{-v(x_N)}}{\Gamma(1-v(x_N))}\left\{\frac{(x_N-x_1)^{1-v(x_N)}}{1-v(x_N)}(3ax_1^2+2bx_1+c)\right.\right.$$
$$\left.\left.+\frac{(x_N-x_1)^{2-v(x_N)}}{(1-v(x_N))(2-v(x_N))}(6ax_1+2b)+\frac{6a(x_N-x_1)^{3-v(x_N)}}{(1-v(x_N))(2-v(x_N))(3-v(x_N))}\right\}\right\}. \tag{8.10}$$

On substituting $n = N-1$, the new interpolant $\tilde{y}_N$ at the knot $x = x_N$ is obtained as,

$$\tilde{y}_N = \sum_{k=1}^{N-1}\left\{f_{k,v}(x_N) + \frac{a_k^{-v(x_N)}}{\Gamma(1-v(x_N))}\left\{\frac{(x_N-x_1)^{1-v(x_N)}}{1-v(x_N)}(3ax_1^2+2bx_1+c) + \frac{(x_N-x_1)^{2-v(x_N)}}{(1-v(x_N))(2-v(x_N))}\right.\right.$$
$$\left.\left.\times(6ax_1+2b)+\frac{6a(x_N-x_1)^{3-v(x_N)}}{(1-v(x_N))(2-v(x_N))(3-v(x_N))}\right\}\right\}\Big/\left\{1-\sum_{k=1}^{N-1}\alpha_k a_k^{-v(x)}\right\}. \tag{8.11}$$

It is clear that the newly obtained variable order Caputo-fractional derivative of a C^1-cubic Hermite FIF, $D_{x_1}^{v(x)} f$ agrees the functional equation of a fractal interpolation function with the new set of interpolation data $\{(x_n, \tilde{y}_n) : n = 1, 2, \ldots, N\}$. □

Corollary 8.1 *Suppose the cubic polynomial q_n takes the form $q_n(x) = A_n\theta^3 + B_n\theta^2 + C_n\theta + D_n$, where $\theta = \frac{x-x_1}{x_N-x_1}$ with*

$$q_n'(x) = \frac{3A_n\theta^2 + 2B_n\theta + C_n}{x_N - x_1},$$

then Theorem 8.5 provides the following

$$\begin{aligned} D_{x_1}^{v(x)} f(L_n(x)) &= \tilde{y}_n + f_{n,v}(x) + \alpha_n a_n^{-v(x)} D_{x_1}^{v(x)} f(x) + \frac{a_n^{-v(x)}}{\Gamma(1-v(x))}\int_{x_1}^{x}(x-u)^{-v(x)} \\ &\quad \times \left(\frac{3A_n\theta^2 + 2B_n\theta + C_n}{x_N - x_1}\right) du \\ &= \tilde{y}_n + f_{n,v(x)}(t) + \alpha_n a_n^{-v(x)} D_{x_1}^{v(x)} f(x) + J_1 + J_2 + J_3, \end{aligned} \tag{8.12}$$

where

$$J_1 = \frac{a_n^{-v(x)}}{(x_N - x_1)\Gamma(1-v(x))}\int_{x_1}^{x} 3A_n(x-u)^{-v(x)}\left(\frac{u-x_1}{x_N-x_1}\right)^2 du,$$
$$J_2 = \frac{a_n^{-v(x)}}{(x_N - x_1)\Gamma(1-v(x))}\int_{x_1}^{x} 2B_n(x-u)^{-v(x)}\left(\frac{u-x_1}{x_N-x_1}\right) du,$$
$$J_3 = \frac{a_n^{-v(x)}}{(x_N - x_1)\Gamma(1-v(x))}\int_{x_1}^{x} C_n(x-u)^{-v(x)} du.$$

Simplifying the above integrals, one can have

$$J_1 = \frac{6A_n a_n^{-v(x)}(x-x_1)^{3-v(x)}}{(x_N-x_1)^3(1-v(x))(2-v(x))(3-v(x))\Gamma(1-v(x))},$$
$$J_2 = \frac{2B_n a_n^{-v(x)}(x-x_1)^{2-v(x)}}{(x_N-x_1)^2(1-v(x))(2-v(x))\Gamma(1-v(x))},$$
$$J_3 = \frac{C_n a_n^{-v(x)}(x-x_1)^{1-v(x)}}{(x_N-x_1)(1-v(x))\Gamma(1-v(x))}.$$

Substituting J_1, J_2, and J_3 in Eq. (8.12), it follows that

$$\begin{aligned} D_{x_1}^{v(x)} f(L_n(x)) &= \tilde{y}_n + f_{n,v}(x) + \alpha_n a_n^{-v(x)} D_{x_1}^{v(x)} f(x) \\ &\quad + \frac{6A_n a_n^{-v(x)}(x-x_1)^{3-v(x)}}{(x_N-x_1)^3(1-v(x))(2-v(x))(3-v(x))\Gamma(1-v(x))} \\ &\quad + \frac{2B_n a_n^{-v(x)}(x-x_1)^{2-v(x)}}{(x_N-x_1)^2(1-v(x))(2-v(x))\Gamma(1-v(x))} + \frac{C_n a_n^{-v(x)}(x-x_1)^{1-v(x)}}{(x_N-x_1)(1-v(x))\Gamma(1-v(x))}. \end{aligned} \tag{8.13}$$

The new interpolation points associated with the function $D_{x_1}^{v(x)} f$ in Eq. (8.13) can be determined using the Eqs. (8.10) and (8.11).

8.3.1 Numerical Examples

In this subsection, a numerical example of a C^1-cubic Hermite fractal interpolation function is illustrated with graphical representation. Let $\{(-1, 0.5), (-0.5, -1), (0, 0), (0.5, -0.9), (1, 0.7)\}$ be the given set of data points. The arithmetic mean method is used for finding the derivative values to generate the required C^1-cubic Hermite FIF:

$$\Delta_n = \frac{y_{n+1} - y_n}{h_n}, \; n = 1, 2, \ldots, N-1, \; \Delta_{3,1} = \frac{x_3 - x_1}{x_3 - x_1}, \; \Delta_{N,N-2} = \frac{y_N - y_{N-2}}{x_N - x_{N-2}},$$

$$d_n = \frac{h_n \Delta_{n-1} + h_{n-1}\Delta_n}{h_{n-1} + f_n}, \; n = 2, 3, \ldots, N-1,$$

$$d_1 = \left(1 + \frac{h_1}{h_2}\right)\Delta_1 - \frac{h_1}{h_2}\Delta_{3,1},$$

$$d_N = \left(1 + \frac{h_{N-1}}{h_{N-2}}\right)\Delta_{N-1} - \frac{h_{N-1}}{h_{N-2}}\Delta_{N,N-2}.$$

To ensure the C^1-continuity of cubic fractal interpolation function, three different set of scaling factors are chosen as

(a) $\alpha_n = 0$, for all $n = 1, 2, 3, 4$
(b) $\alpha = (0.15, -0.15, 0.15, -0.15)$
(c) $\alpha = (0.2, -0.24, 0.23, -0.24)$

such that $|\alpha_n| < a_n$, for all $n = 1, 2, 3, 4$. Consider the cubic polynomial $q_n(x) = A_n(1-\theta)^3 + B_n\theta(1-\theta)^2 + C_n\theta^2(1-\theta) + D_n\theta^3$, where $\theta = \frac{x - x_1}{x_N - x_1}$ and the parameters A_n, B_n, C_n, D_n are determined using the Eq. (8.8), for $n = 1, 2, 3, 4$. The graph of C^1-cubic Hermite FIFs and their derivative corresponding to the given data set and the cubic polynomial q_n are represented in Figs. 8.2, 8.3 and 8.4. Figure 8.2a, b illustrate the classical version of C^1-cubic Hermite FIF and its derivative, respectively with the choice of all the scaling factors to be zero, i.e., $\alpha_n = 0, \; n = 1, 2, 3, 4$. The C^1-cubic Hermite FIFs and their derivative with $\alpha = (0.15, -0.15, 0.15, -0.15)$ and $\alpha = (0.2, -0.24, 0.23, -0.24)$, are demonstrated in Figs. 8.3a, b, and 8.4a, b.

8.4 α-Fractal Function for Caputo-Fractional Derivative of a C^1-Function

This section explores the α-fractal interpolation function for the Caputo-fractional derivative of a continuously differentiable function with constant fractional order $v \in (0, 1)$. On confirming the continuity of Caputo-fractional derivative of the continuously differentiable function, its α-fractal function is defined via the new self-referential equation. The uniform distance of the proposed α-fractal function is

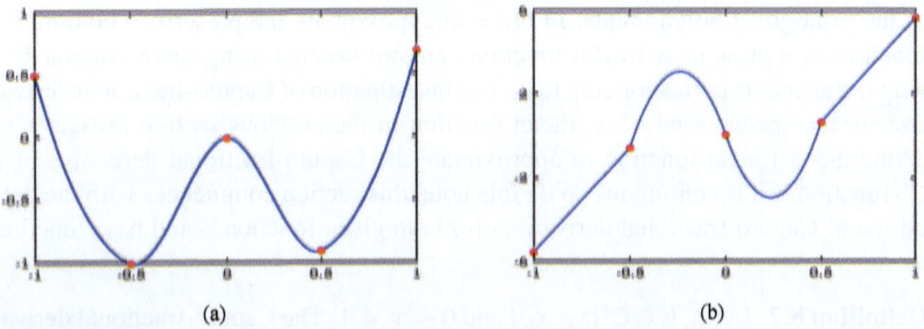

Fig. 8.2 Graphs of **a** C^1-classical cubic Hermite interpolation function and **b** its derivative, with $\alpha = (0, 0, 0, 0)$

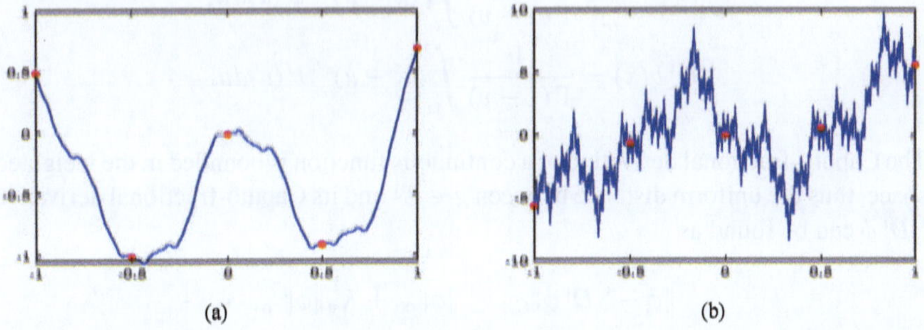

Fig. 8.3 Graphs of **a** C^1-cubic Hermite FIF and **b** its derivative with $\alpha = (0.15, -0.15, 0.15, -0.15)$

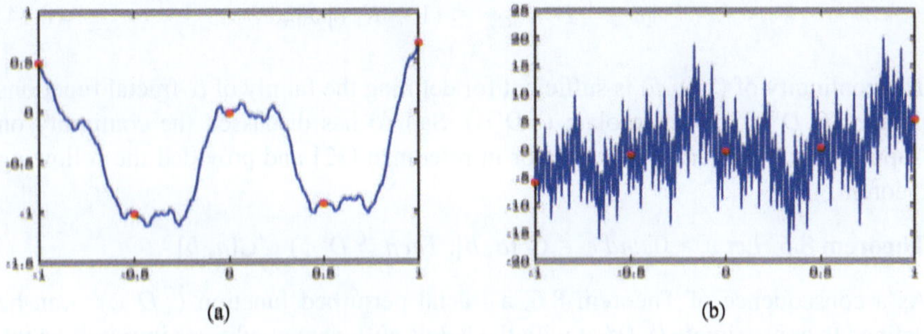

Fig. 8.4 Graphs of **a** C^1-cubic Hermite FIF and **b** its derivative with $\alpha = (0.2, -0.24, 0.23, -0.24)$

compared with the other functions involved in the fractal perturbation process by considering the weighted space of continuous functions. Further, to interpolate the Caputo-fractional derivative of a polynomial, an α-fractal polynomial is generated. With the help of Legendre polynomials, an approximate expansion for the fractional derivative of fractal interpolation function is provided in terms of fractional derivative of the orthogonal polynomials. In order to approximate the prescribed continuous function ϕ, a class of α-fractal functions are constructed using the particular IFS with the elements corresponding to ϕ. The investigation of Caputo-fractional derivative for the special kind of α-fractal function in the previous section instigates to define the α-fractal function to approximate the Caputo-fractional derivative of a C^1-function if it is continuous. With this note, this section commences with the definition of Caputo-fractional derivatives for both given function ϕ and base function b.

Definition 8.2 Let $\phi, b \in C^1[x_1, x_N]$ and $0 < v < 1$. The Caputo-fractional derivatives of ϕ and b are expressed as follows.

$$^C D^v \phi(x) = \frac{1}{\Gamma(1-v)} \int_{x_1}^{x} (x-u)^{-v} \phi'(u) du,$$
$$^C D^v b(x) = \frac{1}{\Gamma(1-v)} \int_{x_1}^{x} (x-u)^{-v} b'(u) du.$$

The Caputo-fractional derivative of a continuous function is bounded in the weighted space, thus the uniform distance between $g \in C^1$ and its Caputo-fractional derivative $^C D^v \phi$ can be found as

$$\|\phi - {}^C D^v \phi\|_{C_{x_1}} \le \|\phi\|_\infty + K_v \|\phi\|_\infty,$$

since $\|^C D^v \phi\|_{C_{x_1}} \le K_v \|\phi\|_\infty$, where $K_v = \frac{(x_N - x_1)^{1-v}}{|\Gamma(1-v)|(2-v)}$. Hence,

$$\|\phi - {}^C D^v \phi\|_{C_{x_1}} \le (1 + K_v)\|\phi\|_\infty. \tag{8.14}$$

The continuity of $(^C D^v \phi)$ is sufficient for defining the family of α-fractal functions namely, $(^C D^v \phi)^\alpha$ to interpolate $(^C D^v \phi)$. Samko has discussed the continuity on Caputo-fractional derivative operator in reference [12] and provided the following theorem.

Theorem 8.6 *Let $v \ge 0$ and $\phi \in C^1[a, b]$. Then $(^C D^v \phi) \in C[a, b]$.*

As a consequence of Theorem 8.6, a fractal perturbed function $(^C D^v \phi)^\alpha$ can be defined to approximate $(^C D^v \phi)$ with the below sufficient conditions imposed on the base function $(^C D^v b)$,

$$(^C D^v b)(x_1) = (^C D^v \phi)(x_1),$$
$$(^C D^v b)(x_N) = (^C D^v \phi)(x_N).$$

Now changing the arguments of the RB operator $\mathcal{T} : C(I) \to C(I)$ which is defined above yields

$$(\mathcal{T}(^C D^v f))(x) = F_n(L_n^{-1}(x), (^C D^v f) \circ L_n^{-1}(x)), \ x \in I_n, \ n = 1, 2, \ldots, N-1.$$

Further, $\mathbb{T}$ admits a unique fixed point, say $({}^C D^\upsilon \phi)^\alpha$ such that

$$(\mathcal{T}({}^C D^\upsilon \phi)^\alpha)(x) = ({}^C D^\upsilon \phi)^\alpha(x) = \alpha_n({}^C D^\upsilon \phi)^\alpha \circ L_n^{-1}(x) + {}^C D^\upsilon \phi(x) - \alpha_n {}^C D^\upsilon b \circ L_n^{-1}(x),$$

$$(i.e.,)\ ({}^C D^\upsilon \phi)^\alpha(L_n(x)) = \alpha_n({}^C D^\upsilon \phi)^\alpha(x) + {}^C D^\upsilon \phi \circ L_n(x) - \alpha_n {}^C D^\upsilon b(x),\ x \in I,\ n = 1, 2, \ldots, N-1. \quad (8.15)$$

Definition 8.3 The function $({}^C D^\upsilon \phi)^\alpha$ satisfying the self-referential equation (8.15) can be referred as the α-fractal function of $({}^C D^\upsilon \phi)$ with respect to Δ and $({}^C D^\upsilon b)$.

Note that the scaling factor employed in Eq. (8.15) is similar to the one employed in Eq. (7.6). The fractal interpolation operator $\mathcal{D}_{\mathcal{F}}$ is defined as a self-map on the weighted space $C_{x_1}(I)$ to provide $({}^C D^\upsilon \phi)^\alpha$ for every continuous function $({}^C D^\upsilon \phi)$ such that

$$\mathcal{D}_{\mathcal{F}}({}^C D^\upsilon \phi) = ({}^C D^\upsilon \phi)^\alpha.$$

The linearity of the operator $\mathcal{D}_{\mathcal{F}}$ can be easily verified. So far, we have discussed the C^1-function g, its fractional derivative $({}^C D^\upsilon \phi)$ and their corresponding fractal perturbations namely, ϕ^α and $({}^C D^\upsilon \phi)^\alpha$. Moreover, the fractal operators for the aforementioned fractal functions have also been investigated. The following subsection estimates the uniform distance of the fractal functions from their respective given functions.

8.4.1 *Estimation of Errors in the Fractal Perturbed Process*

The uniform bound between the functions $({}^C D^\upsilon \phi)^\alpha$ and $({}^C D^\upsilon \phi)$ is presented in the following theorem.

Theorem 8.7 *Let $g \in C^1(I)$ and $({}^C D^\upsilon \phi)$ be its Caputo-fractional derivative. Then the uniform distance between $({}^C D^\upsilon \phi)$ and its associated α-fractal function $({}^C D^\upsilon \phi)^\alpha$ is given by*

$$\|({}^C D^\upsilon \phi)^\alpha - ({}^C D^\upsilon \phi)\|_{C_{x_1}} \leq \frac{2K_\upsilon |\alpha|_\infty}{1 - |\alpha|_\infty} \|\phi\|_\infty. \quad (8.16)$$

Proof One has

$$\begin{aligned} \|({}^C D^\upsilon \phi)^\alpha - ({}^C D^\upsilon \phi)\|_{C_{x_1}} &\leq \frac{|\alpha|_\infty}{1 - |\alpha|_\infty} \|({}^C D^\upsilon \phi) - ({}^C D^\upsilon b)\|_{C_{x_1}} \\ &\leq \frac{|\alpha|_\infty}{1 - |\alpha|_\infty} \{\|({}^C D^\upsilon \phi)\|_{C_{x_1}} + \|({}^C D^\upsilon b)\|_{C_{x_1}}\}. \end{aligned}$$

As $\|b\|_\infty = \|\phi\|_\infty$, then $\|({}^C D^v b)\|_{C_{x_1}} = \|({}^C D^v \phi)\|_{C_{x_1}}$,

$$\begin{aligned}\|({}^C D^v \phi)^\alpha - ({}^C D^v \phi)\|_{C_{x_1}} &\le \frac{|\alpha|_\infty}{1-|\alpha|_\infty}\{K_v\|\phi\|_\infty + K_v\|b\|_\infty\} \\ &= \frac{|\alpha|_\infty}{1-|\alpha|_\infty} K_v\{\|\phi\|_\infty + \|b\|_\infty\},\end{aligned}$$

where $K_v = \frac{(x_N - x_1)^{1-v}}{|\Gamma(1-v)|(2-v)}$. Hence,

$$\|({}^C D^v \phi)^\alpha - ({}^C D^v \phi)\|_{C_{x_1}} \le \frac{2K_v|\alpha|_\infty}{1-|\alpha|_\infty}\|\phi\|_\infty.$$

Corollary 8.2 *$\mathcal{D}_{\mathcal{F}}$ is bounded. Further,*

$$\|\mathcal{D}_{\mathcal{F}}({}^C D^v \phi)\|_{C_{x_1}} \le \frac{K_v(1+|\alpha|_\infty)}{1-|\alpha|_\infty}\|\phi\|_\infty.$$

Proposition 8.1 *If ϕ^α and $({}^C D^v \phi)^\alpha$ are the α-fractal functions corresponding to $\phi \in C^1(I)$ and $({}^C D^v \phi)$, respectively, then the following inequality holds*

$$\|\phi^\alpha - ({}^C D^v \phi)^\alpha\|_{C_{x_1}} \le \left(\frac{2|\alpha|_\infty}{1-|\alpha|_\infty} + 1\right)(1+K_v)\|\phi\|_\infty. \tag{8.17}$$

Proof Consider

$$\begin{aligned}\|\phi^\alpha - ({}^C D^v \phi)^\alpha\|_{C_{x_1}} &= \|\phi^\alpha - ({}^C D^v \phi) + ({}^C D^v \phi) - ({}^C D^v \phi)^\alpha\|_{C_{x_1}} \\ &\le \|\phi^\alpha - ({}^C D^v \phi)\|_{C_{x_1}} + \|({}^C D^v \phi) - ({}^C D^v \phi)^\alpha\|_{C_{x_1}} \\ &\le \|\phi^\alpha - \phi\|_\infty + \|\phi - ({}^C D^v \phi)\|_{C_{x_1}} + \|({}^C D^v \phi) - ({}^C D^v \phi)^\alpha\|_{C_{x_1}}.\end{aligned}$$

Using Theorem 8.7, it follows that

$$\|\phi^\alpha - ({}^C D^v \phi)^\alpha\|_{C_{x_1}} \le \frac{2|\alpha|_\infty}{1-|\alpha|_\infty}\|\phi\|_\infty + (1+K_v)\|\phi\|_\infty + \frac{2K_v|\alpha|_\infty}{1-|\alpha|_\infty}\|\phi\|_\infty.$$

On simplification,

$$\|\phi^\alpha - ({}^C D^v \phi)^\alpha\|_{C_{x_1}} \le \left(\frac{2|\alpha|_\infty}{1-|\alpha|_\infty} + 1\right)(1+K_v)\|\phi\|_\infty.$$

The uniform distance of $({}^C D^v \phi)^\alpha$ from the given function g is evaluated in the following theorem.

Theorem 8.8 *Let $({}^C D^v \phi)^\alpha$ be the α-fractal function for the Caputo-fractional derivative of $\phi \in C^1$. Then the below inequality holds*

$$\|\phi - ({}^C D^\upsilon \phi)^\alpha\|_{C_{x_1}} \leq \left(1 + K_\upsilon + \frac{2K_\upsilon|\alpha|_\infty}{1 - |\alpha|_\infty}\right)\|\phi\|_\infty. \tag{8.18}$$

Proof One can have

$$\begin{aligned}\|\phi - ({}^C D^\upsilon \phi)^\alpha\|_{C_{x_1}} &= \|\phi - ({}^C D^\upsilon \phi) + ({}^C D^\upsilon \phi) - ({}^C D^\upsilon \phi)^\alpha\|_{C_{x_1}} \\ &\leq \|g - ({}^C D^\upsilon \phi)\|_{C_{x_1}} + \|({}^C D^\upsilon \phi) - ({}^C D^\upsilon \phi)^\alpha\|_{C_{x_1}}.\end{aligned}$$

From Theorem 8.7, one can get

$$\|\phi - ({}^C D^\upsilon \phi)^\alpha\|_{C_{x_1}} \leq (1 + K_\upsilon)\|\phi\|_\infty + \frac{2K_\upsilon|\alpha|_\infty}{1 - |\alpha|_\infty}\|\phi\|_\infty.$$

Hence, the required inequality follows

$$\|\phi - ({}^C D^\upsilon \phi)^\alpha\|_{C_{x_1}} \leq \left(1 + K_\upsilon + \frac{2K_\upsilon|\alpha|_\infty}{1 - |\alpha|_\infty}\right)\|\phi\|_\infty.$$

Exploring the construction of α-fractal function to approximate the fractional derivative of a C^1-function, the error bounds of fractal functions are estimated in this section. In the next subsection, a fractal operator is generated to provide the α-fractal approximation to any polynomial of degree n.

8.4.2 Fractal Polynomial

Let $\mathcal{P}_n[a, b]$ denote the set of all polynomials of degree less than or equal to n and $({}^C D^\upsilon \mathcal{P}_{n-\upsilon})[a, b]$ be the set of all Caputo-fractional derivatives of $\mathcal{P}_n[a, b]$ of degree less than or equal to $n - \upsilon$. The standard basis $\{1, x, x^2, \ldots, x^{n-\upsilon}\}$ spans $({}^C D^\upsilon \mathcal{P}_{n-\upsilon})[a, b]$ and

$$\dim({}^C D^\upsilon \mathcal{P}_{n-\upsilon})[a, b] < +\infty.$$

As a result $({}^C D^\upsilon \mathcal{P}_{n-\upsilon})[a, b]$ is closed in $C[a, b]$. Set $({}^C D^\upsilon \mathcal{P})[a, b] = \cup_{n=1}^{\infty}({}^C D^\upsilon \mathcal{P}_{n-\upsilon})[a, b]$.

Remark 8.3 In [9], a α-fractal polynomial is illustrated namely, p^α to approximate $p \in \mathcal{P}[a, b]$ for the same Δ, α and c discussed prior with $\mathcal{F}(p) = p^\alpha$.

In sequel to the above remark, the existence of $({}^C D^\upsilon p)[a, b]$ for the polynomial $p \in \mathcal{P}[a, b]$ put forth the generation of $({}^C D^\upsilon p)^\alpha[a, b]$.

Definition 8.4 The function $({}^C D^\upsilon p)^\alpha[a, b] \in C(I)$ is an α-fractal polynomial to approximate $({}^C D^\upsilon p)[a, b] \in ({}^C D\mathcal{P})[a, b]$ with the partition Δ and scale vector α and the continuous function c obeying the hypothesis in Remark 2.1.

Note 8.1 $\mathcal{F} : C_{x_1}(I) \to C_{x_1}(I)$. Moreover, $\mathcal{F}(({}^C D^\upsilon p)[a, b]) = ({}^C D^\upsilon p)^\alpha[a, b]$.

In [9], using the Weierstrass theorem of approximation, it is demonstrated that for all $\epsilon > 0$ and $\alpha \neq 0$ there exists an α-fractal polynomial p^α for a function $\phi \in C[a, b]$ such that $|\phi(x) - p^\alpha(x)| < \epsilon, \ \forall\, x \in [a, b]$. That is, every continuous function can be approximated by means of a non-smooth continuous function so called the α-fractal function.

Since the fractional derivative of a polynomial p of degree n is again a polynomial of fractional order. Hence, for the function $\phi \in C[a, b]$, there exists a polynomial $({}^C D^v p)(x)$ such that

$$|\phi(x) - ({}^C D^v p)(x)| < \epsilon/2, \ \forall\, x \in [a, b]. \tag{8.19}$$

Similarly, one can have

$$\frac{2K_v|\alpha|_\infty}{1 - |\alpha|_\infty}\|\phi\|_\infty < \epsilon/2, \ \forall\, x \in [a, b]. \tag{8.20}$$

From inequalities (8.19) and (8.20), it is concluded that

$$|\phi(x) - ({}^C D^v p)^\alpha(x)| < \epsilon, \ \forall\, x \in [a, b]. \tag{8.21}$$

The above discussion on the α-fractal polynomial is summarized in the following theorem.

Theorem 8.9 *If $\phi \in C[a, b]$, then for all $\epsilon > 0$ and $\alpha \neq 0 \in \mathbb{R}$, there exists an α-fractal polynomial $({}^C D^v p)^\alpha$ such that*

$$|\phi(x) - ({}^C D^v p)^\alpha(x)| < \epsilon, \ \forall\, x \in [a, b].$$

Note 8.2 A linear subspace $({}^C D^v \mathcal{P}_{n-v})[a, b]$ is dense in $C[a, b]$.

8.5 Numerical Procedure to Obtain the Caputo-Fractional Derivative of a Fractal Function

In this section, an approximate expansion of a fractal interpolation function $\tilde{f}$ in terms of orthogonal polynomials is computed. Let $J = [-1, 1]$ and $\{p_n\}_{n=0}^\infty$ be the system of normalized polynomials of Legendre and are orthonormal with respect to the inner product

$$(f, g) = \int_J f(x)g(x)\, dx. \tag{8.22}$$

The fractal function $\tilde{f}$ is approximated using the shifted Legendre polynomials as

$$\tilde{f} \approxeq \sum_{m=0}^{\mu} c_m \cdot p_m \tag{8.23}$$

and

$$p_m(x) = \sum_{r=0}^{m} d_r^m .x^r,$$

where

$$d_r^m = \frac{(-1)^{m+1}\Gamma(m+r+1)}{\Gamma(m-r+1)\Gamma^2(r+1)}, \tag{8.24}$$

then the Legendre coefficient [14] is given by

$$\begin{aligned} c_m = (2m+1)\int_0^1 \tilde{f}(x)p_m(x)dt &= (2m+1)\sum_{r=0}^{m} d_r^m \int_0^1 x^r \tilde{f}(x)dx \\ &= (2m+1)\sum_{r=0}^{m} d_r^m M_r, \end{aligned} \tag{8.25}$$

where $M_r = \int_{-1}^{1} x^r \tilde{f}(x)dt$ is the moment integral. In the article [15], a recurrence formula for the computation of the moments M_r,

$$M_r = \int_J x^r \tilde{f}(x)\, dx \tag{8.26}$$

is given, for a function $\tilde{f}$ defined by the general iterated function system

$$L_k(x) = a_n x + b_k, \;\; F_n(x, y) = \alpha_n y + q_n(x).$$

The formula is expressed as

$$M_r = \frac{1}{(1 - \sum_{n=1}^{n} a_n^{r+1}\alpha_n)} \sum_{j=0}^{r-1} \binom{r}{j} M_j \sum_{n=1}^{n} a_n^{j+1}\alpha_n b_n^{r-j} + Q_r, \tag{8.27}$$

where

$$\begin{aligned} Q_r &= \int_J x^r Q(x)\, dx \\ &= \sum_{n=1}^{n-1} \int_{I_n} x^r q_n \circ L_n^{-1}(x)dx, \end{aligned}$$

here $Q(x) = q_n \circ L_n^{-1}(x)$ if $x \in I_n$. Without loss of generality, the Fourier-Legendre coefficients of a FIF $\tilde{f}$ with respect to this complete system can be computed with coefficients,

$$c_m = (2m+1)\sum_{r=0}^{m} d_r^m M_r, \tag{8.28}$$

where M_r are the moments defined in (8.27). With the computed coefficients c_m, the exact Caputo-fractional derivative of the polynomial

$$p(x) = \sum_{m=0}^{\mu} c_m . p_m(x)$$

is expressed in terms of ${}^C D^v p_m(x)$ as

$${}^C D^v p_m(x) = \sum_{r=0}^{m} d_r^m . {}^C D^v x^r.$$

It is explicitly handled using

$${}^C D^v p_m(x) = \sum_{r=\lceil v \rceil}^{m} d_r^m \frac{\Gamma(r+1)}{\Gamma(r+1-v)} x^{r-v}.$$

To obtain the Caputo-fractional derivative of $\tilde{f}$ at the nodes, the matrix product is computed as

$$\begin{pmatrix} {}^C D^v p_0(x_1) & {}^C D^v p_1(x_1) & \dots & {}^C D^v p_v(x_1) \\ {}^C D^v p_0(x_2) & {}^C D^v p_1(x_2) & \dots & {}^C D^v p_v(x_2) \\ \dots & \dots & \dots & \dots \\ {}^C D^v p_0(x_N) & {}^C D^v p_1(x_N) & \dots & {}^C D^v p_v(x_N) \end{pmatrix} \begin{pmatrix} c_0 \\ c_1 \\ \dots \\ c_n \end{pmatrix} = \begin{pmatrix} {}^C D^v \tilde{f}(x_1) \\ {}^C D^v \tilde{f}(x_2) \\ \dots \\ {}^C D^v \tilde{f}(x_N) \end{pmatrix} \tag{8.29}$$

the last matrix contains the approximated Caputo-fractional derivatives of the fractal function $\tilde{f}$ at the nodes of the partition Δ. In general, for any FIF $\tilde{f}$,

$${}^C D^v \tilde{f}(x) \approxeq \sum_{m=0}^{\mu} c_m . {}^C D^v p_m(x), \; x \in I.$$

The following is an numerical example of computing the Caputo-fractional derivative of an α-fractal function ϕ^α interpolating the smooth function $g(x) = x^2$. For instance, consider the continuous function c as an identity function, then $\phi^\alpha = g$.

Taking the fractional order $v = 0.5$, the following approximation is verified.

$${}^C D^{0.5}(x^2) \approxeq \sum_{m=0}^{2} c_m . {}^C D^{0.5} p_m(x), \; x \in [0, 1].$$

Using the Eq. (8.25), the Legendre coefficients are computed as

$$c_0 = 1/3,\ c_1 = 1/2,\ c_2 = 1/6.$$

Also, one can get

$${}^{C}D^{0.5} p_0(x) = 0,\ {}^{C}D^{0.5} p_0(x) = \frac{2}{\Gamma(\frac{1}{2})} x^{0.5},\ {}^{C}D^{0.5} p_1(x) = \frac{-6}{\Gamma(\frac{1}{2})} x^{0.5} + \frac{12}{\Gamma(\frac{3}{2})} x^{1.5}.$$

Hence,

$$\sum_{m=0}^{2} c_m . {}^{C}D^{0.5} p_m(x) = \frac{2}{\Gamma(\frac{5}{2})} x^{1.5}$$

which equals the Caputo-fractional derivative of the prescribed α-fractal function ϕ^α.

8.6 Summary

This chapter has presented the investigation of one of the most familiar fractional calculus methods: Caputo-fractional derivative on the fractal interpolation functions. The advantage of studying the Caputo-fractional derivatives is that it allows to deal the non-zero boundary condition problems in various engineering fields, therefore, aids in solving many real world problems. The Caputo-fractional derivative of an α-fractal function is investigated and proved that the resultant function is again an α-fractal function under certain conditions. A fractal operator corresponding to the aforementioned fractal function is proposed and its analytical as well as the algebraic properties are explored. The Caputo-fractional derivative of Hermite fractal function is also studied and its resultant function is verified for fractal nature. Further, the chapter has presented a family of fractal perturbations to interpolate the Caputo-fractional derivative of a C^1-function and discussed the bounds of the functions involved in the perturbation process. It is also demonstrated that the space of fractional order polynomials is dense in the space of continuous functions. Finally, a numerical procedure is presented to verify the approximation of exact Caputo-fractional derivative of an α-fractal function using the Caputo-fractional derivative of the Legendre polynomial.

References

1. M.F. Barnsley, J. Elton, D. Hardin, P. Massopust, Hidden variable fractal interpolation functions. SIAM J. Math. Anal. **20**(5), 1218–1242 (1989)
2. M.A. Navascués, M.V. Sebastián, Fractal splines. Monografías del Seminario Matemático García de Galdeano **33**, 161–168 (2006)
3. M.A. Navascués, Non-smooth polynomials. Int. J. Math. Anal. **1**(4), 159–174 (2007)

4. M.A. Navascués, M.V. Sebastián, Smooth fractal interpolation. J. Inequalities Appl. **1–20**, 2006 (2006)
5. H.-J. Ruan, W.-Y. Su, K. Yao, Box dimension and fractional integral of linear fractal interpolation functions. J. Approx. Theory **161**(1), 187–197 (2009)
6. X. Pan, Fractional calculus of fractal interpolation function on $[0, b](b>0)$, in *Abstract and Applied Analysis*, vol. 2014 (Hindawi, 2014)
7. W.L. Peng, K. Yao, X. Zhang, J. Yao, Box dimension of weyl-marchaud fractional derivative of linear fractal interpolation functions. Fractals **27**(04), 1950058 (2019)
8. A. Gowrisankar, M.G.P. Prasad, Riemann-liouville calculus on quadratic fractal interpolation function with variable scaling factors. J. Anal. **27**(2), 347–363 (2019)
9. M.A. Navascués, Fractal polynomial interpolation. Zeitschrift für Analysis und ihre Anwendungen **24**(2), 401–418 (2005)
10. T.M.C. Priyanka, A. Gowrisankar, Analysis on wely-marchaud fractional derivative for types of fractal interpolation function with fractal dimension. Fractals **29**(07), 2150215 (2021)
11. N. Vijender, Bernstein fractal trigonometric approximation. Acta Applicandae Mathematicae **159**(1), 11–27 (2019)
12. A.A. Kilbas, H.M. Srivastava, J.J Trujillo, *Theory and Applications of Fractional Differential Equations*, vol. 204 (Elsevier, 2006)
13. A.K.B. Chand, P. Viswanathan, A constructive approach to cubic hermite fractal interpolation function and its constrained aspects. BIT Numer. Math. **53**, 841–865 (2013)
14. Z. Odibat, On legendre polynomial approximation with the vim or ham for numerical treatment of nonlinear fractional differential equations. J. Comput. Appl. Math. **235**(9), 2956–2968 (2011)
15. M.F. Barnsley, Fractal functions and interpolation. Constr. Approx. **2**, 303–329 (1986)

Chapter 9
Applications of Fractal Interpolation Functions

9.1 Fractal Interpolation in COVID-19

On December 31, 2019, the World Health Organization (WHO) received the first report of Corona Virus Disease 2019 (COVID-19) from Wuhan city, China. Despite the fact that Chinese cities implemented lockdowns to restrict the spread of infection, the measures were ineffective, and the disease has since spread around the world. As a result, the WHO announced COVID-19, a global pandemic on the month of March 2020. Initially, this unforeseen global pandemic caused chaos on almost every sphere of human life. Since the first reported case of COVID-19, the world was experiencing successive waves. On 24 November 2021, a new SARS-CoV-2 variant was reported to WHO from South Africa. The new variant (B.1.1.529) was officially named *Omicron* (the fifteenth Greek alphabet), as a Variant of Concern. The rapid spread of this highly mutated strain across six continents in less than a week sparked a global health alarm. Since its discovery, Omicron was identified in many countries in six continents. When compared to the previous variants of SARS-CoV-2: Alpha, Beta, Gamma, Delta, Omicron was found to be the most prominent variant and distinct from the millions of SARS-CoV-2 genomes. Comparing to the Wuhan variant, it has 6 mutations which led to its increased transmissibility and vaccine resistance. The aggressive surge of Omicron made it the major corona virus strain, displacing the worldwide prevalent delta form. However, omicron was expected to be highly transmissible, it was unclear whether it was really higher transmissible than delta. Furthermore, emerging research suggested that it was spreading like wildfire in the context of the continued spread of the delta variant and the greater natural immunity to the delta variant. If the trend had been continuing, Omicron had expected to make the delta a more common variant. Refer [1, 2] for more information on Omicron. For analyzing and predicting the nature as well as the successive waves of COVID-19, fractal based methods were used, see for instance [3–5].

A. Gowrisankar et al., *Fractal Functions Communication with Fractional Calculus*,
https://doi.org/10.1007/978-3-031-96993-5_9

In this analysis, the seven days moving average of daily positive cases of the six countries namely Denmark, Germany, India, Netherlands, South Africa and United Kingdom is investigated. The fractal interpolation function is used to construct the graph of seven days moving average. By considering the time (in days) and the value of seven days moving average as the data points, the fractal curve is obtained. In addition, the seven days moving average for the six countries is predicted for the upcoming three months from 13th December 2021.

9.1.1 Method

Moving average is a statistical measure, evaluated to examine the data by calculating series of averages of the different subsets of the entire data set. It is usually calculated to reduce the randomness, short term fluctuations of the data and to identify the trend direction of the data. As it is dependent on the past data, it is considered as a trend following indicator. In this letter, seven days moving average of daily positive cases is calculated by taking the average of past 7 days data. The graph of the seven days moving average is drawn using the fractal interpolation. The data set $\{(t_j, x_j) : j = 0, 1, \ldots, N\}$ is taken where t_j and x_j denote the time in days and seven days moving average of the daily positive cases starting from 27 January 2020, respectively. The fractal interpolation function is constructed using the following affine transformations

$$w_j \begin{pmatrix} t \\ x \end{pmatrix} = \begin{pmatrix} a_j & 0 \\ c_j & d_j \end{pmatrix} \begin{pmatrix} t \\ x \end{pmatrix} + \begin{pmatrix} e_j \\ f_j \end{pmatrix} \tag{9.1}$$

satisfying the join-up conditions $w_j(t_0, x_0) = (t_{j-1}, x_{j-1}),\ w_j(t_N, x_N) = (t_j, x_j)$, $j = 1, 2, \ldots, N$, where a_j, c_j, d_j, e_j and f_j all real parameters. If d_j is a predefined variable in the system (9.1) with join-up conditions, then it provides a unique solution for (9.1). Therefore, for $j = 1, 2. \ldots, N$, the constants a_j, c_j, e_j, f_j can be obtained uniquely as follows

$$\begin{aligned} a_j &= \frac{t_j - t_{j-1}}{t_N - t_0} \\ e_j &= \frac{t_N t_{j-1} - t_0 t_j}{t_N - t_0} \\ c_j &= \frac{(x_j - x_{j-1}) - d_j(x_N - x_0)}{t_N - t_0} \\ f_j &= \frac{(t_N x_{j-1} - t_0 x_j) - d_j(t_N x_0 - t_0 x_N)}{t_N - t_0}. \end{aligned}$$

The free parameter $|d_j| < 1$ is called the vertical scaling factor. In this study, we optimally selected d_j as $x_j - x_{j-1}/\sqrt{(\max_j(x_j) - \min_j(x_j))^2 + (x_j - x_{j-1})^2}$ for $j = 1, 2, \ldots, N$.

9.1.1.1 Data Description

The data for this letter has been taken from the World Health Organization, COVID-19 Data Repository by the Center for Systems Science and Engineering and Our World in Data [6–8].

9.1.2 Results and Discussions

Figure 9.1 compares the 7 days moving average of daily positive cases of the countries Denmark, Germany, India, Netherlands, South Africa and United Kingdom. After the first reporting case of Omicron variant, the following results can be clearly visualized from the subgfigure: there was no major change in the graphs of Denmark and India, the graph of Germany and South Africa increases steeply, whereas, the curves of Netherlands and United Kingdom seemed to have slight fluctuations.

On February 27, 2020, it was announced that the virus has spread to Denmark. The country immediately started its lockdown measures on the month of March. While most of the European countries were on high alert as of beginning of May, due to the COVID-19 problem, Denmark has slowly begun rebuilding its society and industry in the middle of last year April. The Delta form, which was estimated to be twice as contagious as the original Alpha variant, was transmitted rapidly among Denmark's young people who have not been vaccinated. Figure 9.2a clearly shows that, the count of daily positive cases started steeply increasing from October 2021

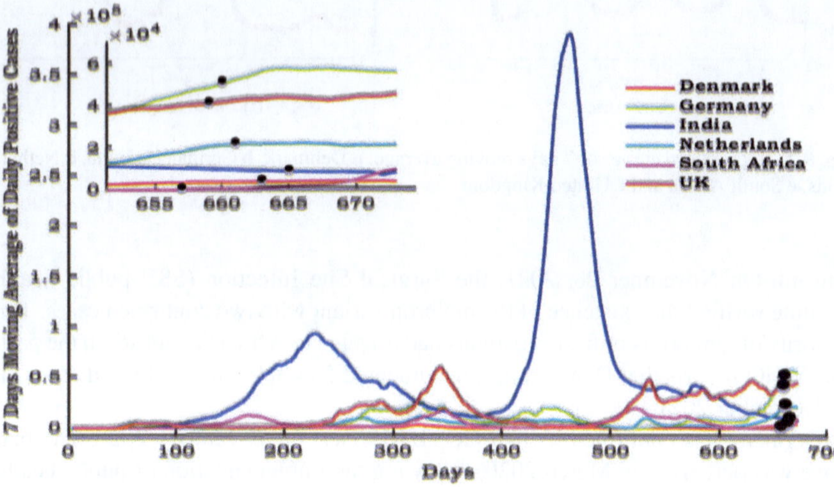

Fig. 9.1 Comparison of 7 days moving average of daily positive cases of Denmark, Germany, India, Netherlands, South Africa and United Kingdom. Black dot indicates the occurrence of first Omicron case

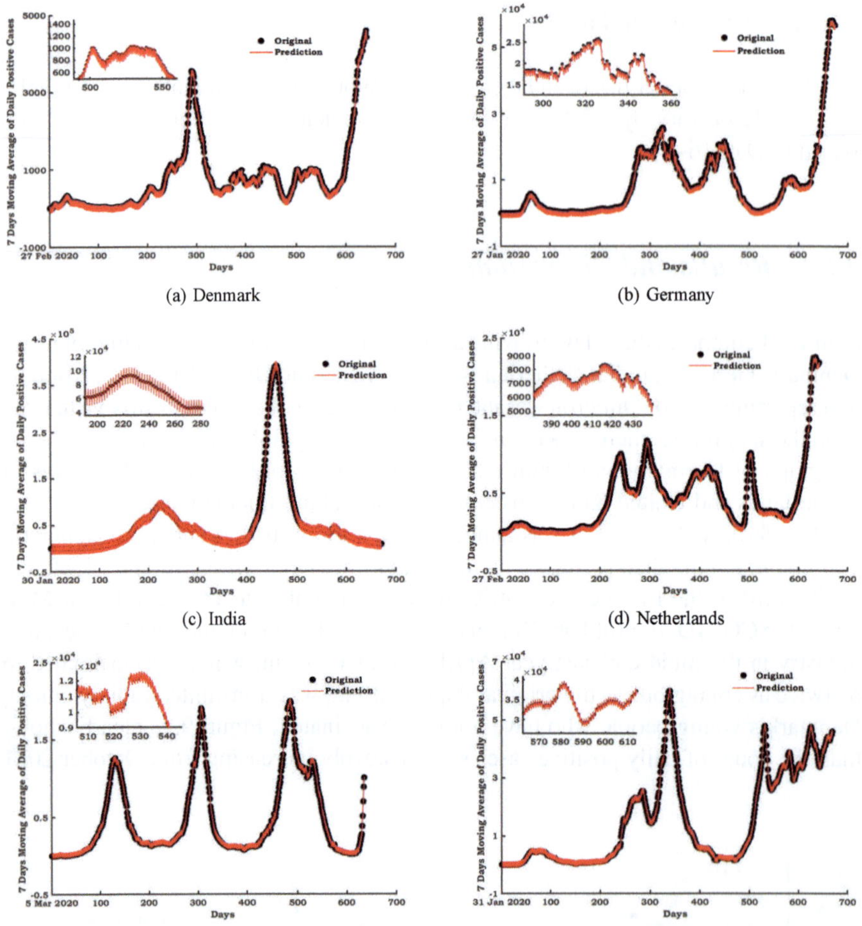

(a) Denmark (b) Germany

(c) India (d) Netherlands

(e) South Africa (f) United Kingdom

Fig. 9.2 Reproduced curves of 7 days moving average: **a** Denmark, **b** Germany, **c** India, **d** Netherlands, **e** South Africa and **f** United Kingdom

onward. On November 28, 2021, the Surgical Site Infection (SSI) public health institute verified the existence of the omicron variant with two confirmed cases. The majority of Omicron confirmed patients had traveled to African countries in the past. The WHO reports that Denmark has documented 554,389 total confirmed cases on 13 December 2021.

Germany has confirmed its first COVID-19 case on 27 January 2020. The first wave was perceived in March 2020, following the implementation of public health measures. Figure 9.2b illustrates that the second wave had began in late September 2020 and peaked in December 2020, with 34,000 new cases each day. The Alpha variant had triggered a third wave of epidemic in March 2021. Infection numbers had began to steadily decline in late April, and the third wave was declared terminated by early May 2021. By the end of June, the Delta form had taken over among new infections, and cases began to rise again. As a consequence, the country entered the fourth wave of the epidemic on August 2021, with the majority of infections coming from the younger age groups. Since about November 25, 2021, 68.1 percent of the overall population had been vaccinated, with significant regional diversity across states. Germany detected its first omicron variant on 27 November 2021. As of December 13, 2021, the total number of confirmed cases and Omicron cases were found to be 6,531,606 and 101, respectively.

The first confirmed case of SARS-CoV-2 was reported in India on 30 January 2020. As the positive cases started to increase day by day, the government declared a national lockdown on 25 March 2020. During the first wave, India, like other parts of the world, seen a sharp increase in COVID-19 cases and deaths. For the first time on the month of June 2020, India's recoveries outnumber current cases. The number of daily positive cases, as well as infection rates, began to decline in September 2020. From the Fig. 9.2c, it is observed that the daily positive cases peaked in mid-September, with over 90,000 recorded each day, have decreased to less than 15,000 in January 2021. Further, Fig. 9.2c indicates that the second wave, which started in March 2021, was observed to be more devastating than the first wave. From 16 January, 2021 onward, vaccination program had been conducted and administered more than 500 million doses as of August 2021. However, there had been no statewide shutdown in India during the second wave of the pandemic. The B.1.617 (dominant in Maharashtra), B.1.1.7 (state of Delhi and Punjab), B.1.618 (West Bengal) variants are the different types of strains reported during the two waves. On December 2, 2021, the first case of Omicron was identified in India. As of 13 December 2021, the country had confirmed 35 Omicron cases.

Netherlands was exposed to the COVID-19 virus when the first case was identified in Tilburg, on February 27, 2020. All the four variants of concern: Alpha variant (B.1.1.7), Beta variant (B.1.351), Gamma variant (P.1) and Delta variant (B.1.617.2) had been detected in Netherlands. Figure 9.2d shows that the first wave was ended at the beginning of July 2020, with no gap, the second wave was started from the end of July 2020, since then there had been a huge rise in the number of daily positive cases. During the second wave, to reduce the risk of transmission of Lineage B.1.1.7, which was thought to be more infectious, the government had proposed strict laws in January 2021. From the Fig. 9.2d, it was observed that the number of daily positive cases in the country had reached its maximum during the months of October and December, 2021. Further, we can notice that the number of daily positive cases had been steadily increasing after the confirmation of first Omicron variant (B.1.1.529) in November 2021. Over 24.4 million vaccinations were estimated to have been provided in the Netherlands by Sunday, December 5, 2021. It was expected that 88.8% of adults over the age of 18 have had at least one vaccination, with 85.5 percent having received all of them. A total of 2,880,426 confirmed cases had been identified in the country on 13 December 2021. In South Africa, the first case had been confirmed on March 1, 2020 and the President of the country, Ramaphosa declared a 21-day national lockdown. From Fig. 9.2a, the first wave can be observed from March to November 2020. At the end of November 2020, 790,004 confirmed cases were documented. From December

2020 onwards, the second wave was commenced and 501.V2 variant was discovered by the scientists. During the second wave, the number of confirmed cases had crossed one million and it has ended in April 2021 with 1,581,210 confirmed cases. With no delay, third wave had began from May 2021 onwards with B.1.1.7 variant of concern, C.1.2 lineage variant and the number of confirmed cases observed at the end of October 2021 were 2,921,886. On November 24, 2021, the B.1.1.529 variant was initially reported to WHO by South Africa. The epidemiological condition of the country had been marked by three different peaks in documented cases, the most recent of which was dominated by the Delta strain. The spread of infection rouse sharply, corresponding with the discovery of the B.1.1.529 strain. A specimen collected on November 9, 2021, was the first reported B.1.1.529 infection. On November 26, 2021, the WHO designed the name Omicron for the new variant. Figure 9.2e shows that as a result of Omicron, South Africa had entered the fourth wave of COVID-19, and the rise in daily positive cases is substantially faster than the previous three waves in almost all provinces. The Omicron variation spread twice as quickly as the Delta variant and was the most contagious of the SARS-CoV-2 strains. On 29 November 2021, 2,858 daily positive cases were reported, while 16,055 daily positive cases were reported on December 4, 2021. This record clearly depicts how severe and fast was the Omicron variant. The country had confirmed 900 Omicron cases as of 13th December.

United Kingdom has reported its first case in the month of January 2020. Prime Minister Boris Johnson declared a state of emergency on March 23, 2020. The first wave of COVID-19 had come to the end at June 2020. In December 2020, vaccination program was implemented in the country. From Fig. 9.2f, it was noticed that the second wave had started at the beginning of September 2020; there was an initial peak in January 2021 and the number of daily positive cases started gradually decreasing during March 2021. With the Delta variant, the third wave emerged in United Kingdom at the end of July 2021 since then the curve was continually increasing. The Omicron variant had reached United Kingdom in late November 2021. United Kingdom stands in the first place with highest number of confirmed Omicron cases in the world. As of 13 December 2021, a total of 10,819,519 confirmed cases and more than 1500 Omicron cases had been documented. In Fig. 9.2, the pattern of the epidemic is reproduced by the fractal interpolation function given in Eq. 9.1, hence each reproduced curve has a self-similar which is elucidated in subfigures.

Figure 9.3 illustrates the prediction of 7 days moving average of Denmark, Germany, India, Netherlands, South Africa and United Kingdom for the months of December 2021, January 2022 and February 2022. From Fig. 9.3a, one can observe that the graph of Denmark shows a rapid growth until the end of January 2022 then the graph begins to fall down. In Fig. 9.3b, Germany's graph exhibits variations until the end of January 2022 and the graph starts to decline steadily. In Fig. 9.3c, India's seven days moving average of daily positive cases graph drops until mid-February 2022, then rises slightly near the end of the month. Figure 9.3d demonstrates that the steep increase in the graph of Netherlands begins to fluctuate for the next several days as of January 2022. Figure 9.3e indicates a health alert to South Africa since the curve shows progressive rise from December 2021 to the end of February 2022

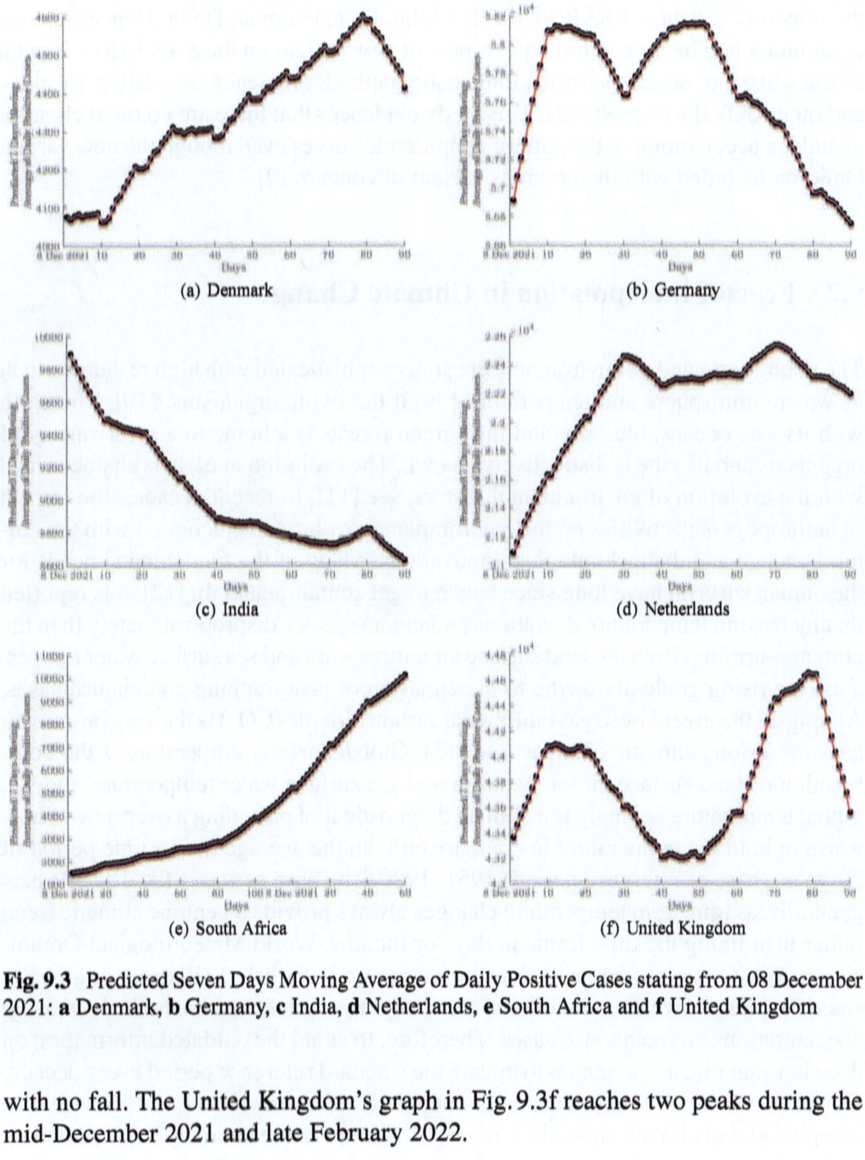

Fig. 9.3 Predicted Seven Days Moving Average of Daily Positive Cases stating from 08 December 2021: **a** Denmark, **b** Germany, **c** India, **d** Netherlands, **e** South Africa and **f** United Kingdom

with no fall. The United Kingdom's graph in Fig. 9.3f reaches two peaks during the mid-December 2021 and late February 2022.

9.1.3 Summary

The evolution of SARS-CoV-2 virus had been changing with respect to time which was the natural character of any virus. In general, while virus changing one mutant to another, its properties remains same or create little impact. In 2021, a new SARS-CoV-2 variant Omicron was reported to the WHO which is distinct from the previous variants of SARS-CoV-2: Alpha, Beta, Gamma, Delta. Hence, science community had been searching the impact of new variant on the SARS-CoV-2, such as transmission, disease severity, diagnostic method, efficiency of existing vaccines and other COVID-19 protocols. This study evidences that there are no rapid changes or sudden acceleration in the pattern of epidemic curves even though the new variant Omicron included with the previous variant of concern [9].

9.2 Fractal Interpolation in Climate Change

The young earth and its environment are super sophisticated with high resources such as water, atmosphere and temperature for all the living organisms, [10]. The earth with its vast oceans, blue sky and lush green forests is a home to a great variety of organisms and its vibe is distinctively its own. The evolution of earth is characterized with the evolution of environmental factors, see [11]. In recent decades, the impact of anthropogenic activities on the environmental evolution is increased with the cultural change and the technological innovations. Many of the fundamental needs for the human survival have long since began to get contaminated. In [12], it is reported that increasing temperature deviations endangers species disproportionately than the climate warming. Both the land surface air temperature and sea surface water temperature are rising gradually owing to the emission of heat trapping greenhouse gases. Among all the greenhouse gas emissions, carbon dioxide (CO_2) is the very prominent causing serious climatic changes, see [13]. Global surface temperature is the combination of land surface air temperature and sea surface water temperature. Indeed, global temperature anomaly is a climate diagnostic tool providing a overview of how warm or cold the temperature in comparison with the average over a time period of 30-years (base or reference period) 1951–1980. The base period is fixed as 30 years generally, as long term temperature changes always provide a genuine climatic trend rather than fixing the time frame in days or months. World Meteorological Organization (WMO) has recommended a new base period of 1991–2020 to adopt globally since the planet's weather conditions are changing much faster due to the increasing concentrations of greenhouse gases. Therefore, to avoid the outdated information on decision-making, it is essential to update the standard reference period every decade. However, WMO still recommends the standard reference period of 1961–1990 to compute global climate anomalies, refer [14]. Temperature anomaly is calculated by taking the difference between the reference temperature and the measured temperature at a particular place and time. The negative (positive) anomaly illustrates that the measured is cooler (warmer) than the reference temperature.

During the last few decades, benefited from the development of nonlinear techniques, several models have been introduced to better diagnose the continuously varying global climate behaviors. In this regard, it is essential to investigate the contemporary as well as the future climatic changes with respect to temperature rise. According to [15], the glaciers will experience losses of 25% to nearly 50% of their mass by 2100 under increases in global temperatures of 1.5 to 4 degree Celsius. In

[16], artificial neural networks, deep learning neural networks, and multiple linear regression have been used to estimate the CO2 rise from soil to atmosphere under greenhouse conditions. In [17], sources of atmospheric carbon dioxide emissions and recent developments in the implementation of carbon capture and storage process to alleviate greenhouse gas emissions are reviewed. A variety of prediction models have been advanced since 21st century to explore several time-series problems on climate changes, for instance, the reader can visit [18–23]. Nevertheless the data used in the aforementioned studies is completely raw and may contain errors leading to inaccurate results. Therefore, whenever a time-series problem is addressed, preprocessing is absolutely essential to eliminate such errors caused by missing data as well as to convert the provided raw data set into a structured format for further analysis. The fractal interpolation functions indeed are the most suitable for interpolating and preprocessing the real-time discrete sequences which share complexity nature, refer [24]. A study on greenhouse gas emissions reported in [25] has preprocessed the raw data to improve the accuracy using the method of linear fractal interpolation. Further, an autoregressive integrated moving average model is employed to forecast the future climatic changes and the results have alarmed the worlds's rapid submerge. Yet the study has concentrated only on the annual and per capita greenhouse gas emissions, and does not considered the global temperature data which has a greater influence in climate behaviours. Hence, by considering both the temperature data and the GHG emissions, this application is an extension of the study in reference [25].

In this investigation, the hidden variable fractal interpolation function is utilized to preprocess the temperature anomaly data by considering the CO_2 emissions as hidden variables. The preprocessed data is further used for predicting the future trends of global, land and sea surface temperatures by indulging the autoregressive fractional integrated moving average model (ARFIMA). It is a superior model comparing to other moving average models in light of the presence of fractional differencing operator, see [26]. The trends of global temperature always show complexity nature, so designing an appropriate model to interpret and predict its uncertain patterns is a bit tedious process. To overcome this issue, an algorithm is constructed comprising hidden variable fractal interpolation function and ARFIMA model to identify the model parameters for expert prediction. Moreover, the proposed model is tested with the evaluation metrics such as mean absolute percentage error, normalized mean absolute percentage error, symmetric mean absolute percentage error, mean absolute error and root mean square error to validate its accuracy. With the proposed Fractal-ARFIMA approach, the global temperature anomalies during the time period 2021–2030 is predicted with respect to the increasing rate of CO_2 emissions to alter the prevailing and future serious environmental changes. In this application, the method of hidden variable fractal interpolation and ARFIMA are utilized. Then a forecast model is constructed using the Fractal-ARFIMA algorithm to identify the model parameters and results are analyzed.

9.2.1 Model Design

Here, construction procedure of hidden variable fractal interpolation function and autoregressive fractional integrated moving average model is briefly discussed to develop a fractal-ARFIMA algorithm with optimized scalings.

9.2.2 Brief on Hidden Variable Fractal Interpolants

Here, let us see the recap of hidden variable fractal function which is discussed in Chap. 2. Hidden variable fractal interpolation function is a fractal function on the higher dimensional space, however it is used to approximate the provided single-valued function efficiently by projecting the fractal function from $\mathbb{R}^3$ to $\mathbb{R}^2$. In modelling the real-time discrete data, generally, fractal functions operate in the $\mathbb{R}^2$-plane to produce self-similar functions interpolating given data set. In contrast, the HVFIF works in the $\mathbb{R}^3$ space to yield a fractal function in the $\mathbb{R}^3$ space and it is projected on the $\mathbb{R}^2$ plane to model the given discrete data. The added dimension makes the HVFIF more flexible thereby producing non self-affine functions.

The given set of data points $\{(t_j, x_j) : j = 1, 2, \ldots, N,\ t_j < t_{j+1}\}$ is extended from $\mathbb{R}^2$ to $\mathbb{R}^3$ by adding the hidden variables $\{y_j : j = 1, 2, \ldots, N\}$ to form a new data set in $\mathbb{R}^3$ such that $\{(t_j, x_j, y_j) : j = 1, 2, \ldots, N\}$. To interpolate the new data set, an IFS $\{\mathbb{R}^3; w_j : j = 1, 2, \ldots, N-1\}$ is generated using following $N-1$ transformations:

$$
\begin{aligned}
w_j \begin{pmatrix} t \\ x \\ y \end{pmatrix} &= \begin{pmatrix} a_j & 0 & 0 \\ b_j & \alpha_j & \beta_j \\ c_j & 0 & \gamma_j \end{pmatrix} \begin{pmatrix} t \\ x \\ y \end{pmatrix} + \begin{pmatrix} k_j \\ l_j \\ m_j \end{pmatrix} \\
&= \begin{pmatrix} L_j(t) \\ F_j(t, x, y) \end{pmatrix}.
\end{aligned}
$$

where

$$
\begin{aligned}
L_j(t) =& a_j t + k_j, \\
F_j(t, x, y) =& (\alpha_j x + \beta_j y + p_j(t), \gamma_j y + q_j(t)),
\end{aligned}
$$

for $j = 1, 2, \ldots, N$, such that the following join-up conditions

$$
w_j \begin{pmatrix} t_1 \\ x_1 \\ y_1 \end{pmatrix} = \begin{pmatrix} t_j \\ x_j \\ y_j \end{pmatrix} \text{ and } w_j \begin{pmatrix} t_N \\ x_N \\ y_N \end{pmatrix} = \begin{pmatrix} t_{j+1} \\ x_{j+1} \\ y_{j+1} \end{pmatrix} \tag{9.2}
$$

are satisfied. The 2 × 2 matrix $A_j = \begin{pmatrix} \alpha_j & \beta_j \\ 0 & \gamma_j \end{pmatrix}$ is called the contraction matrix. By choosing each parameter in the matrix satisfying the constraint $\|A_j\|_\infty < 1$, the remaining parameters a_j, b_j, c_j, k_j, l_j and m_j of the transformation w_j can be determined using the endpoint conditions in Eq. (9.2). The attractor of the IFS defined in $\mathbb{R}^3$ is projected onto the $\mathbb{R}^2$-plane, and it is the graph of a continuous function $f : [t_1, t_N] \to \mathbb{R}^2$ satisfying $f(t_j) = (x_j, y_j)$, $\forall\, j = 1, 2, \ldots, N$. The hidden variable fractal interpolation function for the given data set $\{(t_j, x_j) : j = 1, 2, \ldots, N\}$ is defined as the continuous function $f_1 : [t_1, t_N] \to \mathbb{R}$ which is the first component of $f = (f_1, f_2)$ and the second component f_2 is referred as the fractal function interpolating the data set $\{(t_j, y_j) : j = 1, 2, \ldots, N\}$.

The attractor of the IFS defined in $\mathbb{R}^3$ is self-affine whereas its projection in the $\mathbb{R}^2$ plane may not be a self-affine function. Thus, the attractor is projected onto the $\mathbb{R}^2$ plane exhibiting a symmetrical relationship between the variables x and y corresponding to the t-variable.. However, if the scalings are taken such that $\alpha_j + \beta_j = \gamma_j$ and $y_j = x_j$ for $j = 1, 2, \ldots, N$ in the extended data set, then the function f_1 coincides with the function f_2 and in such case, the graph of f_1 is self-affine as the case of linear fractal interpolation function.

Though the description of HVFIF seems to be expressed succinctly, the generated fractal functions are quiet complicated due to the more number of scaling parameters in the contraction matrix comparing to the scaling parameters in the linear fractal interpolation. Hence, an efficient algorithm is presented in the subsequent section to identify the scaling parameters for the generation of more flexible HVFIF.

9.2.3 Construction of ARFIMA Model

The autoregressive model of order p, AR(p) is expressed by

$$X_t = c + \sum_{k=1}^{p} \phi_k X_{t-k} + \epsilon_t,$$

where $\phi_1, \phi_2, \ldots, \phi_p$ are the parameters of autoregression, ϵ_t is a random variable, called the white noise and c is a constant. The moving average model of order q, MA(q) is expressed by

$$X_t = \mu + \sum_{k=1}^{q} \theta_k \epsilon_{t-k} + \epsilon_t,$$

where $\theta_1, \theta_2, \ldots, \theta_q$ are the parameters of moving average, $\epsilon_t, \epsilon_{t-1}, \ldots$ are the white noise terms and μ is the expectation of X_t. The generalized form of the autoregressive (AR) and moving average (MA) models is given by

$$(1 - \sum_{k=1}^{p} \phi_k B^k)(1 - B)^d (X_t - \mu) = (1 + \sum_{k=1}^{q} \theta_k B^k)\epsilon_t,$$

here, $(1 - B)^d$ is the difference operator. In case of the autoregressive moving average (ARMA) or autoregressive integrated moving average (ARIMA) models, the range of the order p is confined to an integer. The binomial expansion of the operator for any real d with Gamma function is defined as

$$(1 - B)^d = \sum_{k=1}^{\infty} \binom{d}{k} (-B)^k = \sum_{k=1}^{\infty} \frac{\Gamma(d+1)}{\Gamma(k+1)\Gamma(d+1-k)(-B)^k}. \tag{9.3}$$

To deal the fractional systems, it is always better to use fractionally differenced operator rather than using an integer difference operator, hence the ARIMA model is generalized to the autoregressive fractional integrated moving average (ARFIMA) model. The ARFIMA (p, d, q) is defined by

$$\Phi(B)(1 - B)^d X_t = \Theta(B)\epsilon_t,$$

where $0 < |d| < 0.5$ and $(1 - B)^d$ in Eq. (9.3) is the fractional difference operator, p is the autoregressive order, d is the level of fractional differencing, q is the moving average order. For the larger values of d, the model closely approximates the simple integrated series better than the ARIMA model. For more information on ARFIMA, refer [27].

9.2.4 *Fractal-ARFIMA Algorithm*

The algorithm involving HVFIF and ARFIMA methods is constructed as follows.

1. Consider the data $\{(t_j, x_j, y_j) : j = 1, 2, \ldots, N\}$ in $\mathbb{R}^3$, where t_j denotes the time period (in years), x_j indicates the surface temperature (in °C) and y_j represents the per capita CO_2 emissions (in tonnes).
2. Fix the time period t_j. Take initial point and next point in the given data sets x_j and y_j as the first set of interpolation points.
3. The scaling parameters α_j, β_j and γ_j in the contraction matrix A_j are identified using the below formulae,

$$\alpha_j = \frac{x_j - x_{j-1}}{\frac{\Delta x}{\Delta t}\sqrt{(x_{\max} - x_{\min})^2 + (x_j - x_{j-1})^2}},$$
$$\gamma_j = \frac{y_j - y_{j-1}}{\frac{\Delta y}{\Delta t}\sqrt{(y_{\max} - y_{\min})^2 + (y_j - y_{j-1})^2}}, \tag{9.4}$$

   where $\Delta t = t_{\max} - t_{\min}$, $\Delta x = x_{\max} - x_{\min}$, $\Delta y = y_{\max} - y_{\min}$, x_j and y_j are the global temperature and CO_2 emissions with respect to time t_j, respectively for $j = 1, 2, \ldots, N$. The formula for the constrained parameter β_j is chosen same as γ_j in Eq. (9.4) such that $|\beta_j| + |\gamma_j| < 1$.

4. If $|\beta_j| + |\gamma_j| < 1$ go to next step, otherwise β_j is equal to $rand(1, N)$, where $rand(1, N)$ generates N uniformly distributed random numbers between 0 and 1 obeying the constraint.
5. Employing the contraction parameters $\alpha_j, \beta_j, \gamma_j$, determined in the above steps, and endpoint conditions of w_j in Eq. (9.2), the other map parameters a_j, b_j, c_j, k_j, l_j and m_j can be estimated.
6. For the second set of data, the right side end point of the first set of interpolation and its immediate next point are chosen.
7. Repeat steps (1) to (6) to obtain the iterated function system, whose attractor is the required fractal function's graph to preprocess the interpolation data.
8. The ARFIMA (p, d, q) model, which is described in the Sect. 9.2.3 with the chosen values $p = 1$, $d = 0.38$, $q = 1$, is applied on the preprocessed data for prediction.

The degree of fractional differencing d can be any number between 0 and 0.5. In this algorithm, the value of d is taken to be 0.38, as considered in [27].

9.2.5 *Data Description*

The data on global temperature anomalies (in degree Celsius) and per capita CO_2 emissions (in tonnes) are taken from the official websites of *National Aeronautics and Space Administration* (NASA), [28] and *Our World in Data*, [29] time period varying from 1880 to 2020. For estimating the temperature anomalies, the base temperature is considered over 30 years from 1951 to 1980. Excel-19 is used to create a time-series database separately for each of the following temperature data: global temperature anomaly, land surface air temperature anomaly, sea surface water temperature anomaly, northern and southern hemisphere temperature anomalies. The built time-series database for 40 years is preprocessed using the HVFIF and its future trends are predicted using ARFIMA model. Global heat maps demonstrating the relationship between temperature anomaly and chosen time length are generated with the aid of NASA website, [28] by providing required input details such as map type, time interval, base period, mean period and map projection.

9.2.6 *Evaluation Metrics*

This section presents the formulations of evaluations metrics such as MAPE, NMAPE, sMAPE, MAE and RMSE in order to validate our proposed forecast model. With an aim of testing the accuracy of the forecast models, various performance evaluation measures have been employed in the literature based on both percentage and relative errors of the models. For the present investigation, some of the familiar statistical metrics are discussed as follows. Let $x(t_i)$ and $\hat{x}(t_i)$ represent the original and forecast data, for $i = 1, 2, \ldots, N$. Here the original data are preprocessed data and the forecast data are values predicted using the preprocessed data. The mean absolute percentage error (MAPE) is used to validate the forecast model's accuracy. It is actually a mean of the absolute percentage errors of forecast values and is formulated by

$$\text{MAPE} = \frac{1}{N}\sum_{i=1}^{N}\left|\frac{x(t_i) - \hat{x}(t_i)}{x(t_i)}\right| \times 100\%. \tag{9.5}$$

Suppose the MAPE returns values greater than 50% then the model is not a good forecast model, on the other hand, if it returns values less than 10% then the constructed forecast model is a well-efficient one. Normalized mean absolute error (NMAPE) is one more frequently used metric for evaluating the model accuracy and it is expressed by

$$\text{NMAPE} = \frac{1}{N}\sum_{i=1}^{N}\left|\frac{x(t_i) - \hat{x}(t_i)}{\frac{1}{N}\sum_{i=1}^{N} x(t_i)}\right| \times 100\%. \tag{9.6}$$

Symmetric mean absolute percentage error (sMAPE) is a relative error measure, which combines the errors determined for different time-series into a single number and its mathematical formulation is provided by

$$\text{sMAPE} = \frac{1}{N}\sum_{i=1}^{N}\frac{\left|x(t_i) - \hat{x}(t_i)\right|}{(|x(t_i)| + |\hat{x}(t_i)|)/2} \times 100\%. \tag{9.7}$$

The forecasts are better, the closer the sMAPE value is to zero. Mean absolute error (MAE) is a another evaluation metric to compare the observed and predicted time series. It is the average of sum of the absolute errors for the two observations exhibiting same phenomena and it is given by

$$\text{MAE} = \frac{\sum_{i=1}^{N}\left|x(t_i) - \hat{x}(t_i)\right|}{N}. \tag{9.8}$$

Root mean square error (RMSE) is employed to verify the experimental outcomes of the forecast models, particularly when the model exhibits undesirable large errors. The RMSE for two time-series (original and predicted) is expressed by

$$\text{RMSE} = \sqrt{\frac{\sum_{i=1}^{N}(x(t_i) - \hat{x}(t_i))^2}{N}}. \tag{9.9}$$

RMSE provides greater weight to large errors, since before taking the average, the errors are squared. Both MAE and RMSE are negatively-oriented and lower values always suggest the best fit.

9.2.7 *Results and Discussion*

The unprocessed (raw) data of temperature anomaly (in deg Celsius) and per capita CO_2 emissions (in tonnes) taken from the official websites of *NASA* [28] and *Our World in data* [29] are represented graphically in Fig. 9.4a, b, respectively. From Fig. 9.4a, it is seen that the global temperature attains a peak at the year 1940 and from the beginning of 1950, there is a constant fluctuation in the temperature patterns. Figure 9.4b shows that there is a gradual increase in the per capita CO_2 emissions from the year 1945 to 1980, then the curve starts to fluctuate and shows a decline trend for the two years 2019 and 2020.

The worldwide scenarios of environmental changes are portrayed in terms of temperature anomalies through the global heat maps by choosing two different base periods for several set of ten years time interval in Figs. 9.5 and 9.6. The maps are developed using Robinson map projections over the mean period-January to December with anomalies map type. Figure 9.5a–d represent the global maps of annual land and ocean temperature anomaly with common base period 1951–1980 for the time intervals 1981–2020, 1981–1990, 1991–2000 and 2001–2010, respectively. Although the two 10 years time intervals 1981–1990 and 1991–2000 show less temperature rise 0.26 and 0.38 °C, the third 10 year time interval from 2001 to 2010 has showed a greater rise 0.63 °C and which leads to the increased average temperature rise during the time period of 40 years from 1981–2020 with 0.52 °C. Fixing the base period 1981–2010, global heat maps of annual land surface temperature anomaly and ocean temperature anomaly are separately provided for the time interval 2011–2020 in Fig. 9.6a, b, respectively. Note that the land temperature with 0.52 °C is greater than the ocean temperature with 0.31 °C. Global heat maps of annual global temperature anomaly for the time interval 2011–2020 with base period 1981–2010 is depicted in Fig. 9.6c showing 0.41 °C.

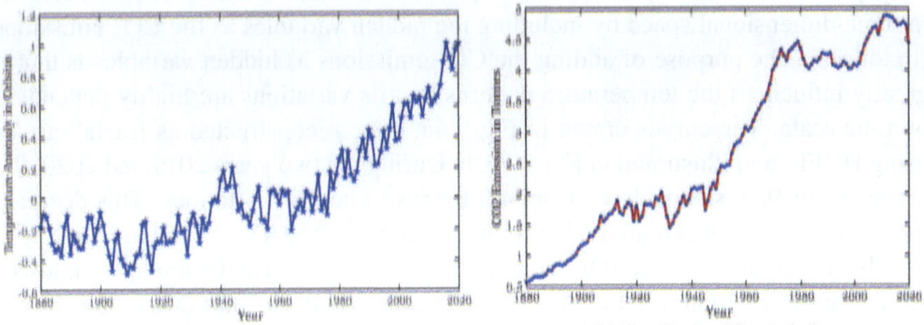

(a) Global Temperature

(b) Per Capita CO_2 Emissions

Fig. 9.4 Graphical representation of **a** global temperature anomaly and **b** per capita CO_2 emissions, during the time period 1880–2020

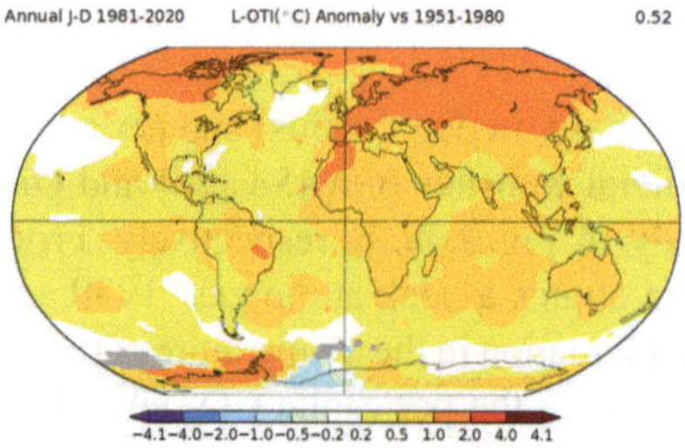

(a) Global Temperature Anomaly: 1981-2020

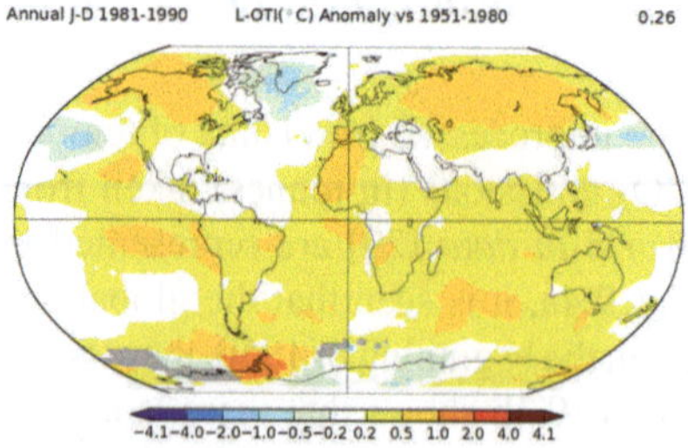

(b) Global Temperature Anomaly: 1981-1990

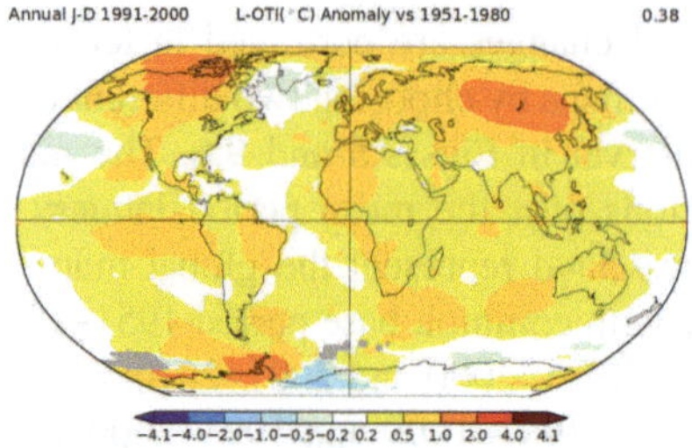

(c) Global Temperature Anomaly: 1991-2000

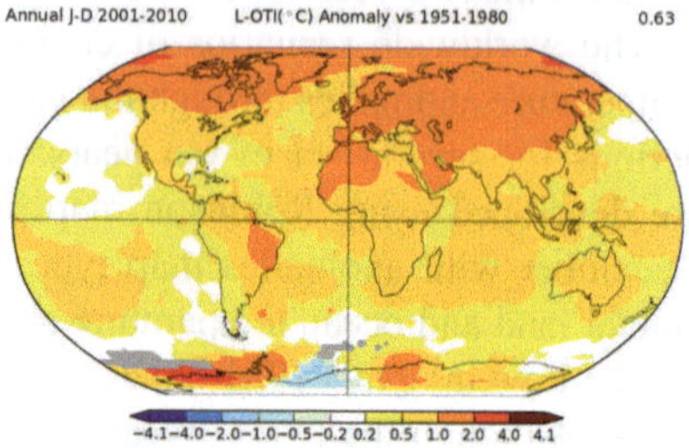

(d) Global Temperature Anomaly: 2001-2010

Fig. 9.5 Global heat maps showing annual land and ocean temperature anomaly across the planet with base period 1951–1980 for the time intervals **a** 1981–2020, **b** 1981–1990, **c** 1991–2000 and **d** 2001–2010 with anomaly estimations of 0.52, 0.26, 0.38 and 0.63 °C, respectively. Annual J-D represents annual mean period of January-December, L-OTI indicates Land-Ocean Temperature Index (Map projection: Robinson, Map type: Anomalies)

Fractal function is an excellent choice for preprocessing the temperature data as it exhibits irregular patterns at any time range. Method of hidden variable fractal interpolation is employed for preprocessing the raw data. Initially, the temperature data (in °C) collected with respect to time (in years) is extended to the generalized data set in three-dimensional space by including the hidden variables as the CO_2 emissions (in tonnes). The purpose of adding the CO_2 emissions as hidden variables is that it greatly influences the temperature patterns and its variations are highly dependent on time scale. The curves drawn in Fig. 9.4a, b are reconstructed as fractal curves using HVFIF and illustrated in Fig. 9.7a, b. During the two years 2019 and 2020, the curve in Fig. 9.7b steeps down from 4.9 tonnes to nearly 4.4 tonnes. This decreasing trend of CO_2 emissions is due to the impact of COVID-19 pandemic and it is clearly investigated in [30]. International borders are closed and national lockdowns are implemented, both public and private organizations are shut down, transportation and energy consumption rates have been greatly diminished, the entire aviation industry has been affected severely. Governments around the world have established progressively stringent policies in an effort to isolate cases, restrict the spread of the virus, and gradually slow it down. As a consequence of all this forced confinements,

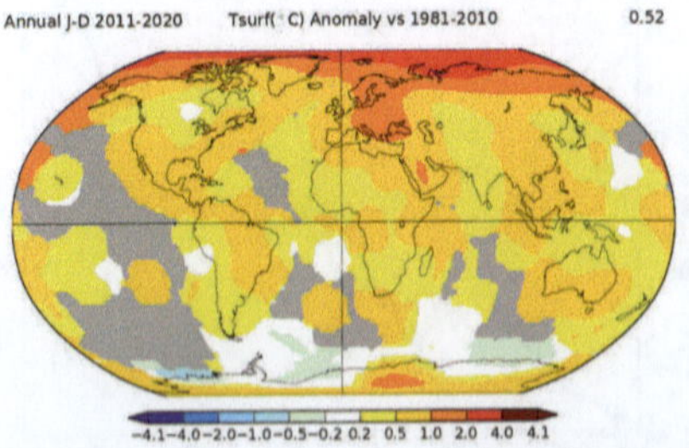

(a) Land Temperature Anomaly: 2011-2020

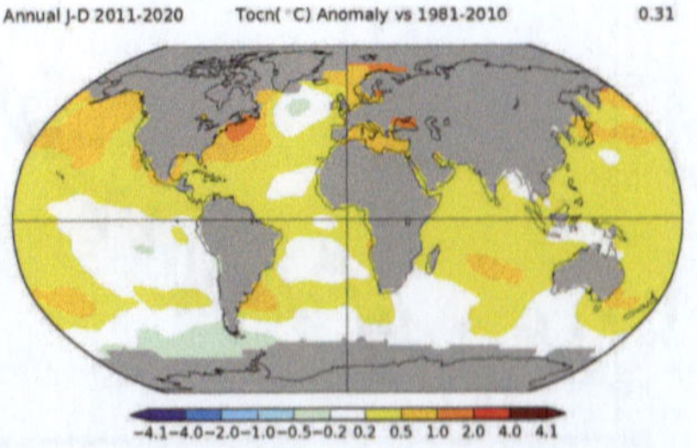

(b) Ocean Temperature Anomaly: 2011-2020

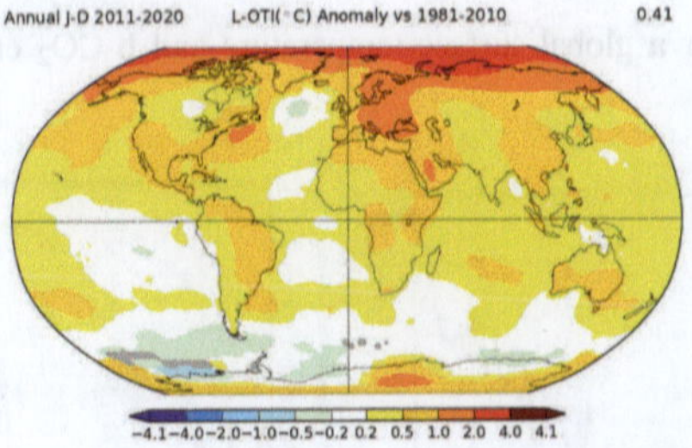

(c) Global Temperature Anomaly: 2011-2020

Fig. 9.6 Global heat maps showing temperature anomaly across the planet with base period 1981–2010 and the time interval 2011–2020 for **a** Land Surface, **b** Ocean Surface and **c** Land and Ocean Surface with anomaly estimations of 0.52, 0.31 and 0.41 °C, respectively. Annual J-D represents annual mean period of January-December, Tsurf denotes Land-surface, Tocn represents Ocean-surface, L-OTI indicates Land-Ocean Temperature Index (Map projection: Robinson, Map type: Anomalies)

the concentration of global atmospheric CO_2 has decreased significantly. Nevertheless, the rate of worldwide CO_2 emissions from the residential sectors is about the same or higher than it was prior to 2019. Overall, the COVID-19 crisis has contributed to a temporary decline in the global CO_2 emissions. Figure 9.8a, b demonstrate the fractal curves of land surface air temperature and sea surface water temperature in deg Celsius, respectively. At the end of 2020, the land and sea temperature anomalies are found out to be 1.5 and 0.6 °C. For broader analysis, the global surface temperature is estimated hemisphere-wise. Figure 9.9a, b show the temperature anomalies of northern and southern hemispheres, respectively. The curves in Fig. 9.9a, b show a fluctuating trend throughout the time period from 1980 to 2020.

This study aims to predict the future temperature anomalies for global, land and sea surfaces during the time period 2021–2030 based on the rate of per capita CO_2 emissions. In prior to predict the future temperature anomaly trend, the proposed model is tested for accuracy. The global, land and sea surface temperature anomalies are preprocessed using the HVFIF during the time period 1980–2010. The prepro-

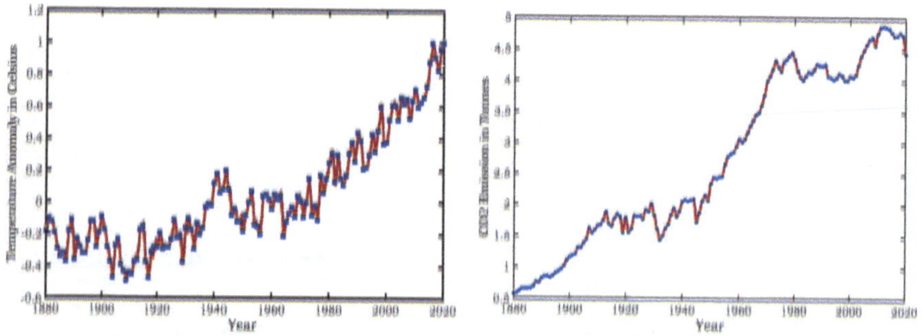

(a) Land and Sea Temperature (b) Per Capita CO_2 Emissions

Fig. 9.7 Fractal curves for **a** global surface temperature and **b** CO_2 emissions during the time period 1880–2020

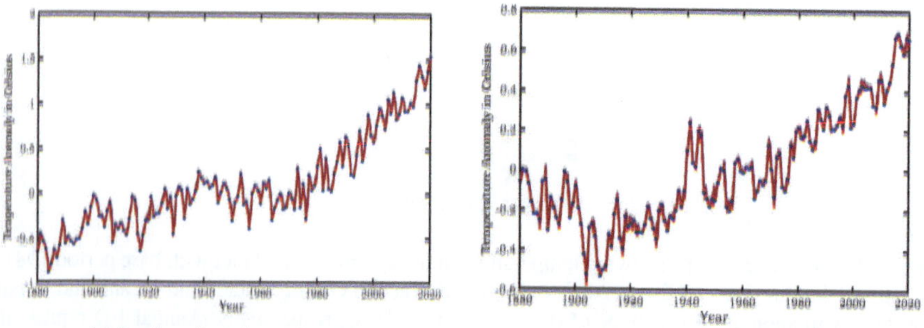

(a) Land-surface air temperature (b) Sea-surface water temperature

Fig. 9.8 Fractal curves for **a** land-surface and **b** sea-surface temperature anomalies during the time period 1880–2020

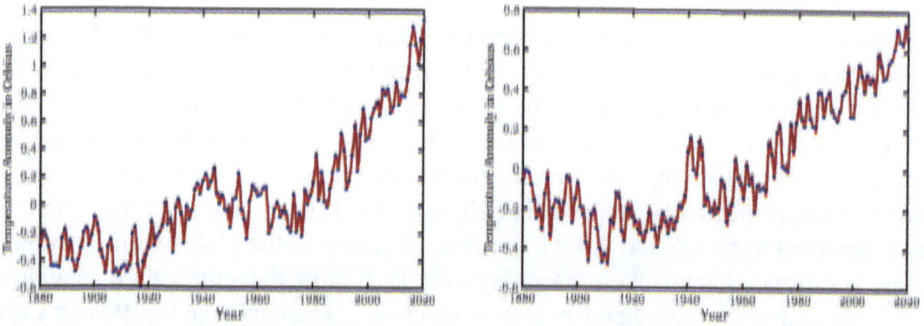

(a) Northern hemisphere (b) Southern hemisphere

Fig. 9.9 Fractal curves for **a** northern hemisphere and **b** southern hemisphere temperature anomalies during the time period 1880–2020

cessed data is then utilized to predict the changes of temperature anomaly during the time period 2011–2020 and compared with the original temperature anomaly which

Table 9.1 Estimation of statistical measures during the time period 2010–2020

Error estimation (in °C)	MAPE	NMAPE	sMAPE	MAE	RMSE
Global temperature	0.0017	0.0016	0.0017	0.1369	0.1490
Land temperature	0.0016	0.0014	0.0014	0.1734	0.2024
Sea temperature	0.0020	0.0017	0.0017	0.1162	0.0949

is previously available. For the evaluation of Fractal-ARFIMA model, a sample data during the time period 2010–2020 is statistically tested. The statistical measures such as MAPE, NMAPE, sMAPE, MAE and RMSE are estimated for the time period 2010–2020 and the results are tabulated in Table 9.1. In a similar fashion, data for any time period can be evaluated using the aforementioned measures. The evaluation metrics are computed by taking the actual values as raw data and the predicted values as data predicted using the proposed model. The estimated values of all the measures in Table 9.1 are approaching zero ensuring that the proposed model is an efficient one for forecasting the future evolutions. In [23], Hosking has recommended an ARFIMA $(1, d, 1)$ model with any fractional differencing $d \in (0, 0.5)$ can be applied for better forecasting of temperature data. In this study, in particular, an ARFIMA $(1, 0.38, 1)$ model with $d = 0.38$ is employed for predicting the future temperature anomalies. The prediction of preprocessed temperature anomaly for the time period 2011–2020 is graphically illustrated in Fig. 9.10. The predictions are performed by taking the base period as 1981–2010 for global, land and sea temperature anomalies and are respectively represented in Fig. 9.10a–c.

Taking the base period as 1991–2020, future patterns of global, land and sea temperature anomalies are predicted for the 10 years time period 2021 to 2030 and are graphically portrayed in Fig. 9.11a–c. The temperature anomaly for the year 2021 already exists, hence, the original data for the year 2021 can be used for the evaluation of model's accuracy with the predicted data. The actual and predicted temperature anomalies to global, land and sea surfaces for the year 2021 are estimated and the results are presented in Table 9.2. As a sample evaluation, year 2021 is chosen nevertheless the comparison of temperature anomaly can be made for any particular year. This comparison further guarantees the validity of the proposed model since the predicted values are not so far from the actual values. From Fig. 9.11, it is seen that the global surface tends to reach a temperature anomaly greater than 1 °C, the land surface attains a temperature anomaly of 1.7 °C and the sea surface is expected to attain more than 0.7 °C. In Figs. 9.10 and 9.11, the blue curves indicate the fractal

Table 9.2 Comparison of temperature anomalies for the year 2021

Comparison (in °C)	Actual anomaly	Predicted anomaly
Global temperature	0.84	0.9556
Land temperature	1.29	1.5163
Sea temperature	0.56	0.6634

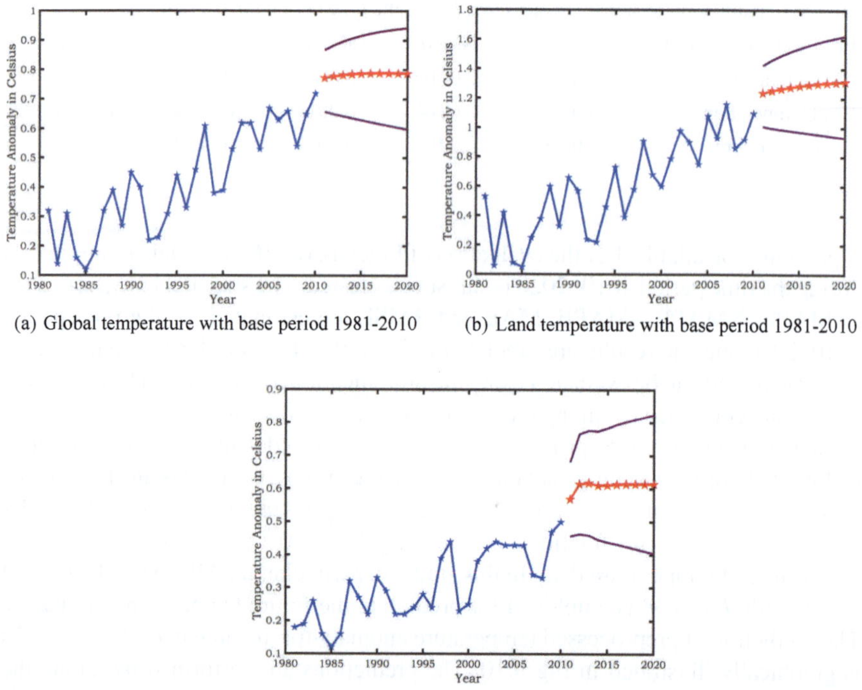

(a) Global temperature with base period 1981-2010

(b) Land temperature with base period 1981-2010

(c) Sea temperature with base period 1981-2010

Fig. 9.10 ARFIMA prediction of **a** global temperature, **b** land temperature and **c** sea temperature anomalies, with base period 1981–2010 for the time interval 2011–2020

curves of preprocessed data for time period 1981–2010, the red curves denote the fractal curves of predicted data for the time period 2011–2020 and the purple curves are 95% of prediction interval.

9.2.8 Summary

This application has estimated the future temperature anomalies of the planet during the time period 2021–2030, where the forecasts are mainly based on the per capita CO_2 emissions. The results of the study provide a clear scientific picture of our climate system influenced by the CO_2 emissions and it is the right time to investigate the standard CO_2 emission scenarios as they affect many of the environmental factors rigorously. To correlate the CO_2 emissions with the global temperature changes, hidden variable fractal interpolation function is utilized. On applying the Fractal-ARFIMA algorithm, the available raw data on temperature and CO_2 emissions are preprocessed and predicted. The predictions suggest that the temperatures of global, land and sea surfaces are expected to reach 1, 1.7 and 0.7 °C, respectively, at the end

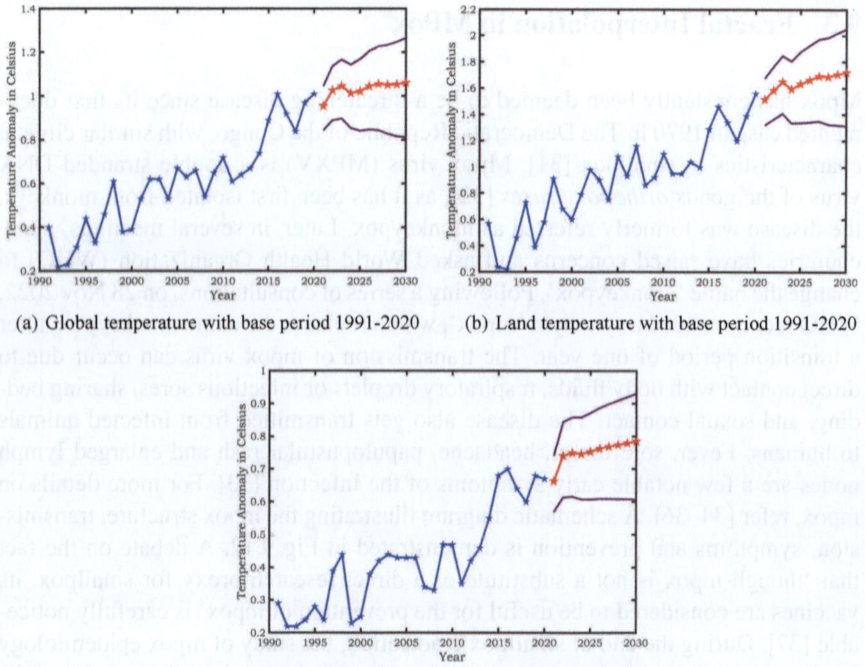

(a) Global temperature with base period 1991-2020 (b) Land temperature with base period 1991-2020

(c) Sea temperature with base period 1991-2020

Fig. 9.11 ARFIMA prediction of **a** global temperature, **b** land temperature and **c** sea temperature anomalies with base period 1991–2020 for the time interval 2021–2030

of 2030. The model is validated for accuracy using the evaluation metrics: MAPE, NMAPE, sMAPE, MAE and RMSE. The statistical evaluation pinpoints that the proposed model, with its prediction outcomes, remarkably contributes to maintain the greenhouse emissions within the permissible limits and in the habitat safety of almost all lifeforms.

The choice of hidden variable fractal interpolation function has connected global temperature changes and CO_2 emissions by considering the hidden variables as CO_2 emissions. Indeed, a short term prediction for ten years is presented, the study is limited within the concentration of CO_2 emissions and has uncovered the effect of other greenhouse gases and environmental factors like cutting down trees, burning fossil fuels, freshwater withdrawals and so on. It inevitably leads to more open problems of considering other greenhouse gases as hidden variables individually for the prediction of climate scenarios. Unsurprisingly, all the aforementioned environmental factors are interdependent on each other with respect to time scale and therefore, can be considered together to make predictions much better. However, the literature does not report any such multivariate fractal interpolation functions to investigate several variables simultaneously. Henceforth, in order to concentrate on numerous climate-affecting factors in forecasting, any alternative multivariate technique should be purchased.

9.3 Fractal Interpolation in MPox

Mpox has constantly been deemed to be a threatening disease since its first documented case in 1970 in The Democratic Republic of the Congo, with similar clinical characteristics to smallpox [31]. Mpox virus (MPXV) is a double stranded DNA virus of the genus *orthopoxviruses* [32], as it has been first isolated from monkeys, the disease was formerly referred as monkeypox. Later, in several meetings, many countries have raised concerns and asked World Health Organization (WHO) to change the name 'monkeypox'. Following a series of consultations, on 28 Nov 2022, WHO recommended a synonym 'mpox', which replaces the name monkeypox, after a transition period of one year. The transmission of mpox virus can occur due to direct contact with body fluids, respiratory droplets or infectious sores, sharing bedding, and sexual contact. The disease also gets transmitted from infected animals to humans. Fever, sore throat, headache, papulopustular rash and enlarged lymph nodes are a few notable early symptoms of the infection [33]. For more details on mpox, refer [34–36]. A schematic diagram illustrating the mpox structure, transmission, symptoms and prevention is demonstrated in Fig. 9.12. A debate on the fact that 'though mpox is not a substitute or a direct research-proxy for smallpox, its vaccines are considered to be useful for the prevention of mpox' is carefully noticeable [37]. During the end of smallpox eradication, the study of mpox epidemiology has commenced, however new assessments are needed to analyze the severity of the disease.

On 14 Aug 2024, WHO determined that the upsurge of mpox in the Democratic Republic of the Congo (DRC) and declared a public health emergency of international concern (PHEIC). This PHEIC declaration is the second time in two years

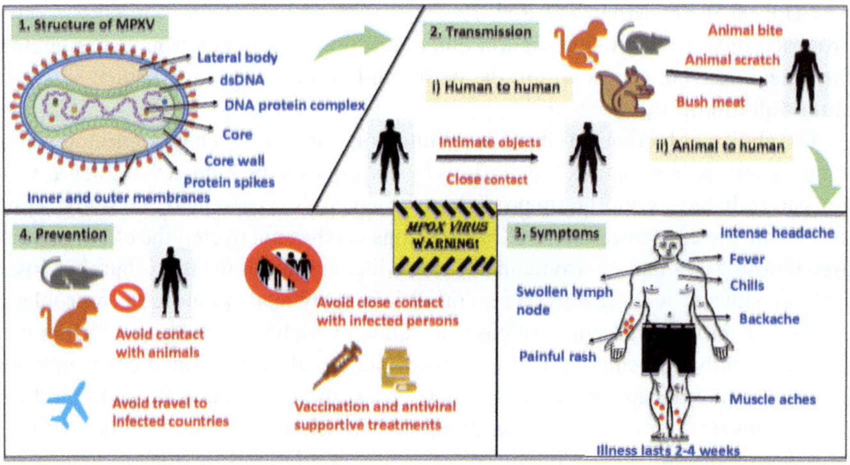

Fig. 9.12 Schematic diagram showcasing the structure, transmission, symptoms and prevention of mpox virus

relating to the spread of mpox. In July 2022, the PHEIC was declared due to the outbreak of a new mpox virus strain in the DRC, *clade 1b*, as its spread is rapid via sexual transmission. Later, in May 2023, PHEIC was declared over on observing a sustained decline in the global mpox cases. There have been 116 member states from all 6 WHO regions that have reported mpox cases since Jan 1, 2022. In the month of Jun 2024, four neighbouring countries of the DRC had documented over 100 laboratory confirmed cases of *clade 1b* which were not reported mpox before. Different clades of mpox have had multiple outbreaks in various countries, each with its own routes of transmission and risk levels. There had been a total of 99176 laboratory confirmed cases and 535 probable cases, including 208 deaths documented to WHO as of Jun 30, 2024 [38]. A variety of research is performed on the mpox outbreak, for some informative research on mpox spread, refer [39–48]. A single confirmed case of mpox in a country is also regarded as an outbreak. In order to better understand and regulate the outbreak, this work concentrates on the mpox spread by considering the data of confirmed daily cases which underscores the necessity for increased surveillance. The mpox dynamics is analyzed through pre-processesing and further explored for heterogeneity with the aid of fractal interpolation and multifractal measure, respectively. In addition, to forecast the future mpox scenario in the DRC, a bidirectional long-short term memory (BiLSTM) neural network is employed on the pre-processed mpox cases.

9.3.1 *Focused Regions*

WHO reported that at the global level, 866 mpox cases had been confirmed monthly in the past six months (01 Jan 2024–30 Jun 2024) on an average. The African region was the most affected region with 1854 confirmed cases and 13 deaths. This was followed by the Region of the Americas with 1812 confirmed cases and 3 deaths, and the European Region, where 826 cases were confirmed with 3 deaths [38]. Therefore, this application has focused on the confirmed mpox cases of three primarily affected regions namely, Regions of Africa, Americas (North America and South America) and Europe. Further, The Democratic Republic of the Congo was given a special focus, where a significant increase in mpox cases and deaths are continuously being reported from 2023. It was documented that from Jan 2024 to Jun 2024, 10337 suspected cases with 1754 laboratory-confirmed and 433 deaths were reported in the DRC. Due to the increasing mpox cases in the DRC and its expansion to other regions, a PHEIC was constituted. As DRC contributed 96 per cent of the confirmed new cases within the African region in Jun 2024, it was essential to shed a special focus to control the spread in DRC.

9.3.2 Materials and Methods

9.3.2.1 Fractal Interpolation

Here is a recap of linear fractal interpolation. It is developed as an attractor of a special kind of Iterated Function System (IFS), hence the provided data set can be enlarged to increase its size by increasing the iteration. Let $\{(x_j, y_j) \in \mathbb{R}^2 : j = 0, 1, \ldots, N\}$, $N \geq 2$ be the given set of interpolation data with $\Delta x_j (= x_j - x_{j-1}) > 0$, for $j = 1, 2, \ldots, N$. With the given data set and a scaling factor s_j obeying $|s_j| < 1$ for all $j = 1, 2, \ldots, N$, a hyperbolic IFS $\{\mathbb{R}^2; W_j : j = 1, 2, \ldots, N\}$ can be associated, where W_j are affine maps provided by

$$W_j(\mathbf{x}) = \begin{bmatrix} a_j & 0 \\ c_j & s_j \end{bmatrix} \mathbf{x} + \begin{bmatrix} e_j \\ f_j \end{bmatrix}, \ \mathbf{x} = \begin{bmatrix} x \\ y \end{bmatrix}. \tag{9.10}$$

Here, the parameters a_j, c_j, e_j and f_j can be determined using the below formulas [49],

$$\begin{aligned} a_j &= \frac{x_j - x_{j-1}}{x_N - x_0}, \\ e_j &= \frac{x_N x_{j-1} - x_0 x_j}{x_N - x_0} \\ c_j &= \frac{(y_j - y_{j-1}) - s_j(y_N - y_0)}{x_N - x_0} \\ f_j &= \frac{(x_N y_{j-1} - x_0 y_j) - s_j(x_N y_0 - x_0 y_N)}{x_N - x_0}. \end{aligned}$$

By the IFS theory, the graph of *linear fractal interpolation function* $f : [x_0, x_N] \times \mathbb{R}$ interpolating the given data set $\{(x_j, y_j) \in \mathbb{R}^2 : j = 0, 1, \ldots, N\}$ can be obtained as an attractor of the IFS. While investigating mpox cases, the x_j's denote the time (in days) and y_j's indicate the confirmed daily new cases of mpox in respective regions.

9.3.2.2 Multifractal Measure

Multifractal measure reveals the non-uniform distribution of the considered data set, which is captured by a spectrum of singularities. Multifractal spectrum is a characteristic curve generated by the singularity exponent α and its function $f(\alpha)$. If the data set is covered with boxes of size ϵ, then the number of boxes having same α, denoted by $N_\alpha(\epsilon)$, obeys the power law relation $N_\alpha(\epsilon) \sim \epsilon^{-f(\alpha)}$. Here, $f(\alpha)$ is defined as the fractal dimension of the subsets of the boxes with Hölder exponent α [50] and is expressed by

$$f(\alpha) = \alpha q + \tau(q), \tag{9.11}$$

where $\tau(q)$ is the mass exponent with moment of order q. Multifractality is basically an idea to discuss the distribution of a (self-similar) measure on a support in a heterogeneous rather than uniform way [50]. In other words, the concentration of the measure at different points along the support ranges from very high to high concentrations, as well as low to very low concentrations. At this scenario, a spectrum of dimensions is required rather than a single dimension, thus for the complete characterization of the given object, the spectrum of dimensions $f(\alpha)$ is employed. Multifractal spectrum is basically a bell shaped curve, larger the spectrum width stronger the multifractality that the considered object has.

9.3.2.3 Bidirectional Long-Short Term Memory (BiLSTM) Neural Network

The bidirectional recurrent neural networks have been introduced as unique recurrent neural networks (RNNs) to process the data sequence in both directions [51]. They carry over all of the attributes of RNNs and expand their functionality to situations where information from the past and future influences the result. Thus, an LSTM layer and a bidirectional neural layer are combined to form a BiLSTM network. In other words, the Bidirectional layer receives the LSTM application, creating two hidden layers—one for the Forward LSTM and one for the Backward LSTM [22]. Due to the two-way inputs in this neural network, the interaction between the inputs is improved, resulting in boosting the quantity of information available in the network.

With the provided current value x_t, previous hidden state h_{t-1} and previous state C_{t-1}, the following mathematical operations [52] are performed,

$$\begin{aligned} f_t &= \sigma(W_f \cdot [h_{t-1}, x_{t-1}] + b_f), \\ i_t &= \sigma(W_i \cdot [h_{t-1}, x_{t-1}] + b_i), \\ \hat{C}_t &= \tanh(W_c \cdot [h_{t-1}, x_{t-1}] + b_c), \\ C_t &= f_t \cdot C_{t-1} + i_t \cdot \hat{C}_t, \\ o_t &= \sigma(W_o \cdot [h_{t-1}, x_{t-1}] + b_o). \end{aligned} \tag{9.12}$$

where, σ and tanh are the sigmoid and hyperbolic tangent functions, respectively. f_t, i_t and o_t are the forget gate, input gate and output gate, respectively. W's and b's are the weight matrices and bias vectors. An LSTM layer is made up of a sequence of LSTM cells, and the sequence data are fed forward, on the other hand, BiLSTM has an additional LSTM layer that has data fed backward. A sample BiLSTM architecture is shown in Fig. 9.12.

9.3.2.4 Evaluation Metrics

The evaluation metrics used during the training and testing of the BiLSTM model are described in the following. Three classical evaluation statistical metrics that this work use are the Root Mean Squared Error, the Mean Absolute Error and the Mean

Absolute Percentage Error. All the formulae of RMSE, MAE and MAPE are defined in Eqs. (9.9), (9.8) and (9.5). Employing more than one metric is mainly to evaluate different characteristics and performance of the considered BiLSTM model before forecasting the future mpox spread.

9.3.3 Data Description

For this study, the data on mpox has been taken from the official site of Our World in Data [53]. Confirmed mpox cases in the regions of Africa, North America, South America, Europe and the Democratic Republic of the Congo have been considered for the analysis. The considered data is a seven days rolling average of confirmed daily cases and the rolling average is considered to spot trends in the specified time interval. The seven days rolling average is estimated by dividing the sum of past seven days data by seven. Different time intervals are considered for the above-mentioned four different regions based on the severity of the infection. For Africa and the DRC, confirmed daily cases were taken from 1 Aug 2023 to 28 Jul 2024, for North America from 1 May 2022 to 24 Feb 2023, for South America from 31 May 2022 to 20 Mar 2023 and for Europe from 7 May 2022 to 22 Nov 2022.

9.3.4 Multifractality in Mpox Transmission

This section is devoted to analyze the multifractal patterns in the transmission of mpox disease. Initially, the confirmed daily cases in the most affected regions—Africa, Americas, Europe and the Democratic Republic of the Congo—are pre-processed through the technique of fractal interpolation, then the pre-processed cases are investigated for multifractality. Data pre-processing is not only important in machine learning but also it is essential whenever a real data is considered for the investigation. It refers to the cleaning or transforming the data to improve the quality of raw data set making ready for the specific task. In this study, fractal interpolation function is utilized for pre-processing the raw data of confirmed daily mpox cases to improve its quality. How the quality is increased? In the case of fractal interpolation, the size of the data set is increased on every successive iteration, since new data points generation specifically depends on the iteration. If $n + 1$ number of sample points are given, in the first iteration, $n^2 + 1$ new points are generated. At the end of second iteration, $n^3 + 1$ distinct points are generated. When the iteration is increased to N, then the fractal interpolation function passes through a total number of $n^{N+1} + 1$ new points. As a consequence of increasing the size of the data set, the provided raw data set becomes finer with more data points and any missing information will be corrected with more arbitrary accuracy. With the above-described advantages on the fractal interpolation function, the confirmed daily mpox cases are pre-processed with the choice of scaling factor as $s_j = 0.01$ for all $j = 1, 2, \ldots, n$ and fixing the

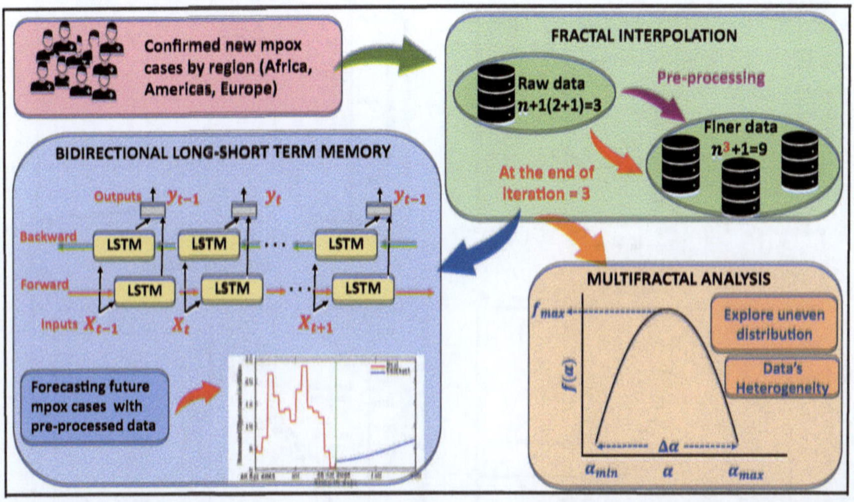

Fig. 9.13 Graphical theme of application of fractal interpolation in the study of mpox transmission

iteration as 2. The pre-processed curves of confirmed daily mpox cases in the aforementioned four regions are demonstrated in Fig. 9.14 (the blue boxes highlighted are discussed in Sect. 9.2.6). By studying the pre-processed data using fractal interpolation, the fractal characteristics like self-similarity in the trend of mpox spread can be perceived, to further analyze the uneven distribution of mpox cases, multifractal measure is employed. Multifractal measure effectively captures the irregular distribution of mpox cases in each region. It generally refers to the self-similar measure and it is stronger than the fractal dimension. It gives quantitative information about the self-similar measure associated with the considered object/function rather than merely the discussing irregularity of the object in its structure. In this study, multifractal measure is tailored to discuss how irregular and heterogeneous the distribution of mpox cases. The graphical abstract of the proposed study is demonstrated in Fig. 9.13.

9.3.4.1 Focus on Africa

The seven day average of daily mpox cases of Africa from 1 Aug 2023 to 28 Jul 2024 are pre-processed through fractal interpolation and its graph is illustrated in Fig. 9.14a. In Fig. 9.14a, the black dots represent the raw data set and orange curve denotes the fractal interpolated mpox cases of Africa. The mpox cases in Africa was analyzed the month of August 2023 till mid June 2024. From Fig. 9.14a, though the graph seems to be decreasing after June 2024, there is a possibility for the hike when we observe the past pattern of the curve. The trend of mpox cases in Africa follows a self-similar pattern, that is, the pattern from each small portion of the

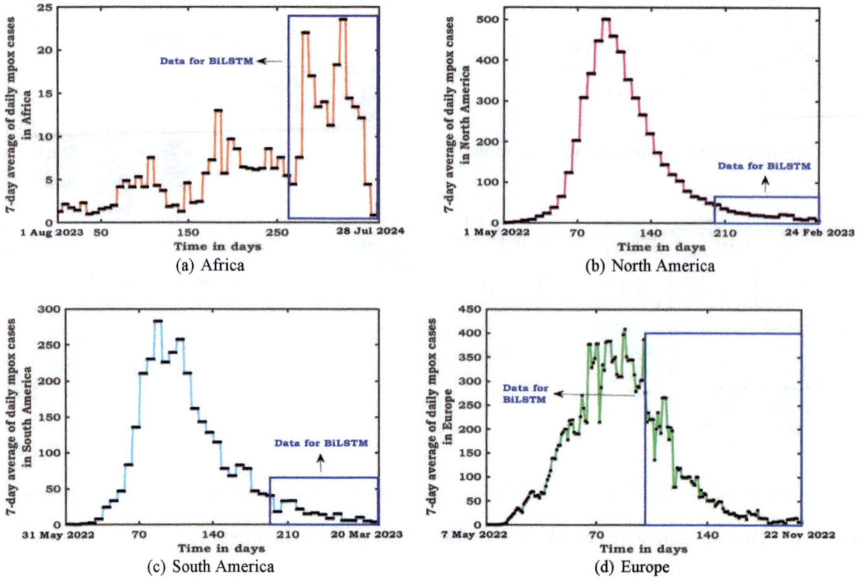

Fig. 9.14 Seven day average of daily mpox cases of **a** Africa (1 Aug 2023–28 Jul 2024), **b** North America (1 May 2022–24 Feb 2023), **c** South America (31 May 2022–20 Mar 2023) and **d** Europe (7 May 2022–22 Nov 2022) pre-processed through fractal interpolation

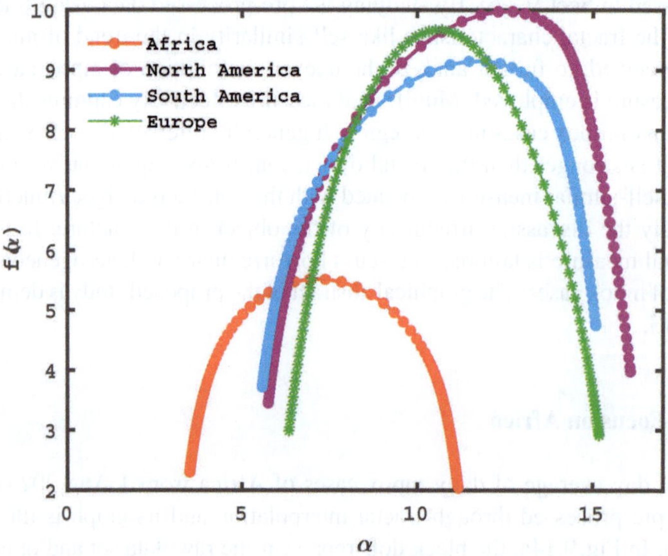

Fig. 9.15 Multifractal spectrum for analyzing the randomness of pre-processed mpox cases in the regions of Africa, Americas and Europe

Table 9.3 Estimated parameter values of multifractal spectrum for analyzing the irregular distribution of mpox spread

	Africa	North America	South America	Europe
$\Delta\alpha$	7.6583	10.3277	9.5194	8.8505
f_{max}	5.5181	9.9758	9.1522	9.6719
α_0	6.9204	12.7255	11.9552	10.7058

graph is approximately similar to the whole, which is a unique property of *fractals*. This self-similar patterns suggest a hike in the transmission of mpox in the upcoming days. In addition, the graph throughout the considered time interval is irregular consisting of both peaks and downs, this uneven distribution is measured using the multifractal measure. The orange curve demonstrated in Fig. 9.15 represents the multifractal spectrum of the mpox cases in Africa and its corresponding parameter values are estimated and shown in Table 9.3. The multifractal spectrum of mpox cases in Africa is not symmetric with slight increase in the right side of the spectrum. The tabulated values and this $f(\alpha)$ spectrum confirms the multifractality in the mpox spread. Further, the asymmetric nature of the spectrum indicates the distribution of singularities is not balanced, conveying that the mpox spread in Africa is unpredictable.

9.3.4.2 Focus on Americas

The confirmed mpox cases of both North America and South America are discussed in this subsection. In Fig. 9.14a, b, black dots represents the raw data set, and the magenta and blue curves represent the graphs of pre-processed mpox cases of North America and South America, respectively. The time intervals focused are 1 May 2022 to 24 Feb 2023 for North America and 31 May 2022 to 20 Mar 2023 for South America. This particular choice is made in order to analyze the infection spread during its peak in the considered regions. The time intervals for North America and South America are indeed different, since it is not interesting or beneficial to study the disease when its spread is less or lack. From Fig. 9.14a, b, both the curves of North America and South America look like a mountain, however the mpox cases in South America shows more irregularity with peaks and downs. In case of North America, with a maximum of 500 daily mpox cases the spread has touched its peak during Aug 2022, later, the spread has begun to gradually decrease till Feb 2023. In South America, during the mid of Sep 2022, the mpox spread has attained its peak with around 258 daily mpox cases and the spread is almost nil during the Feb 2023. This kind of highly irregular curves can only be fitted using fractal interpolation rather than the classical interpolation. For further analysis, the multifractal spectrum of pre-processed mpox cases of North America and South America are depicted in Fig. 9.15 using magenta and blue curves. Here, the two spectra can be compared since the time intervals considered for both the regions are almost same. Both the spectrum are asymmetric as the spectra are slightly tilted right. Comparing the values of their multifractal spectrum tabulated in Table 9.3, the multifractality of North America is stronger, since larger the width ($\Delta\alpha$) of the spectrum stronger the multifractality. Also, the other values $f_{\max}$ and α_0 of North America are also higher than that the South America indicating the distribution and the spread of mpox is highly irregular in North America with more number of confirmed cases than in the South America with less confirmed cases.

9.3.4.3 Focus on Europe

The third most affected mpox region is the Europe, its confirmed mpox cases during the time interval 7 May 2022 to 22 Nov 2022 are pre-processed through fractal interpolation and the respective graphical illustration is portrayed in Fig. 9.14d. In Fig. 9.14d, the black dots represent the original mpox cases and the green curve represents the pre-processed mpox cases in Europe. The mpox spread is high in Europe during the period of Jul 2022 to mid Sep 2022. The point here to be noted is that at the same time the transmission is highly irregular which is clearly illustrated using the fractal interpolation. The gradually decreasing curve from Sep 2022 indicates that the spread has began to reduce and it is almost nil at the end of Nov 2022. To further analyze the multifractal nature of the mpox spread, the multifractal spectrum of the pre-processed mpox cases of Europe is demonstrated in Fig. 9.15 using the green curve. The spectrum's associated values are estimated and shown in Table 9.3. Unlike the spectra of Africa, North America and South America, the multifractal spectrum of Europe is less asymmetric, meaning that the underlying dynamics may have slight differences across scales, but the spread is not drastically different. The estimated width value of the spectrum suggests there is multifractal pattern in the mpox transmission in Europe.

9.3.4.4 Situation in the DRC

While discussing the mpox transmission in the most affected regions globally, a special focus is shed on the Democratic Republic of the Congo, the one country in African region continues to report mpox cases over time. In the current year, from Jan 2024 to Jun 2024, there are 30711 suspected mpox cases and 1326 deaths have been reported in the DRC. Also, this country is the major reason behind the Africa's spread and making Africa to retain in the first place of most affected mpox disease. This can be clearly visualized on comparing the graphs demonstrated in Fig. 9.14a and Fig. 9.16. Figure 9.16a portrays the pre-processed mpox cases in DRC from 1 Aug 2023 to 28 Jul 2024. The irregular pattern of both Africa and the DRC matches closely evidencing the impact of the DRC mpox cases on the African mpox cases. One could not conclude by simply analyzing the raw data of both the DRC and Africa, fractal interpolation has qualitatively revealed their relation besides exhibiting the fractal characteristics of self-similarity. To analyze the randomness and heterogeneity of the

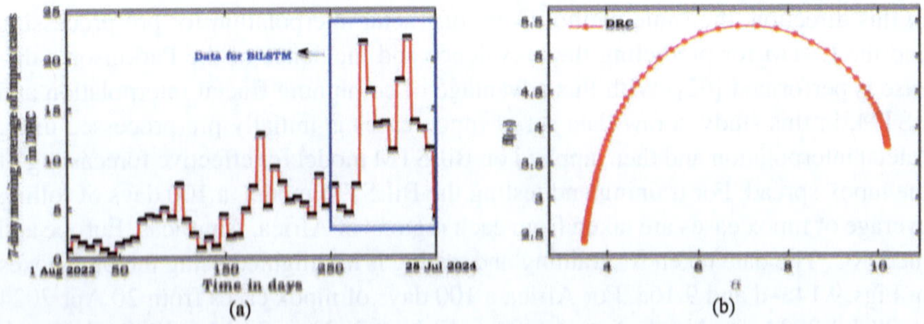

Fig. 9.16 **a** Seven day average of daily mpox cases in the DRC pre-processed through fractal interpolation and **b** its multifractal spectrum

mpox spread in the DRC, multifractal spectrum is drawn for the pre-processed mpox cases of DRC. Figure 9.16b represents the multifractal spectrum of pre-processed mpox cases of DRC. The width of this spectrum is closely related to the width of the spectrum of Africa in Fig. 9.15, confirming that the spread of mpox disease is highly irregular with non-uniform distribution across scales.

9.3.5 *Forecasting Mpox Spread Using BiLSTM*

This section discusses the applications of a deep learning model in short term forecasting of the mpox spread in the most affecting region, Africa. Multiple artificial intelligence and machine learning models have been designed for precise and timely diagnosis of mpox, for instance [48, 54–57]. Approaches based on deep learning are being used more and more to increase research productivity and create countermeasures for mpox. Several noteworthy studies have thoroughly examined and reviewed different AI applications for drug research, mpox diagnosis, forecasting and other scenarios of use. For example, Patel et al. [58] examined several AI techniques used for mpox virus identification, viral genome characterization and mortality risk prediction. Gul et al. [59] examined a number of mpox diagnostic techniques including immunodiagnostics, nucleic acid testing. The comprehensive review presented in the article [60] focuses on the automated identification of mpox through the use of machine learning and deep learning techniques. The afore-mentioned literature concretely acknowledges that how significant and suitable the artificial neural network and deep learning models for analyzing and forecasting the mpox spread with biological interpretation. The fractal interpolation technique is being effectively employed to obtain the finer version of provided time series, then the fractal pre-processed finer time series is applied on a deep learning model, namely the long-short term memory neural network to forecasting the future time series [61]. It is observed that when a test fit is performed on unknown data, the predictions obtained from the fractal interpolated time series significantly outperforms predictions from the original data.

In this direction, the same methodology of fractal interpolation for pre-processing and the LSTM for predicting the prevalence and incidence of the Parkinson's disease is performed [62]. With this advantage of combining fractal interpolation and LSTM, in this study, a raw data set of mpox cases is initially pre-processed using fractal interpolation and then, applied on BiLSTM model for effective forecasting of the mpox spread. For training and testing the BiLSTM model, a 100 days of rolling average of mpox cases are taken from each regions of Africa, Americas, Europe and the DRC. The data taken for training and testing is highlighted using the blue boxes in Figs. 9.14a–d and 9.16a. For Africa a 100 days of mpox cases from 20 Apr 2024 to 28 Jul 2024, for North America from 17 Nov 2022 to 20 Mar 2022, for South America 11 Dec 2022 to 20 Mar 2023, for Europe 15 Aug 2022 to 22 Nov 2022 and for the DRC 20 Apr 2024 to 28 Jul 2024 have been taken. However, 100 data points are very less to training, hence the 100 days of rolling average data is initially pre-processed through fractal interpolation which sufficiently enlarges the data set. Then, the corresponding finer preprocessed data points are utilized for training and testing the BiLSTM model.

The deep learning neural network model is built using a Bidirectional Long Short-Term Memory (BiLSTM) layer with 100 units and ReLU activation function. A dropout layer with a dropout rate of 0.2 is included after the BiLSTM layer to mitigate overfitting. The model concludes with a dense (fully connected) output layer containing a single neuron, which is commonly suitable for regression tasks, such as predicting a continuous target variable. The model is compiled using the mean squared error loss function, which measures the average squared difference between predicted and actual values, and the Adam optimizer, a widely-used gradient descent optimization algorithm that adjusts the learning rate adaptively to accelerate convergence. This BiLSTM model is trained for 10 epochs for adjusting the model weights to minimize the loss function. The BiLSTM performances of training and testing for each region are depicted in Figs. 9.17a, c, e, g and 9.18a. The respective losses during training and validation are illustrated in Fig. 9.17b, d, f, h and 9.18b. Fractal interpolated data of 100 mpox cases, (i.e.,) $(n^2 + 1)$ data with $n = 99$ at the end of iteration 2, have been generated as a result of pre-processing. This interpolated data is utilized for training and testing the BiLSTM model as shown in the x axis of Figs. 9.17 and 9.18, and the time steps here is the quotient of number of original data over the number of interpolated data. From Figs. 9.14 and 9.16, it is observed that the fractal interpolated time series exhibits higher levels of persistency and long term memory since the complexities of sub-intervals closely match with the original data. This significant advantage enables the BiLSTM predict the future mpox spread with greater precision rather than the prediction from the original data, which is evident from the Figs. 9.17 and 9.18. Therefore, the purpose of employing fractal interpolation is not only to reveal the hidden fractal characteristics in the mpox spread but also it has enhanced the representation of data by maintaining variability at different scales, enabling the BiLSTM to adapt to the unpredictable nature of disease transmission for effective forecasting. The model performance is

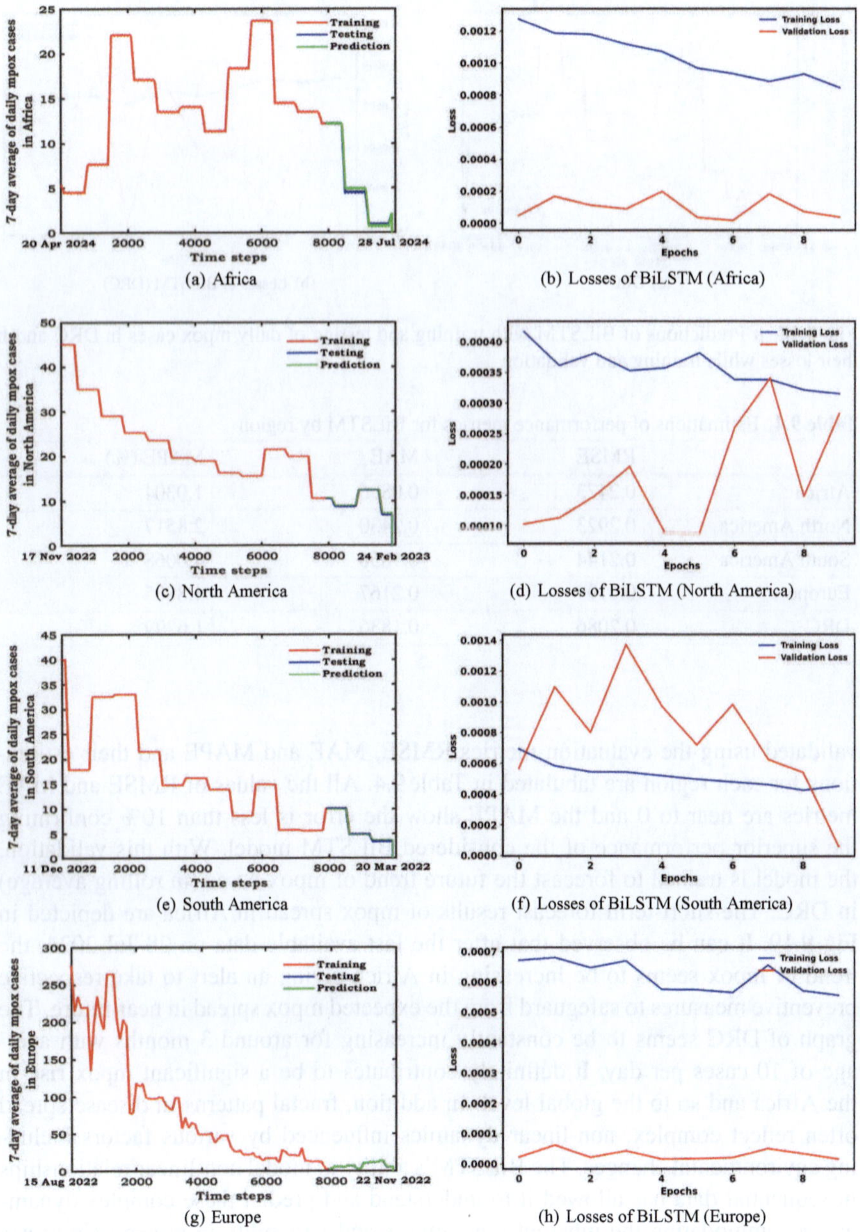

(a) Africa (b) Losses of BiLSTM (Africa)

(c) North America (d) Losses of BiLSTM (North America)

(e) South America (f) Losses of BiLSTM (South America)

(g) Europe (h) Losses of BiLSTM (Europe)

Fig. 9.17 Predictions of BiLSTM with training and testing of daily mpox cases **a**, **c**, **e**, **g** and their losses while training and validation **b**, **d**, **f**, **h**

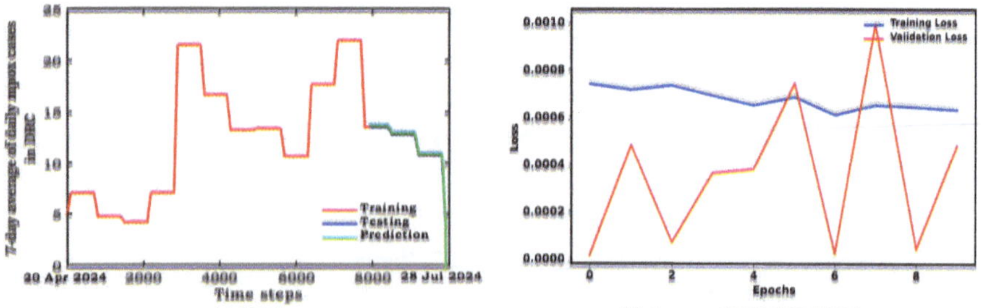

(a) DRC (b) Losses of BiLSTM (DRC)

Fig. 9.18 **a** Predictions of BiLSTM with training and testing of daily mpox cases in DRC and **b** their losses while training and validation

Table 9.4 Estimations of performance metrics for BiLSTM by region

	RMSE	MAE	MAPE (%)
Africa	0.2473	0.0896	1.0304
North America	0.2923	0.2430	2.8517
South America	0.2144	0.1856	4.0068
Europe	0.2199	0.2167	4.8805
DRC	0.2086	0.1836	1.6299

validated using the evaluation metrics RMSE, MAE and MAPE and their estimations for each region are tabulated in Table 9.4. All the values of RMSE and MAE metrics are near to 0 and the MAPE show the error is less than 10% confirming the superior performance of the considered BiLSTM model. With this validation, the model is trained to forecast the future trend of mpox cases (in rolling average) in DRC. The short-term forecast results of mpox spread in Africa are depicted in Fig. 9.19. It can be observed that after the last available data on 28 Jul 2024, the trend of mpox seems to be increasing in Africa giving an alert to take respective preventive measures to safeguard from the expected mpox spread in near future. The graph of DRC seems to be constantly increasing for around 3 months with average of 10 cases per day. It definitely contributes to be a significant mpox rise in the Africa and so to the global level. In addition, fractal patterns in disease spread often reflect complex, non-linear dynamics influenced by various factors including environmental changes. The BiLSTM's ability to model non-linear relationships in sequential data has allowed it to understand and predict these complex dynamics, accommodating the inherent randomness and variability in transmission rates of mpox. Moreover, the model has incorporated stochastic variability by learning from the diverse, fractal-like patterns present in the interpolated data for effectively capturing multi-scale patterns while forecasting future spread under varying conditions.

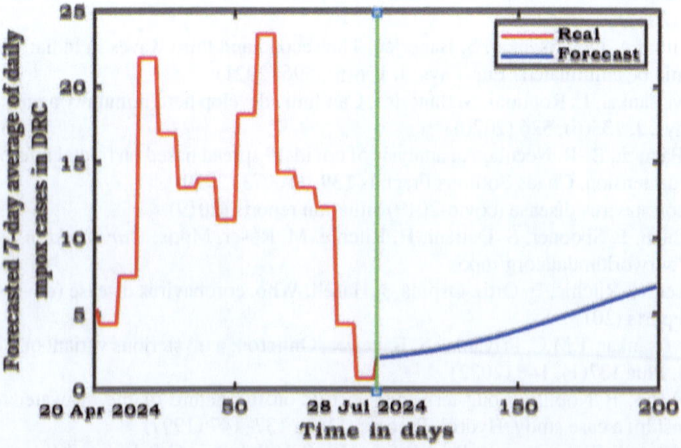

Fig. 9.19 Forecasted 7-day rolling average of daily mpox cases in DRC using BiLSTM model

9.3.6 Summary

In this application, the mpox transmission is explored to analyze its trend in the most affected regions: Africa, Americas and Europe, according to WHO reports. In each region, a specific time period is chosen depending on the severity of the infection and explored how the mpox dynamics has evolved over time. Notably, seven days rolling average of new mpox cases is taken in order to avoid data insufficiency. The results have confirmed the fractal as well as multifractal patterns in the mpox spread, more precisely, an approximate self-similarity and heterogeneity are confirmed in the dynamics of mpox spread [63]. This novel information is expected to be beneficial to scientists and researchers to slow down the spread. Furthermore, a deep learning BiLSTM neural network is employed to forecast the disease spread in the DRC. On confirming the superior performance of the model using the evaluation metrics, a short-term forecast is carried out. Though the employed BiLSTM model is advantageous in forecasting the future mpox spread based on the estimations of statistical metrics, this work is limited within this particular model evaluation, however, other deep learning and AI models can be employed and compared for better forecasting results.

References

1. S.S.A. Karim, Q.A. Karim, Omicron sars-cov-2 variant: a new chapter in the covid-19 pandemic. The Lancet **398**(10317), 2126–2128 (2021)
2. E. Petersen, F. Ntoumi, D.S. Hui, et al., Emergence of new sars-cov-2 variant of concern omicron (b. 1.1. 529)-highlights africa's research capabilities, but exposes major knowledge

gaps, inequities of vaccine distribution, inadequacies in global covid-19 response and control efforts (2022)
3. C. Kavitha, A. Gowrisankar, S. Banerjee, The second and third waves in india: when will the pandemic be culminated? Eur. Phys. J. **136**(5), 596 (2021)
4. A. Gowrisankar, L. Rondoni, S. Banerjee, Can India develop herd immunity against covid-19? Eur. Phys. J. **135**(6), 526 (2020)
5. C.-M. Păcurar, B.-R. Necula, An analysis of covid-19 spread based on fractal interpolation and fractal dimension. Chaos Solitons Fractals **139**, 110073 (2020)
6. Who, coronavirus disease (covid-2019) situation reports (2019)
7. E. Mathieu, F. Spooner, S. Dattani, H. Ritchie, M. Roser, Mpox. *Our World in Data* (2022). https://ourworldindata.org/mpox
8. M. Roser, H. Ritchie, E. Ortiz-Ospina, J. Hasell, Who, coronavirus disease (covid-2019) situation reports (2019)
9. A. Gowrisankar, T.M.C. Priyanka, S. Banerjee, Omicron: a mysterious variant of concern. Eur. Phys. J. Plus **137**(1), 1–8 (2022)
10. B.W. Webb, F. Nobilis, Long-term perspective on the nature of the air-water temperature relationship: a case study. Hydrol. Process. **11**(2), 137–147 (1997)
11. W. Steffen, K. Richardson, J. Rockström, H.J. Schellnhuber, O.P. Dube, S.ébastien Dutreuil, T.M. Lenton, J. Lubchenco, The emergence and evolution of earth system science. Nat. Rev. Earth Environ. **1**(1), 54–63 (2020)
12. D.A. Vasseur, J.P. DeLong, B. Gilbert, H.S. Greig, C.D.G. Harley, K.S. McCann, V. Savage, T.D. Tunney, M.I. O'Connor, Increased temperature variation poses a greater risk to species than climate warming. Proc. R. Soc. B: Biol. Sci. **281**(1779), 20132612 (2014)
13. P. Friedlingstein, R.A. Houghton, G. Marland, J. Hackler, T.A. Boden, T.J. Conway, J.G. Canadell, M.R. Raupach, P. Ciais, C. Le Quéré, Update on co2 emissions. Nat. Geosci. **3**(12), 811–812 (2010)
14. It's warmer than average. but what is average? *World Meteorological Organization*, published 13 July 2022
15. D.R. Rounce, R. Hock, F. Maussion, R. Hugonnet, W. Kochtitzky, M. Huss, E. Berthier, D. Brinkerhoff, L. Compagno, L. Copland, et al., Global glacier change in the 21st century: every increase in temperature matters. Science **379**(6627), 78–83 (2023)
16. S. Altikat, Prediction of co2 emission from greenhouse to atmosphere with artificial neural networks and deep learning neural networks. Int. J. Environ. Sci. Technol. **18**(10), 3169–3178 (2021)
17. K.O. Yoro, M.O. Daramola, Co2 emission sources, greenhouse gases, and the global warming effect, in *Advances in Carbon Capture* (Elsevier, 2020), pp. 3–28
18. M. Isik, S. Devadoss, An analysis of the impact of climate change on crop yields and yield variability. Appl. Econ. **38**(7), 835–844 (2006)
19. S.L. Grotch, M.C. MacCracken, The use of general circulation models to predict regional climatic change. J. Clim. 286–303 (1991)
20. Y. Liu, D. Li, S. Wan, F. Wang, W. Dou, X. Xu, S. Li, R. Ma, L. Qi, A long short-term memory-based model for greenhouse climate prediction. Int. J. Intell. Syst. **37**(1), 135–151 (2022)
21. A.A. Jamali, R.G. Kalkhajeh, T.O. Randhir, S. He, Modeling relationship between land surface temperature anomaly and environmental factors using gee and giovanni. J. Environ. Manag. **302**, 113970 (2022)
22. Y. Lai, D.A. Dzombak, Use of the autoregressive integrated moving average (arima) model to forecast near-term regional temperature and precipitation. Weather. Forecast. **35**(3), 959–976 (2020)
23. R.M. Jonathan, Hosking, modeling persistence in hydrological time series using fractional differencing. Water Resour. Res. **20**(12), 1898–1908 (1984)
24. S. Banerjee, D. Easwaramoorthy, A. Gowrisankar, Fractal interpolation function for countable data. *Fractal Functions, Dimensions and Signal Analysis* (2021), pp. 61–77

25. A. Gowrisankar, T.M.C. Priyanka, A. Saha, L. Rondoni, K. Hassan, S. Banerjee, Greenhouse gas emissions: a rapid submerge of the world. Chaos: Interdiscip. J. Nonlinear Sci. **32**(6) (2022)
26. W.J. Clive, Granger, R. Joyeux, An introduction to long-memory time series models and fractional differencing. J. Time Ser. Anal. **1**(1), 15–29 (1980)
27. J. Xiu, Y. Jin, Empirical study of arfima model based on fractional differencing. Phys. A: Stat. Mech. Appl. **377**(1), 138–154 (2007)
28. accessed 07 July 2022 NASA Goddard Institute for Space Studies. Gistemp team, giss surface temperature analysis (gistemp), version 4,
29. accessed 07 July 2022 Our World in Data. Per capita co−2 emissions
30. C. Le Quéré, R.B. Jackson, M.W. Jones, A.J.P. Smith, S. Abernethy, R.M. Andrew, A.J. De-Gol, D.R. Willis, Y. Shan, J.G. Canadell, et al., Temporary reduction in daily global co2 emissions during the covid-19 forced confinement. Nat. Clim. Chang. **10**(7), 647–653 (2020)
31. A.M. McCollum, I.K. Damon, Human monkeypox. Clin. Infect. Dis. **58**(2), 260–267 (2014)
32. S. Realegeno, A.S. Puschnik et al., Monkeypox virus host factor screen using haploid cells identifies essential role of garp complex in extracellular virus formation. J. Virol. **91**(11), 10–1128 (2017)
33. H. Naji, Monkeypox virus: transmission, signs and symptoms, prevention, and epidemiology. Eur. J. Med. Health Sci. **4**(5), 30–34 (2022)
34. H. Ejaz, K. Junaid et al., Emergence and dissemination of monkeypox, an intimidating global public health problem. J. Infect. Public Health **15**(10), 1156–1165 (2022)
35. Y. Araf, J.F. Nipa et al., Insights into the transmission, host range, genomics, vaccination, and current epidemiology of the monkeypox virus. Vet. Med. Int. **2024**(1), 8839830 (2024)
36. Y. Sun, W. Nie, et al., Human monkeypox virus: epidemiologic review and research progress in diagnosis and treatment. J. Clin. Virol. 105662 (2024)
37. M.G. Reynolds, I.K. Damon, Outbreaks of human monkeypox after cessation of smallpox vaccination. Trends Microbiol. **20**(2), 80–87 (2012)
38. 2022-24 Mpox Outbreak: Global Trends. Geneva: World Health Organization (2024)
39. W. Matsee, S. Charoensakulchai et al., The silent, rapidly unfolding threat of mpox spread via international travel to southeast Asia: are we ready? Travel Med. Infect. Dis. **54**, 102597 (2023)
40. M.I. Paredes, N. Ahmed et al., Underdetected dispersal and extensive local transmission drove the 2022 mpox epidemic. Cell **187**(6), 1374–1386 (2024)
41. K.K. Mercy, B. Tibebu, et al., Mpox continues to spread in Africa and threatens global health security. Nat. Med. **30**(5), 1225–122 (2024)
42. R.K. Mohapatra, P.K. Singh et al., Transmission dynamics, complications and mitigation strategies of the current mpox outbreak: a comprehensive review with bibliometric study. Rev. Med. Virol. **34**(3), e2541 (2024)
43. H. Ajmal, E. Hunter, et al., Agent-based modelling of mpox infection dynamics: simulating disease transmission and control strategies. J. Artif. Soc. Soc. Simul. **27**(2) (2024)
44. N.L. Bragazzi, S.A. Iyaniwura et al., Quantifying the basic reproduction number and underestimated fraction of mpox cases worldwide at the onset of the outbreak. J. R. Soc. Interface **21**(216), 20230637 (2024)
45. G.S. Lira, V.A. Ota et al., Mpox outbreak in Rio de Janeiro, Brazil: a translational approach. J. Med. Virol. **96**(5), e29621 (2024)
46. S.U.R. Khan, S. Asif et al., Deep hybrid model for mpox disease diagnosis from skin lesion images. Int. J. Imaging Syst. Technol. **34**(2), e23044 (2024)
47. L. Zhang, J. Huang et al., Global prediction for mpox epidemic. Environ. Res. **243**, 117748 (2024)
48. A.K. Nayak, S.K. Bisoyi, et al., Mpox classifier: a deep transfer learning model for monkeypox disease detection and classification, in *2024 1st International Conference on Cognitive, Green and Ubiquitous Computing (IC-CGU)*, pp. 1–6. IEEE (2024)
49. M.F. Barnsley, Fractal functions and interpolation. Constr. Approx. **2**, 303–329 (1986)
50. F.R. Dayeen, M.K. Hassan, Multi-multifractality, dynamic scaling and neighbourhood statistics in weighted planar stochastic lattice. Chaos Solitons Fractals **91**, 228–234 (2016)

51. M. Schuster, K.K. Paliwal, Bidirectional recurrent neural networks. IEEE Trans. Signal Process. **45**(11), 2673–2681 (1997)
52. A. Sagheer, M. Kotb, Time series forecasting of petroleum production using deep lstm recurrent networks. Neurocomputing **323**, 203–213 (2019)
53. Covid-19 data repository by the center for systems science and engineering (csse) at johns hopkins university (2019)
54. G. Ou, Y. Tang, et al., Automated robot and artificial intelligence-powered wastewater surveillance for proactive mpox outbreak prediction. Biosaf. Health (2024)
55. T. Nayak, K. Chadaga et al., Deep learning based detection of monkeypox virus using skin lesion images. Med. Nov. Technol. Devices **18**, 100243 (2023)
56. V. Singh, S.A. Khan et al., Modeling global monkeypox infection spread data: a comparative study of time series regression and machine learning models. Curr. Microbiol. **81**(1), 15 (2024)
57. S. Asif, M. Zhao, et al., Ai-based approaches for the diagnosis of mpox: challenges and future prospects. Arch. Comput. Methods Eng. 1–33 (2024)
58. M. Patel, M. Surti, M. Adnan, Artificial intelligence (ai) in monkeypox infection prevention. J. Biomol. Struct. Dyn. **41**(17), 8629–8633 (2023)
59. I. Gul, C. Liu et al., Current and perspective sensing methods for monkeypox virus. Bioengineering **9**(10), 571 (2022)
60. K. Chadaga, S. Prabhu et al., Application of artificial intelligence techniques for monkeypox: a systematic review. Diagnostics **13**(5), 824 (2023)
61. S. Raubitzek, T. Neubauer, A fractal interpolation approach to improve neural network predictions for difficult time series data. Expert. Syst. Appl. **169**, 114474 (2021)
62. Z. Dai, S. Liu, and C. Liu, Predict the prevalence and incidence of parkinson's disease using fractal interpolation-lstm model. Chaos: Interdiscip. J. Nonlinear Sci. **34**(5) (2024)
63. T.M.C. Priyanka, A. Gowrisankar, S. Banerjee, Mpox outbreak: Time series analysis with multifractal and deep learning network. Chaos **34**(10) (2024)

MIX
Papier aus verantwortungsvollen Quellen
Paper from responsible sources
FSC® C105338

If you have any concerns about our products,
you can contact us on
ProductSafety@springernature.com

In case Publisher is established outside the EU,
the EU authorized representative is:
Springer Nature Customer Service Center GmbH
Europaplatz 3, 69115 Heidelberg, Germany

Printed by Libri Plureos GmbH
in Hamburg, Germany